WORLDS OF CHEMISTRY

A TEXT FOR LIBERAL ARTS STUDENTS

WORLDS OF CHEMISTRY

A TEXT FOR LIBERAL ARTS STUDENTS

James N. Lowe
The University of the South

McGraw-Hill Publishing Company

New York St. Louis San Francisco Auckland Bogotá
Caracas Hamburg Lisbon London Madrid Mexico Milan
Montreal New Delhi Oklahoma City Paris San Juan
São Paulo Singapore Sydney Tokyo Toronto

A TEXT FOR LIBERAL ARTS STUDENTS

3456789 DOC DOC 8932109

ISBN 0-07-038846-6

This book was set in Caledonia by Progressive Typographers, Inc.
The editors were Karen Misler and Jack Maisel;
the designer was Jo Jones;
the production supervisor was Denise L. Puryear.
Drawings were done by J & R Services, Inc.
R. R. Donnelley & Sons Company was printer and binder.

Library of Congress Cataloging-in-Publication Data

Lowe, James N.
Worlds of chemistry.

1. Chemistry. I. Title.
QD33.L873 1989 540 88-844
ISBN 0-07-038846-6

ABOUT THE AUTHOR

JAMES N. LOWE

Jim Lowe was born in North Dakota and grew up in Oregon. He received a B.S. in chemistry from Antioch College and a Ph.D. in physical organic chemistry from Stanford University. After teaching for two years at Smith College, he joined the chemistry department at The University of the South (Sewanee) in 1965. He has spent sabbaticals doing research in biochemistry at the University of California at Davis and at the University of Illinois. He has written and coauthored articles on research and chemical education. With Lloyd L. Ingraham, he wrote a short book on biochemical reaction mechanisms.

DEDICATION

This book is dedicated to three teachers: Richard Yalman of Antioch College, Richard Eastman of Stanford University, and Lloyd Ingraham of the University of California, Davis.

CONTENTS

This text is written for use in a one-semester course for nonscience students. In particular, it is written for liberal arts students seeking to satisfy a science requirement or electing to gain some exposure to chemistry at the college level. Attitudes about science and insights into science that students get from an introductory course may be more important than a specific knowledge of any topic studied. This text is written in the hope that students using it will gain insight into the processes of science and will become aware of impacts that science and technology have on their lives and world.

Liberal arts students can better understand and appreciate science by knowing some of its history and methods. In the early chapters, some of the experimental evidence and reasoning from experiment that marked the development of ideas concerning atomic structure and chemical bonding are presented. Many of the experiments are simple in concept. Scientists proposed theories about atoms and molecules to explain small whole-number ratios determined by experiment. This text presents some of the history of chemistry and makes some of the philosophy of science explicit. In doing so, it offers liberal arts students an opportunity to relate their study of science to their study of other disciplines.

In the middle chapters, basic ideas of bonding, acid-base chemistry, and redox chemistry are presented and illustrated. In developing basic chemical ideas, a balance of concept and application has been attempted. Material on buffers is developed without using equilibrium calculations, and Lewis acid-base chemistry is included. Throughout, there is an effort to relate the chemistry presented to the periodic table. Applications are given to make concepts come alive for students and to arouse student interest in the underlying chemistry. In contrast with other texts for liberal arts students, I have presented fewer examples and sought to develop more complete connections between concepts and applications.

In presenting material on reaction rates and on thermodynamics, the emphasis is on applications. The use of radioactive isotopes to estimate the age of the earth and the use of carbon-14 dating to trace human history illustrate nuclear decay reactions. The active-site model for enzyme-catalyzed reactions is presented as part of a consideration of the rates of chemical reactions. This leads into a discussion of the mode of action of sulfa and penicillin. To elucidate the concepts of energy and entropy, Le Chatelier's principle is discussed in relation to the control of automobile emissions.

Later chapters emphasize the relationship between the structures of materials and their physical and chemical properties. The history of the introduction of copper, iron, and aluminum is related to differences in the chemistry of these metals and their compounds. The properties of silicates and clays are presented along with a discussion of glass and ceramics. The ion-exchange properties of soils are presented along with the production of chemical fertilizers. A discussion of petroleum and gasoline, structure-property relationships for soaps, detergents and membranes, and environmental issues surrounding the use of halogenated hydrocarbons is used to illustrate some of the chemistry of organic compounds. Structure property relationships of synthetic polymers illustrate the way that scientists can design molecules with a wide variety of desired properties. A discussion of the important biochemical polymers provides a background for understanding the growing field of genetic engineering.

Special Chapters

Two chapters in the text focus on topics important for the consideration of technology and the environment and for science and society. Problems associated with the growth of the use of energy and with extensive reliance on petroleum, coal, or nuclear energy are presented and discussed together in Chapter 11. In Chapter 18, I have tried to identify some issues that will be of growing importance to students as they participate in the political process. Some consideration of the roles of science and technology in society is important in a course for liberal arts students, and these chapters are intended to provide background for such consideration.

Asides

Asides are an important feature of this text. They are used to enrich the reading, even though subsequent material does not build on them. The asides go beyond the material under discussion in a variety of directions. They may present additional history, an aspect of the philosophy of science, an application to another scientific discipline, or a short exploration deeper into the topic at hand.

Problems

Problems are an important part of any chemistry text. Many of the problems in this text require students to think about an experiment or an application of chemistry. It is hoped that students will enjoy discovering that they too can understand experiments and can appreciate the scientific basis of some of their everyday experiences. Many of the problems are open-ended and are addressed particularly to liberal arts students.

Acknowledgements

I appreciate the suggestions and support from my students and colleagues. Many here at Sewanee have read portions of the manuscript. I owe a special thanks to David Camp, who carefully read and commented on early drafts of the manuscript. Without his criticisms and suggestions, this book may not have reached publication. I am appreciative of the students who have enrolled in Chemistry 100, a course taken to satisfy part of the math/science requirement at Sewanee. These students have worked with patience and good will in spite of the shortcomings of early drafts of this manuscript. Their interest and occasional enthusiasm have been of great help in bringing this book to completion.

I appreciate the time and thoughtfulness of all the reviewers of this book. I trust that they will see that I had the good sense to heed many of their suggestions. They did much to improve the organization of the book and to make the level of coverage more uniform. They helped make the book more readable and caught an occasional error. They are Ronald Backus, American River College; Neal Bush, South Dakota State University; Gregory Choppin, Florida State University; Kenneth Fornsberg, St. Louis University; Ralph Gable, Davidson College; Arnulf Hagen, The University of Oklahoma; Walter Hamilton, Texas Woman's University; Richard Hanson, University of Arkansas; Lynne Hardin, Tarrant County Junior College; Robert Harris, University of Nebraska–Lincoln; Allan Hovland, St. Mary's College; Colin D. Hubbard, University of New Hampshire; Peggy Hurst, Bowling Green State University; Judith Kelley, University of Lowell; Kenneth Loach, SUNY–Plattsburgh; Lyle Lowry, Brevard Community College; Ralph Powell, Eastern Michigan University; Otis Rothenberger, Illinois State University; Leo Spinar, South Dakota State University; Robert Swindell, Tennessee Technical University; John S. Thompson, Texas A & I University; and Everette Turner, University of Massachusetts.

A special thanks goes to James L. Bills at Brigham Young University who made many helpful suggestions through two rounds of the reviewing process.

Finally, I appreciate the encouragement and support of the people at McGraw-Hill. In particular, my editors Karen Misler and Jack Maisel never waivered in their encouragement and good will. I have also come to appreciate the creative contributions of those involved in graphics and design. They have contributed much to this book.

Jim Lowe

WORLDS OF CHEMISTRY

A TEXT FOR LIBERAL ARTS STUDENTS

CHAPTER

ONE

FROM THE BEGINNINGS OF MODERN CHEMISTRY TO DALTON'S ATOMIC THEORY

The use of chemistry has long been a part of everyday experience. The brewing of tea, the baking of bread, the dyeing of cloth, and the forging of tools are chemical processes. People have chemically changed materials in the preparation of food and in the making of tools throughout human existence. These processes continue, sometimes with new twists. We brew decaffeinated coffee, bake bread using enriched flour, wear coats of colors unknown to Joseph and his brothers, and cut with knives made of new and harder steel alloys.

Chemical processes range from simple separations such as the extraction of tea from leaves to more complex analyses such as the identification of caffeine in tea and coffee and of vitamins in flour and bread. Chemists combine metals in new ways to make harder or tougher alloys, and they synthesize new chemicals such as dyes not found in nature. Chemistry is applied to practical problems such as the preservation of food, and it is studied to enhance our understanding of the complex processes taking place in living organisms.

Chemical processes include not only the old and familiar but also the new and exotic. Chemists have separated and identified compounds called pheromones by which many insects send and receive signals over distances. The chemical bases for many folk medicines have been unraveled; this knowledge has been used to develop new medicines. The ability to purify silicon to previously unheard of levels and to carefully introduce controlled trace amounts of impurities has made possible the manufacture of the tiny chips used to make smaller and more powerful computers.

Scientists, in performing and studying processes like those described above, have sought to gain an understanding of the composition of materials around them and have tried to identify the changes in composition that occur when materials are converted to new substances. Their investigations and discoveries are part of the science called chemistry.

Chemistry is the study of the composition and transformation of matter. Chemists seek to understand the structure of materials and the changes that occur when one material is converted to another.

Even though chemical change has always been a part of human experience, the emergence of modern chemistry as a science is a recent development. Modern chemistry began during the period 1750–1815: it is a young science.

In this chapter, the development of chemistry as a science will be presented to show how people gained an understanding of the composition of matter and its changes. Individuals performed experiments that led to discoveries of regular patterns in nature. Efforts to explain regularities in nature gave rise to the atomic theory that is basic to modern chemistry.

ANTECEDENTS OF CHEMISTRY

The idea that all matter was composed of a small number of *elements* first arose in Greece. Greek philosophers considered air, earth, fire, and water to be the four elements from which all other substances were made. Elements provided their properties, hot or cold, wet or dry, to materials. Changes in materials were explained by trying to account for changes in the quantities of these four elements. For example, a metal obtained by heating an ore was considered a compound of the elements earth and fire. It had substance due to the presence of earth, and it was shiny because it contained fire. This viewpoint was a forerunner for the development of scientific ideas, as explanations were tested against observations.

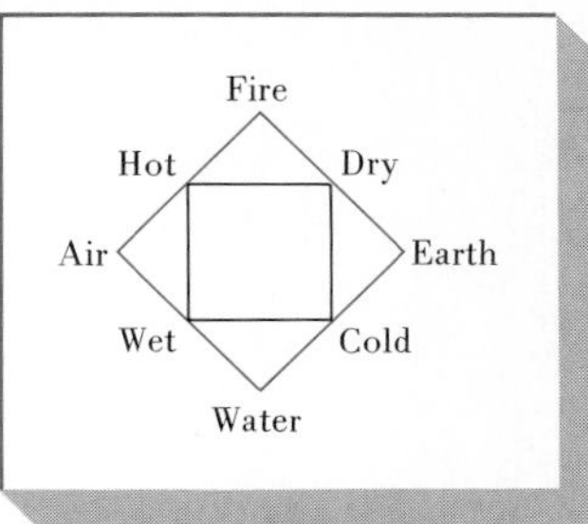

FIGURE 1.1
Early Greek ideas about matter emphasized properties of substances. For example, the element earth was thought to be dry and cold. The properties of real materials were thought to reflect the presence and properties of the elements earth, air, water, and fire.

The name *chemistry* is derived from *alchemy* which describes a wide variety of practices occurring over several centuries and in many places. Alchemy originated in Alexandrian Egypt. It flourished in the Arabic world and entered Europe by way of Spain. Its origins included both practical technology utilized by artisans and speculations about nature devised by the Greeks.

One aim of alchemy was *transmutation*, the effort to turn base metals into gold. From a modern chemical standpoint such an effort is doomed to failure, but to some alchemists it made much sense. Gold was considered to be the most noble and incorruptible metal. According to Islamic records from the tenth century, the following doctrine formed a basis for attempting to transmute gold. Just as health is the state of perfection for the body, gold is the state of perfection for metals. Metals strive for the essence of goldness. In practice, alchemists sought to remove that which was imperfect and to add for perfection that which was lacking. It is also fair to say that at some times and in some places alchemy included recipes for making metals, such as silver, appear as gold.

Alchemists made important discoveries in the areas of early metallurgy and pharmacology. They extracted plant materials and they purified volatile materials by distillation. In the course of their work they developed recipes for the production of some acids and bases used in the investigations that established modern chemistry. Among the materials first described by alchemists are phosphorus, arsenic, antimony, bismuth, and zinc, substances that are elements in the modern sense. Both apparatus and vocabulary in chemistry have been passed down from alchemy, and alchemy provided a crude framework of explanations for natural phenomena.

Unfortunately, competing and often confusing explanations for chemical changes abounded. Alchemy had at times been forbidden by civil or religious authorities, and then it was practiced as a secret art. Its vocabulary was obscure, and its symbols carried both chemical and mystical meanings. For modern chemistry to develop, it was necessary to sort the natural explanations for chemical changes from the mystical interpretations of those changes.

SETTING THE STAGE—QUESTIONS FACING EARLY CHEMISTS

What are the properties of a substance? Is a given substance pure or is it a mixture? These are not trivial questions, and answers were slow in coming. For example, even the description of something as familiar as water poses problems. Samples of water differ. Seawater tastes salty, the water from some springs bubbles, and water from other springs smells foul and tastes bad. How can one know if water is pure? Which properties of a sample of water are due to impurities, and which are the properties of water itself?

If the analysis of water poses problems, consider the greater problems posed by a metal such as copper. Early samples of metals usually contained contaminants from the ore, and ores mined in different places often contained different impurities. Which sample of copper might be purer? How would one decide?

Greek philosophers, alchemists, and early chemists sought to explain chemical change in terms of elements and compounds. But what are elements? And what are compounds? Some substances had been observed in a wide variety of chemical reactions. For example, copper was obtained from a variety of ores, and sulfur reacted with most metals. Are copper and sulfur (brimstone) elements, or are fire and earth? The problem was how to decide whether a given substance was an element, a compound, or a mixture.

In asking these questions, a scientific attitude is implied. It was not an attitude found among educated people in the Middle Ages. Western attitudes were influenced by the Greek philosopher Plato. According to Plato's philosophy, real things were imperfect representations of the ideal just as circles drawn in the sand are imperfect representations of the ideal circle. When the goal of study is to understand the ideal or essence, experimental knowledge can only be of lesser value. The decision to study the particular properties of particular samples of materials by observation represented an important change in people's thought.

THE BEGINNINGS OF MODERN CHEMISTRY

The philosopher Francis Bacon (1561 – 1626) set forth a program for experimental science. He called for the direct observation of nature. In his preface to *True Directions Concerning the Interpretation of Nature,* he wrote:

> Those who have taken upon them to lay down the law of nature as a thing already searched out and understood . . . have therein done philosophy and the sciences great injury. For as they have been successful in inducing belief, so they have been effective in quenching and stopping inquiry; and have done more harm by spoiling other men's efforts than good by their own.

In 1661, Robert Boyle (1627 – 1691) published *The Sceptical Chemist*; Boyle wrote that elements were

> certain primitive and simple, or perfectly unmingled bodies; which not being made of any other bodies, or of one another, are the ingredients of which all those called perfectly mixt bodies are immediately compounded, and into which they are ultimately resolved.

Boyle criticized Greek ideas concerning a four-element explanation of matter, and he critically examined and rejected some of the claims of alchemy. Boyle's definition set the stage for the further evolution of ideas concerning compounds and elements and compounds, but it failed to indicate how one could decide if a particular substance was an element.

To find better explanations, additional and careful observations were required. Those qualities most useful for the characterization of substances had to be identified, and ways to prove or disprove the interpretations of chemical changes needed to be found. In this process, there was a shift from qualitative observations to quantitative measurements. The qualities hot and cold were replaced by thermometer readings, and light and heavy by balance readings.

Modern chemistry began with discoveries that disproved plausible explanations of natural phenomena based on four elements. Air was shown to be a mixture, and water, a compound. The important measurements and the logic with which the data are examined are simple. We will examine the intimate relation of early chemical explanations to experimental observations.

THE STUDY OF GASES

During the period 1760–1780, scientists prepared, collected, and described a number of pure gases. This work contributed to the development of a new understanding of the nature of gases. Gases were collected by the displacement of water from an inverted container. Properties of each gas were studied. Did it burn? Did it support the combustion of a candle? Did it dissolve in water? (Gases that were highly soluble in water were collected over mercury.) Similarities and differences of gases were catalogued.

Because a gas separates from the liquid or solid from which it is generated, gases prepared by synthesis were generally very pure. This is important, for one can conclude that the properties of these gases were not altered by the presence of impurities. One outcome of the study of gases was the demonstration that air, one of the four elements listed by Greek philosophers, is a mixture of gases.

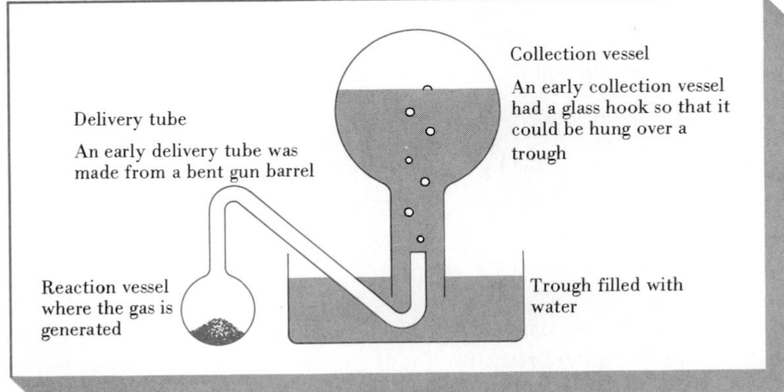
Collection vessel
An early collection vessel had a glass hook so that it could be hung over a trough
Delivery tube
An early delivery tube was made from a bent gun barrel
Reaction vessel where the gas is generated
Trough filled with water

FIGURE 1.2
Early scientists collected gases by the displacement of water from an inverted container.

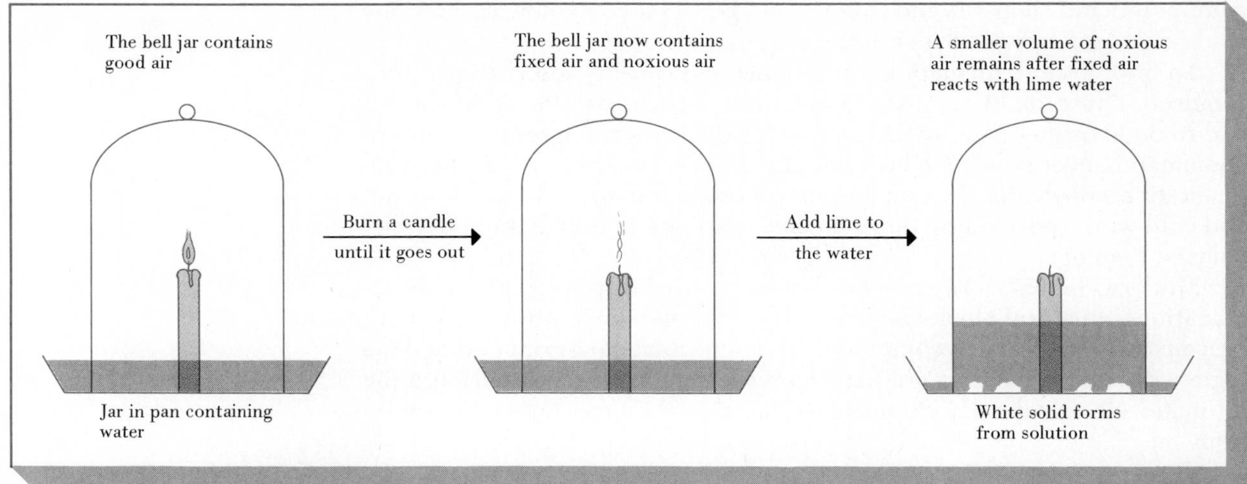
The bell jar contains
good air
The bell jar now contains
fixed air and noxious air
A smaller volume of noxious
air remains after fixed air
reacts with lime water
Burn a candle
until it goes out
Add lime to
the water
Jar in pan containing
water
White solid forms
from solution

FIGURE 1.3
In the eighteenth century, scientists studied the changes in air that accompanied the burning of a candle. In the course of their studies, they prepared fixed air now known as carbon dioxide and noxious air now known as nitrogen.

A S I D E

ACIDS AND BASES DESCRIBED BY THE ALCHEMISTS WERE USED TO PREPARE GASES

Acids and bases, substances known to the alchemists, were used to produce some gases. *Acids* can be defined as substances that taste sour, that fizz with soda, and that cause some vegetable dyes to change color. *Bases* taste bitter and reverse the color changes caused by acids.

For the studies concerning the generation and testing of gases we need to consider three acids and two bases. The common acids are:

Sulfuric acid, prepared by heating an iron salt (hydrated ferrous sulfate)

Nitric acid, prepared by adding sulfuric acid to saltpeter (potassium nitrate) and distilling the acid from the mixture

Hydrochloric acid, prepared by adding sulfuric acid to common salt and distilling a gas into water

Common bases include:

Quicklime, produced by heating limestone

Potash, obtained from the extract of wood ashes

AIR AND COMBUSTION

Early scientists were interested in combustion. Air would support the burning of a candle or the respiration of a mouse under a bell jar. When the flame went out or the mouse died, the remaining "air" had different properties. A part of this gas called "fixed air" readily dissolved in limewater (a solution

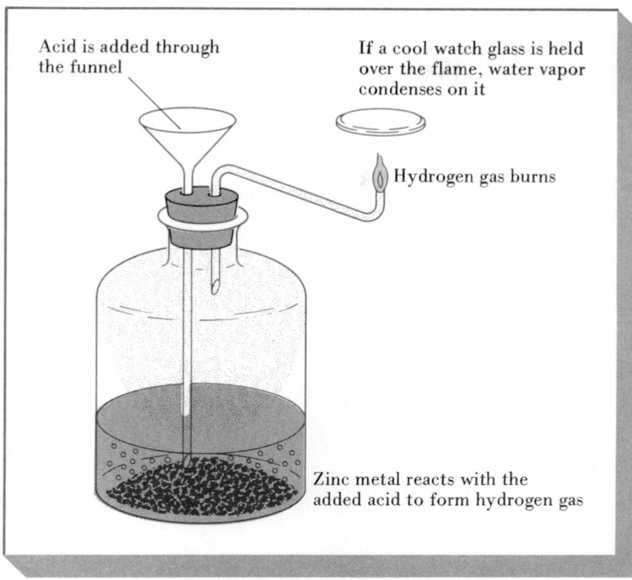
Acid is added through the funnel
If a cool watch glass is held over the flame, water vapor condenses on it
Hydrogen gas burns
Zinc metal reacts with the added acid to form hydrogen gas

FIGURE 1.4
Hydrogen gas is produced by the reaction of zinc metal with dilute sulfuric acid. Water is produced when the hydrogen gas burns in air.

made by dissolving quicklime in water) to produce a white precipitate. (A *precipitate* is a solid formed by a reaction occurring in solution.) Another part, the "noxious air" remaining after the fixed air was dissolved in base, did not support combustion or respiration.

The carbon dioxide produced by combustion and the nitrogen remaining after combustion were two of the first gases to be described and characterized. Fixed air, now known as *carbon dioxide,* was produced by respiration or combustion. It could be prepared in a pure form and studied by adding acid to the white solid formed by the reaction of fixed air with limewater. Noxious air, now known as *nitrogen,* remained after the carbon dioxide was dissolved in base.

Early scientists also produced *hydrogen* or "inflammable air," by the action of sulfuric acid or hydrochloric acid on zinc or iron. Hydrogen gas burns in air. By condensing the vapor above the flame on a cold surface, it was shown that water is produced by the burning of hydrogen.

JOSEPH PRIESTLEY DISCOVERS OXYGEN

Joseph Priestley (1733–1804) prepared and described a number of previously unknown gases. In the course of his investigations, he developed a chemical test for the "goodness" of air, its ability to support respiration. (Modern names rather than the names given by Priestley are used to describe his test.) When 2.0 L of common air was mixed with 1.0 L of nitric oxide over water, orange fumes formed; the volume decreased to 1.8 L as the orange gas dissolved. The remaining gas, being mostly nitrogen, would

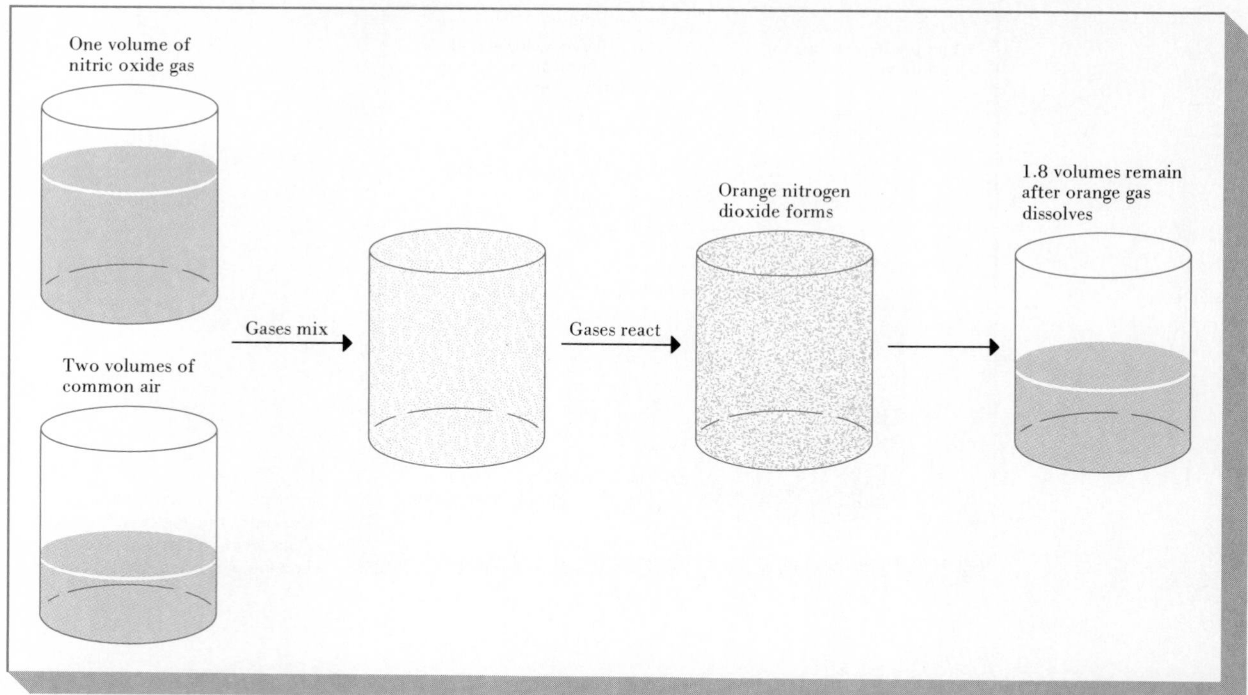
One volume of
nitric oxide gas
Two volumes of
common air
Gases mix
Gases react
Orange nitrogen
dioxide forms
1.8 volumes remain
after orange gas
dissolves

FIGURE 1.5
Joseph Priestley devised a test for the goodness of air that involved the reaction of air with nitric oxide. A larger final volume of gas remained if the air had been spoiled by burning or breathing.

not support combustion. If the common air had been spoiled by breathing or burning, a larger final volume of gas would have remained.

Experiments using this test led to Priestley's discovery of oxygen. He found that heating an orange solid known as mercury calx produced liquid mercury and a gas. He collected and studied this gas. Glowing charcoal placed in the gas burst into flame. He then tested the gas for its goodness as air by mixing two volumes of the gas with one volume of nitric oxide. Orange nitrogen dioxide formed and then dissolved in water. The next day, he happened to put a glowing candle in the gas remaining after the test. The candle burst into flame. Unlike common air, all of the gas obtained by heating the mercury calx would support combustion.

Priestley had discovered oxygen. The mercury calx that Priestley heated is now known to be mercury oxide. Priestley had decomposed mercury oxide into mercury and oxygen.

$$\text{Mercury oxide} \rightarrow \text{mercury} + \text{oxygen}$$

Priestley's test for the goodness of air can now be interpreted. Nitric oxide reacts with oxygen in the air to form orange nitrogen dioxide.

$$\text{Nitric oxide} + \text{oxygen} \rightarrow \text{nitrogen dioxide}$$

When the reaction is run using air as the source of oxygen, the nitrogen in the air does not react. It is present both before and after the reaction with oxygen.

$$\text{Nitric oxide} + \text{air} \rightarrow \text{nitrogen dioxide} + \text{nitrogen}$$
$$\text{Nitric oxide} + \text{oxygen} + \text{nitrogen} \rightarrow \text{nitrogen dioxide} + \text{nitrogen}$$

When nitrogen dioxide dissolves in water, the total volume of gas decreases, and only nitrogen and excess nitric oxide remain. If the air had been used to support combustion, less oxygen would be present, less nitrogen dioxide would form and then dissolve, and a larger volume of gas would remain.

The discovery of oxygen was critical to the development of a clear distinction between the various pure gases and air. Because oxygen is responsible for the ability of air to support combustion, the identification and description of oxygen was also critical for unraveling the mystery of that most fundamental chemical process, fire.

A S I D E

PRIESTLEY WAS AN AMATEUR SCIENTIST

Priestley was a dissenting churchman in England. He was persecuted for his religious and political views. For his sympathies toward the French Revolution, his home and library were vandalized. He ended his career as an exile in America where he declined appointments to become a Congregational minister or a professor of chemistry at either the University of Pennsylvania founded by Benjamin Franklin or the University of Virginia being planned by Thomas Jefferson.

Joseph Priestley. *(National Portrait Gallery, Smithsonian Institution, Washington, D.C.)*

Was the discovery of oxygen an accident? Priestley explains:

> I cannot, at this distance of time, recollect what it was that I had in view in making this experiment; but I know I had no expectation of the real issue of it. Having acquired a considerable degree of readiness in making experiments of this kind, a very slight and evanescent motive would be sufficient to induce me to do it. If, however, I had not happened for some purpose, to have a lighted candle before me, I should probably never have made the trial; and the whole train of my future experiments relating to this kind of air might have been prevented.

Seemingly chance discoveries frequently have played important roles in the development of science. Joseph Priestley's curiosity and willingness to experiment illustrate important characteristics of many scientists. Louis Pasteur, a later scientist, said, "Chance favors only the mind that is prepared."

THE LAW OF CONSERVATION OF MASS

Antoine Lavoisier (1743–1794), a Frenchman working around the time of the American Revolution, played a most important role in the birth of modern chemistry. Lavoisier systematically set out to bring order to the emerging science of chemistry.

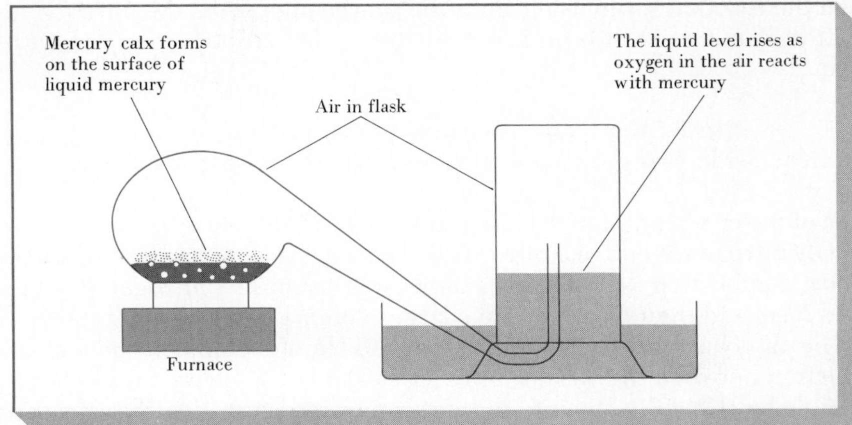
Mercury calx forms
on the surface of
liquid mercury
Air in flask
The liquid level rises as
oxygen in the air reacts
with mercury
Furnace

FIGURE 1.6
Lavoisier heated mercury under air to form mercury calx. He measured the decrease in the volume of air. He also found that the final weight of the mercury plus the mercury calx plus the residual air is the same as the weight of the original mercury plus the original air.

Lavoisier studied the quantities of materials involved in chemical reactions. For example, he studied the formation and decomposition of mercury calx, the compound from which Priestley obtained oxygen. In one set of experiments, Lavoisier prepared mercury calx by heating mercury under a closed atmosphere of air in a furnace. He weighed the red powder that had formed, and he measured the decrease in the volume of air. When this sample of mercury calx was later heated more strongly, the solid lost weight to re-form mercury and oxygen. The volume of oxygen was equal to the decrease in the volume of air during the formation of the solid.

Lavoisier carefully measured the mass of each material reacting and the mass of each product formed for a large number of reactions.

$$\text{10 g sulfur} + \text{10 g oxygen} \rightarrow \text{20 g sulfur oxide}$$
$$\text{7.75 g phosphorus} + \text{10.00 g oxygen} \rightarrow \text{17.75 g phosphorus oxide}$$
$$\text{3.0 g carbon} + \text{8.0 g oxygen} \rightarrow \text{11.0 g carbon dioxide}$$

He found that in each reaction, the total mass of the substances reacting was always equal to the total mass of the products formed.

Lavoisier generalized his observations to formulate the *law of conservation of mass.* In a chemical reaction, the total mass of reactants is always equal to the total mass of the products. Matter is neither created nor destroyed. (*Mass is a fundamental measure of the quantity of matter.* The greater the mass of an object, the greater the force required to give it the same acceleration. The weight of an object that we measure on earth is the gravitational force acting on its mass. Objects placed in orbit around the earth have the same mass as on earth even though they may be weightless in space.)

SAMPLE CALCULATION

When strongly heated, limestone loses carbon dioxide to form lime. What mass of limestone would be needed to produce 100 g of lime if 78.5 g of carbon dioxide is also produced?

$$\underset{?\text{g}}{\text{Limestone}} \rightarrow \underset{100\text{ g}}{\text{lime}} + \underset{78.5\text{ g}}{\text{carbon dioxide}}$$

According to the law of conservation of mass, the mass of the reactant(s) must equal the mass of the products.

$$\text{Mass of limestone} = 100\text{ g} + 78.5\text{ g} = 178.5\text{ g}$$

LAVOISIER AND COMBUSTION

The examples given to illustrate the law of conservation of mass happen to involve combustion or burning. Lavoisier was the first to understand that when burned, substances combine with oxygen. Lavoisier's interpretation of combustion replaced an earlier explanation, the phlogiston theory, that had been widely held during the prior hundred years.

According to the phlogiston theory, things rich in phlogiston burned. In combustion, phlogiston departed, and the remaining ash was "dephlogistinated." This theory had a certain success in explaining observations. For example, carbon burned because it was rich in phlogiston. Some metal calxes (metal oxides) could be converted to the uncombined metals by treating them with carbon. Because phlogiston from the carbon passed into the calx, metals such as iron could burn.

It is instructive to consider why one "scientific" theory replaced another. The phlogiston theory was satisfactory as long as qualitative explanations were required, but it proved less attractive in a quantitative way. If a log burned because phlogiston left the wood, then the mass of the ashes is less than that of the wood because of the loss of phlogiston. However, iron and other metals gain mass when burning. To explain this observation, the departing phlogiston would have to have negative mass. That phlogiston sometimes had positive mass and other times had negative mass provided a far more complicated and less attractive theory than that proposed by Lavoisier.

Lavoisier. *(National Portrait Gallery, Smithsonian Institution, Washington, D.C.)*

Lavoisier extended his research into the field now called biochemistry. Carbon burns in air to give carbon dioxide. Since that same oxide is present in exhaled air but not in inhaled air, Lavoisier recognized that we obtain energy by the burning of foods to form carbon dioxide which we then exhale.

This promising inquiry was cut short by the beheading of Antoine Lavoisier in 1794. Although his services to the government of France had ranged from promoting the metric system to improving gunpowder, his service as a bureaucrat and his association with the collection of taxes on tobacco led to his trial and execution during the French Revolution.

OPERATIONAL DEFINITIONS PERMIT SCIENCE TO PROGRESS

Scientists use *operational definitions* to pose and to resolve questions in science. To illustrate the use of operational definitions, consider a simple example: "Who is the biggest person in a class?" This question is not posed in operational terms, so people may supply their own definitions and may disagree on the answer. Change the question to "Who is the tallest person in a class?" or "Who is the heaviest person in a class?" Now measurements, height or weight, can be used to answer the questions posed.

Lavoisier introduced and used an operational definition to distinguish compounds from elements. Elements had been defined as those simple chemical substances from which all other substances were formed. But what did it mean to say that a substance was simple? Lavoisier assumed that a substance in a chemical reaction is simpler if it has less mass than the substance from which it is formed. Mercury calx cannot be an element, for it was decomposed to mercury and oxygen, each of which has less mass than the original sample of the orange compound. *Chemicals can be shown to be compounds when they can be formed from or decomposed into lighter materials.*

According to Lavoisier, *a substance may be an element if it has not been shown to be decomposed by chemical reactions.* In 1789, Lavoisier published a treatise in which he named 33 substances that he considered to be elements. Many of these substances appear on current tables of elements.

The ability to use measurement to resolve questions speeds the choice between competing explanations. By choosing to use measurement in deciding questions, scientists restrict the domain of scientific inquiry. In particular, many questions of value lie outside the domain of science.

THE LAW OF CONSTANT COMPOSITION

A period of increased activity in analytical chemistry followed Lavoisier's demonstration of the importance of measuring weights in chemical reactions. A large number of compounds were analyzed by decomposing them to their constituents.

About 1800, Joseph Proust (1754–1826) analyzed samples of malachite, an ore of copper, obtained from several different sources. These samples of this dark-green mineral were free of impurities. Proust prepared the same compound present in the mineral from other chemicals. Gentle heating of each sample of malachite produced water. On stronger heating, carbon dioxide evolved. The product of these operations was a black oxide of copper. When heated, the black oxide reacted with hydrogen gas to produce water vapor and copper metal.

When the mass driven off in each transformation was expressed as a percentage of the original mass, Proust found that each sample gave the same analysis. Every sample of each intermediate compound had the same percentage copper as all other samples of that compound.

Proust formulated the *law of constant composition.* He generalized that *every compound contains a fixed ratio of elements by mass.* For example, in

	Noms nouveaux.	*Noms anciens correfpondans.*
Subftances fimples qui appartiennent aux trois règnes tqu'on peut regarder comme les élémensdescorps.	Lumière	Lumière,
	Calorique	Chaleur. Principe de la chaleur. Fluide igné. Feu. Matière du feu & de la chaleur.
	Oxygène	Air déphlogiftiqué. Air empiréal. Air vital. Bafe de l'air vital.
	Azote	Gaz phlogiftiqué. Mofète. Bafe de la mofète.
	Hydrogène	Gaz inflammable. Bafe du gaz inflammable.
Subftances fimples non metalliques oxidables & acidifiables.	Soufre	Soufre.
	Phofphore	Phofphore.
	Carbone	Charbon pur.
	Radical muriatique	Inconnu.
	Radical fluorique	Inconnu.
	Radical boracique	Inconnu.
Subftances fimples metalliques oxidables & acidifiables.	Antimoine	Antimoine.
	Argent	Argent.
	Arfenic	Arfenic.
	Bifmuth	Bifmuth.
	Cobalt	Cobalt.
	Cuivre	Cuivre.
	Etain	Etain.
	Fer	Fer.
	Manganèfe	Manganèfe.
	Mercure	Mercure.
	Molybdène	Molybdène.
	Nickel	Nickel.
	Or	Or.
	Platine	Platine.
	Plomb	Plomb.
	Tungftène	Tungftène.
	Zinc	Zinc.
Subftances fimples falifiables terreufes.	Chaux	Terre calcaire, chaux.
	Magnéfie	Magnéfie, bafe du fel d'epfom.
	Baryte	Barote, terre pefante.
	Alumine	Argile, terre de l'alun, bafe de l'alun.
	Silice	Terre filiceufe, terre vitrifiable.

FIGURE 1.7
Lavoisier's table of the elements. *(From Lavoisier, Traité élémentaire de chimie [1789].)*

any sample of pure water, the ratio of the mass of oxygen to the mass of hydrogen is 8 : 1. The fraction of the total mass contributed by each element is also constant for every compound. For example, carbon dioxide is always three-elevenths carbon by mass.

THE LAW OF MULTIPLE PROPORTIONS

The most striking evidence for the atomic theory is the *law of multiple proportions*. This law was first proposed by John Dalton (1766–1844) in 1803, about the same time he formulated an atomic theory. The recognition of the mass relations summarized by this law can either be considered

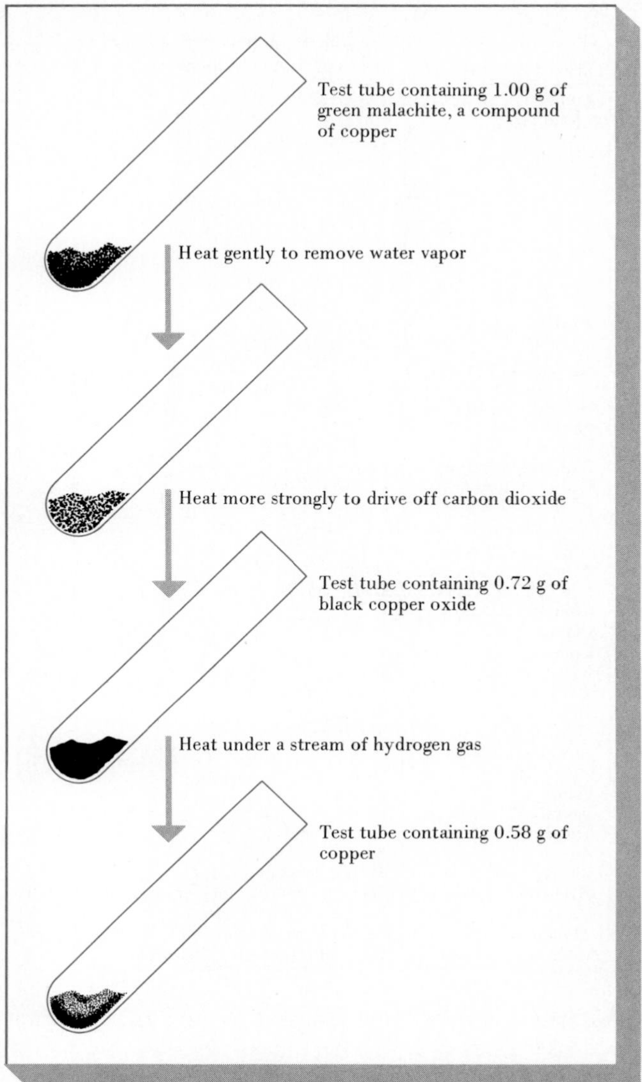
Test tube containing 1.00 g of green malachite, a compound of copper
Heat gently to remove water vapor
Heat more strongly to drive off carbon dioxide
Test tube containing 0.72 g of black copper oxide
Heat under a stream of hydrogen gas
Test tube containing 0.58 g of copper

FIGURE 1.8
Joseph Proust carefully analyzed samples of malachite mined in different places as well as a sample of the same compound prepared in the laboratory. He found that each sample of malachite contained 58% copper and that each sample of copper oxide contained 80% copper. From these and other analyses he formulated the law of constant composition.

powerful evidence for postulating chemical atoms, or it can be seen as evidence confirming a prediction of atomic theory.

Consider an experiment illustrating the law of multiple proportions. Two compounds are formed between the pair of elements, copper and bromine. When one of these copper bromides, a dark shiny compound, is heated, orange bromine gas evolves and a second compound, a pale copper bromide, remains. This second copper bromide can be converted to copper metal (and other products containing the bromine) in a series of chemical operations.

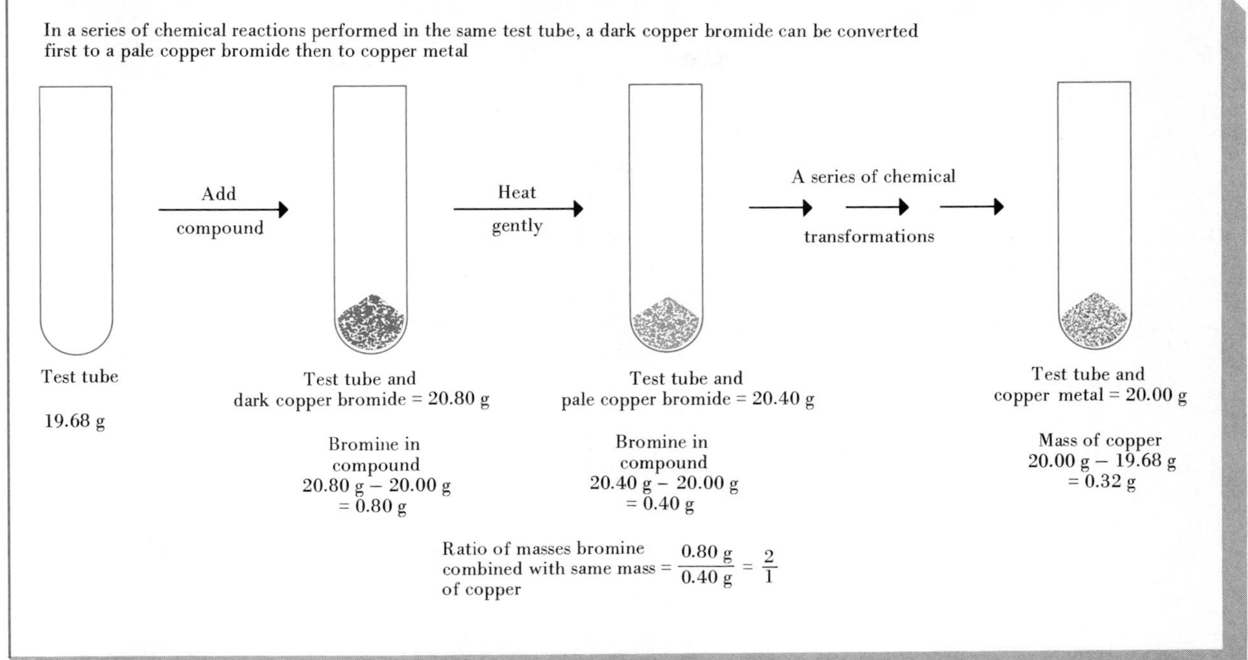

FIGURE 1.9
The law of multiple proportions can be illustrated using the analyses of two compounds composed of the elements copper and bromine.

Consider the quantities involved in this experiment. At successive times, there is a dark shiny copper bromide, a pale copper bromide, and, finally, copper. Both of the copper bromides contain the same quantity of copper. By subtracting the mass of the copper isolated at the end of the experiment from the mass of each compound, the mass of bromine in each compound can be determined. There is a most striking outcome when these masses are compared. The mass of bromine in the dark solid is exactly twice the mass of bromine in the pale copper bromide. (Note that this is a statement about masses. Neither the names nor the formulas of the copper bromides were necessary to show this mass relation.)

The law of multiple proportions generalizes the results of this and similar experiments. *When two compounds are formed between a pair of elements, then with a fixed mass of one element the masses of the second element in the two compounds are in a small whole number ratio.* Examples of such simple integer ratios are 2 : 1, 3 : 1, and 3 : 2.

SAMPLE CALCULATION

Two gaseous compounds of carbon and oxygen are known. From 6.0 g of carbon, 14.0 g of carbon monoxide or 22.0 g of carbon dioxide can be prepared. Show that the composition of these gases is in accord with the law of multiple proportions.

$$\begin{gathered} 6.0 \text{ g carbon} + \text{oxygen} \rightarrow 14.0 \text{ g carbon monoxide} \\ \text{Oxygen in } 14.0 \text{ g carbon monoxide} = 14.0 \text{ g} - 6.0 \text{ g carbon} = 8.0 \text{ g} \\ 6.0 \text{ g carbon} + \text{oxygen} \rightarrow 22.0 \text{ g carbon dioxide} \\ \text{Oxygen in } 22.0 \text{ g carbon dioxide} = 22.0 \text{ g} - 6.0 \text{ g carbon} = 16.0 \text{ g} \end{gathered}$$

The ratio of the masses of oxygen combined with a fixed mass of carbon is 8.0 : 16.0 or 1 : 2, a ratio of small integers.

Scientists use laws to summarize the results of real observations and to describe ways that the physical world can be seen to be regular. They use scientific laws to predict the results of new experiments. By generalizing from several experiments to all similar experiments in formulating or using a law, scientists assert a belief that nature is predictable.

ASIDE

EXPERIMENTAL RESULTS VARY FROM THE IDEAL

What does it mean to say the results are in a simple integer ratio? This is an idealization of experimental results, for measured ratios from individual experiments vary. Consider the experiment performed with compounds of copper and bromine. When a class of students performs the experiment converting samples of the dark copper bromide first to the pale solid then to copper, results vary. Most students find that their measured ratio of masses of bromine in the two compounds lies between 1.90 and 2.10. One student may fail to drive all the readily removable bromine from the first compound. Another may expose hot copper to the air and have some of it react with oxygen, giving too great a mass of copper in each of the compounds. These and other experimental errors would cause the measured ratio of masses of bromine to vary. With small errors, the measured ratios are close to the integer ratio, 2 : 1. Variation is assumed to be due to experimental errors and uncertainties. An assumed, precise regularity of nature is expressed in the scientific law.

DALTON'S ATOMIC THEORY

Three important generalizations about the composition of pure substances have been discussed. First, mass is always conserved when substances react. Second, a pure compound always contains a fixed ratio of elements by mass. Third, if two compounds form between a pair of elements, the masses of the one element found combined with a fixed mass of the other element in the two compounds are in a simple integer ratio. To the scientist, these regularities in nature invite explanation.

In 1803, Dalton postulated an *atomic theory* to account for the observed mass relations in chemical combination. Dalton made the following assumptions:

1. All matter is composed of tiny indivisible, indestructible particles called atoms

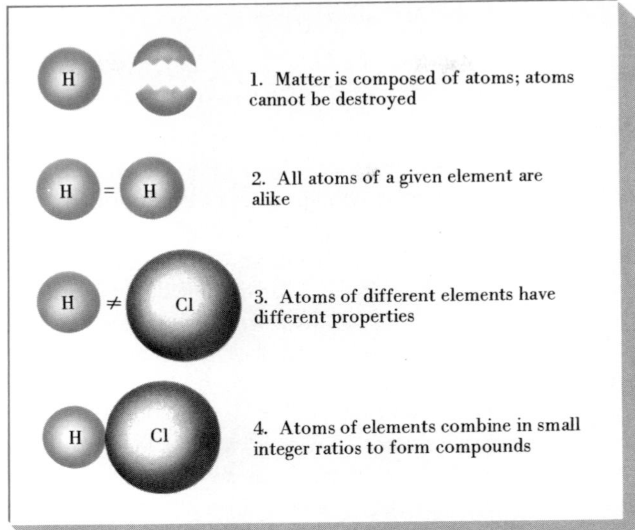
H
1. Matter is composed of atoms; atoms cannot be destroyed
H = H
2. All atoms of a given element are alike
H ≠ Cl
3. Atoms of different elements have different properties
H Cl
4. Atoms of elements combine in small integer ratios to form compounds

FIGURE 1.10
In 1803 John Dalton formulated an atomic theory that incorporated these assumptions.

2. All atoms of a given element are alike. In particular, they have the same mass.
3. Atoms of different elements have different properties. Most important, they have different masses.
4. Atoms of different elements combine in small integer ratios to form compounds.

Because atoms can neither be created nor destroyed, masses are conserved in chemical reactions. All of the atoms that start out in the reactants end up in the products. Hence the mass of the products is equal to the mass of the reactants. In a compound, the atoms of different elements are present in the ratio of small whole numbers, and the masses of the atoms are fixed. Therefore the fraction of the total mass of each element in a compound is fixed. Finally, since chemical combinations of atoms are in a simple integer ratio, the law of multiple proportions follows. For example, if the copper bromides in the experiment had the formulas $CuBr_2$ and CuBr, then with a given number of atoms of copper, one compound would have twice the number of atoms of bromine and twice the mass of bromine as the other.

Dalton's atomic theory was a major milestone in the development of the science of chemistry. His theory has successfully explained observed mass relations in chemical combination, and provided the framework for the subsequent exploration of chemical structures and transformations. The working out of the consequences of atomic theory to generate rules for chemical structure and reactions was to be a major activity in chemistry following Dalton.

A S I D E

DALTON, THE WEATHER, AND ATOMS

John Dalton. *(National Portrait Gallery, Smithsonian Institution, Washington, D.C.)*

John Dalton, a Quaker, was born into a poor English family. Although he had little formal schooling, a wealthy patron helped him acquire some instruction in science and mathematics. He studied newtonian science and began to teach in school at an early age. For most of his life, he supported himself modestly by tutoring.

For more than 50 years, Dalton regularly made and recorded weather observations. Among other things, Dalton studied humidity, the contribution of water vapor to the atmosphere. (Those of you who have experienced hot days on the Gulf Coast or in the desert of the Southwest know that humidity can make a difference.) Dalton measured the relative humidity by measuring the number of degrees the air must be cooled to reach the dew point.

His atmospheric studies convinced him that the gases in air acted independently and were not chemically combined as some others thought. Dalton thought of the gases in air as consisting of tiny solid particles or atoms that repelled one another. He believed that the properties of the mixture of gases were due to the numbers and properties of each kind of atom present.

Dalton applied his early atomic ideas to the consideration of masses in chemical compounds. He pursued his ideas to develop an atomic theory for chemical combination. Dalton is a foremost example of a scientist who brought insights gained in one area of science to the solution of problems and advancement of understanding in another area of science.

QUESTIONS

1. A philosophy student argues that it should be no harder to distinguish an element from a compound than it is to distinguish a planet from a star. Why were ancient astronomers able to distinguish planets from stars at a time that alchemists were unable to distinguish most elements and compounds?
2. Hydrogen, oxygen, and nitrogen were among the first gaseous elements to be studied by early chemists. How was each gas prepared? Which of these gases do you think was least pure when it was first studied? Why?
3. By collecting gases by the displacement of mercury, Priestley was able to identify and describe some previously undiscovered gases. What property of these gases prevented their preparation by the displacement of water?
4. In Priestley's test for the goodness of air using nitric oxide, what gas remains if the sample tested is air? If the sample tested is pure oxygen?
5. How does the burning of hydrogen gas to form water contradict an explanation of matter based on the existence of the elements air, earth, water, and fire?
6. How could the reaction of hydrogen gas with oxygen to form water be

used to show that water is a compound under Lavoisier's operational definition of elements?

7. What mass of carbon dioxide is produced in the following reaction? What law is invoked to answer this question?

$$100 \text{ g calcium carbonate} \rightarrow 56 \text{ g lime} + ?\text{g carbon dioxide}$$

8. What mass of oxygen is required to burn 1000 g of isooctane, a component of gasoline?

$$1000 \text{ g isooctane} + ?\text{g oxygen} \rightarrow 3082 \text{ g } CO_2 + 1419 \text{ g } H_2O$$

9. How does the formation of lime and carbon dioxide from limestone prove that limestone is not an element?
10. How can it be shown that carbon dioxide is not an element?
11. Which of the following analyses were obtained from samples of the same compound?

Sample	Mass of Sample, g	Mass Iron, g	Mass Chlorine, g
A	0.600	0.206	0.394
B	0.750	0.258	0.492
C	0.500	0.220	0.280

12. The composition of air is about 20% oxygen and 80% nitrogen. Why do you believe air to be a mixture rather than a compound of oxygen and nitrogen?
13. A student wakes up from a nightmare in which rivers and streams were aflame. He knows that mixtures of hydrogen gas and oxygen gas burn. How can he reassure himself that water does not burn even though it is composed of hydrogen and oxygen. (Note that the meanings in chemistry, as in all other areas, depend upon context.)
14. Lavoisier considered lime to be an element. Why might his operational definition of an element have led to the inclusion of the compound lime in a list of elements?
15. From 100 g of a red oxide of iron, 70 g of iron can be obtained by reaction with air and charcoal. The same mass of iron can be obtained from 90 g of a black oxide of iron. Show that these results are consistent with the law of multiple proportions.
16. An oxide of sulfur is composed of equal masses of oxygen and sulfur. According to Dalton's postulates, this compound could not have the formula SO. Why?
17. Acetylene and ethane are both compounds of carbon and hydrogen. When acetylene is burned, 4.89 g of carbon dioxide is formed for each gram of water. When ethane is burned, 1.63 g of carbon dioxide is formed per each gram of water. Show that these compounds illustrate the law of multiple proportions.

18. In some texts, an atom is defined as the smallest part of an element that retains the physical properties of the element. Is this an operational definition of an atom? Use the fact that diamond, graphite used in pencil leads, and sootlike carbon black are three forms of the element carbon to criticize this definition.
19. Does the fact that three forms of the element carbon exist contradict Dalton's postulate that all atoms of an element are alike? That they have the same mass?
20. Does Dalton's atomic theory lead to a new operational definition of an element? If so, what would the operation be?
21. If Dalton's atomic theory had been proposed on the basis of the law of conservation of mass and the law of constant composition, the law of multiple proportions would have been a consequence of atomic theory. Describe how the law of multiple proportions would be predicted from atomic theory.
22. Lavoisier proposed his theory of combustion in 1783, only a few years before the convening of the Constitutional Convention in the United States. Which has undergone the greater amount of change in the subsequent two centuries—chemistry or political science?

CHAPTER

TWO

FROM MOLECULES TO THE PERIODIC TABLE

Dalton's atomic theory provided a framework for the exploration of the chemistry of elements and compounds. However two major questions needed to be addressed. First, in what ratio do atoms of elements combine in a compound? For example, is water HO_2 or HO or H_2O? Second, how are the properties of compounds related to their elemental constituents? For example, oxygen combines with carbon and nitrogen to form gases, but it forms high melting solids with iron and aluminum.

These were difficult questions. To provide answers, chemists needed to find independent ways to determine formulas, and they needed to accumulate a body of descriptive chemistry of elements and their compounds. This chapter opens with the successful approach to determining chemical formulas, and it closes with the revelation of patterns of chemical behavior summarized in Mendeleev's periodic table.

DALTON'S ATOMIC THEORY IS INCOMPLETE

To assign formulas to compounds, Dalton made an assumption when formulating his atomic theory. If only one compound is known to occur between two elements, Dalton assumed that the elements combine in a 1 : 1 ratio of atoms. His assumption led him to assign an incorrect formula HO for water.

Consider the consequences of Dalton's assumption for the relative weights of oxygen and hydrogen atoms. It can be demonstrated by experiment that 8.0 g of oxygen combines with 1.0 g of hydrogen to form water. If water has the formula HO, then in every sample of water, no matter how large or small, there would be equal numbers of hydrogen and oxygen atoms. In 8.0 g of oxygen there would be the same number of oxygen atoms as in 1.0 g of hydrogen. Each oxygen atom would weigh 8.0 times as much as each hydrogen atom.

The assumption that atoms combine 1 : 1 when a single compound between a pair of elements is known led Dalton to calculate incorrect relative atomic weights for carbon and some other elements, as well as for oxygen. Confusion about relative atomic weights slowed the acceptance of Dalton's atomic theory. To construct a consistent and correct table of relative atomic weights, scientists needed to discover how to determine formulas of compounds by experiment.

Does this incorrect assumption mean that Dalton's atomic theory is wrong? With hindsight we might note that his other assumptions are also faulty. As you explore chemistry, you will learn of the existence of isotopes, atoms of an element that differ in weight, and that atoms are not indivisible but are composed of subatomic particles. Finally, some compounds are known in which atoms do not combine in simple integer ratios. For each of Dalton's assumptions there are exceptions; yet his atomic theory continues to inform the way we perceive the physical world.

Theories, such as Dalton's, that chart new ground in science are often simple. When nature is found to be more complex than predicted by the simple theory, the theory may be modified. Dalton's theory has been modified as scientists learned more about atoms. But if the complexity of modern atomic theory had been presented in 1810, it would have seemed unbelievable. It is only after a theory has been found to explain phenomena that it is accepted. Later, the inadequacy of the theory for explaining additional phenomena can be seen as a problem that requires modification of the theory.

THE LAW OF GAY-LUSSAC

Measurements of combining volumes of gases in chemical reactions led to a way for determining formulas. Gay-Lussac (1778–1850), a Frenchman, measured the oxygen content of air by adding hydrogen and then sparking the mixture. Oxygen and hydrogen combined to form liquid water and a smaller volume of gas remained. When excess hydrogen is present, the decrease in volume was equal to three times the volume of oxygen present in the air. Gay-Lussac concluded that two volumes of hydrogen gas react with one volume of oxygen gas to make water.

Gay-Lussac then investigated volume relationships in other reactions of gases. Quantities of gases in chemical reactions were studied by measuring volumes of both reactants and products at the same temperature and pressure. For example, in Priestley's chemical test for the goodness of air, nitric

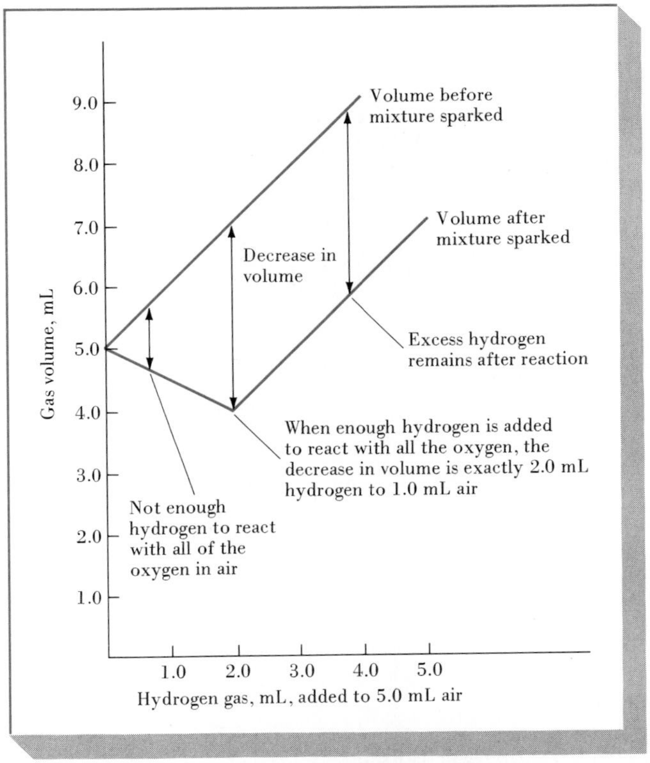

FIGURE 2.1
By adding small amounts of hydrogen gas to air then sparking the mixture and measuring the remaining gas, Gay-Lussac found that in the reaction of hydrogen gas with oxygen gas to form water, the ratio of reacting gases was exactly 2:1.

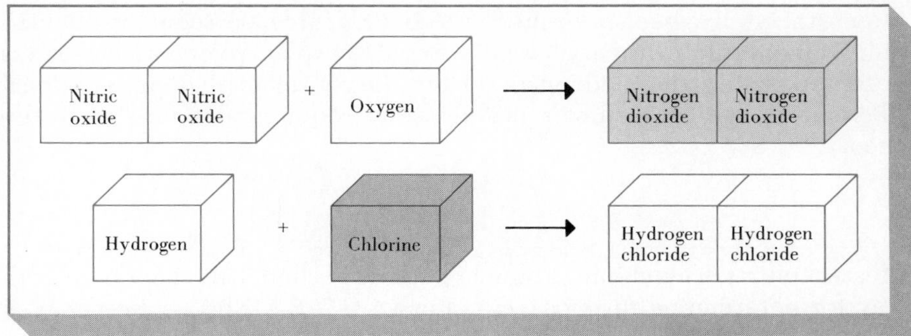

FIGURE 2.2
In reactions of gases, Gay-Lussac found that volumes of reacting gases and of product gases (measured at the same temperature and pressure) were in small whole number ratios. For example, 2.0 L of nitric oxide combine with 1.0 L of oxygen to form 2.0 L of orange nitrogen dioxide. In a second example, 1.0 L of hydrogen reacts with 1.0 L of chlorine to form 2.0 L of hydrogen chloride gas.

oxide reacted with oxygen to give nitrogen dioxide. When volumes of reacting gases and of product gases are measured, it is found that 2.0 L of nitric oxide reacts with 1.0 L of oxygen to give 2.0 L of nitrogen dioxide. In another example, 1.0 L of hydrogen gas reacts with 1.0 L of chlorine gas to form 2.0 L of hydrogen chloride gas.

Gay-Lussac discovered that a very simple relationship described the relationship of volumes observed for reacting gases. The *law of Gay-Lussac,* formulated in 1806, states that in reactions, *volumes of reacting gases measured at the same temperature and pressure can be expressed as small whole number ratios.*

GASES ARE COMPOSED OF MOLECULES

The simplicity of the relationship formulated by Gay-Lussac invites explanation. In 1811, Avogadro (1776–1856) *hypothesized that equal volumes of gases measured at the same pressure and temperature contain equal numbers of particles.* These particles are known as *molecules.* Avogadro's hypothesis together with measurements of volume in the reactions of gases provide the information needed to determine formulas for gaseous elements such as hydrogen and oxygen and gaseous compounds such as hydrogen chloride.

The relation of atoms to molecules can best be explored by considering an example of a reaction of gases. Consider the reaction of nitric oxide with oxygen to form nitrogen dioxide.

$$\underset{2.0\ \text{L}}{\text{Nitric oxide}} + \underset{1.0\ \text{L}}{\text{oxygen}} \rightarrow \underset{2.0\ \text{L}}{\text{nitrogen dioxide}}$$

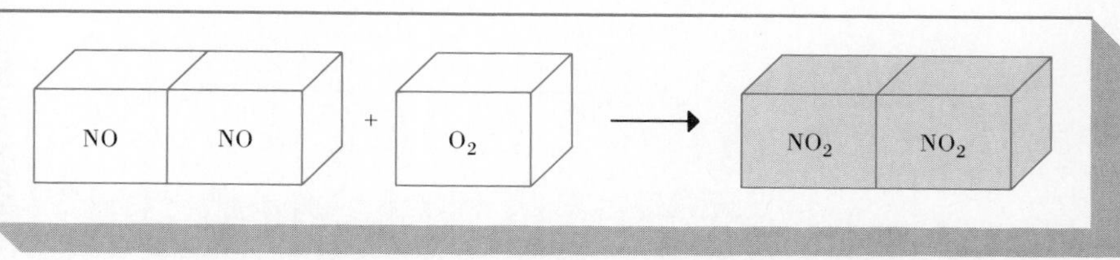

FIGURE 2.3
The formation of 2.0 L of NO_2 from 1.0 L of oxygen is evidence for the existence of oxygen molecules with the formula O_2.

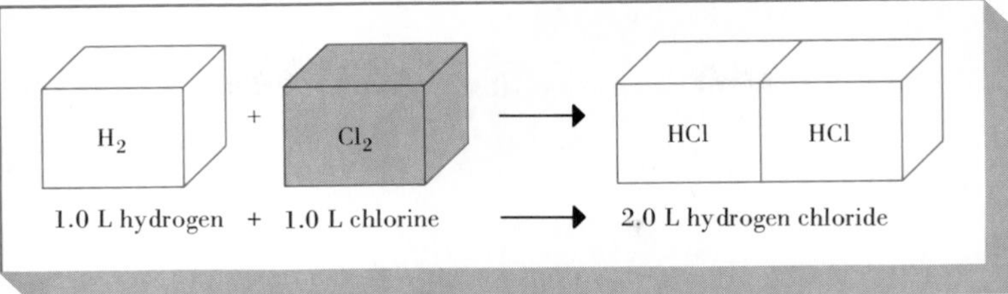

FIGURE 2.4
The formation of 2.0 L of hydrogen chloride gas from 1.0 L of hydrogen and 1.0 L of chlorine is evidence for the diatomic molecules H_2 and Cl_2. Each molecule of hydrogen and chlorine furnishes one atom to a molecule of HCl. Because hydrogen chloride occupies double the volume of hydrogen and chlorine, there are twice as many molecules of the compound as there are of either element.

According to Avogadro's hypothesis, there are equal numbers of nitric oxide and nitrogen dioxide molecules since the volumes of these two gases are equal. The number of molecules of oxygen is only half the number of molecules of either nitric oxide or nitrogen dioxide since the volume of oxygen is only half that of each compound. Each molecule of the reactant nitric oxide must gain at least one atom of oxygen in going to product. The atoms contained in one volume of oxygen are divided equally between two volumes of the product. Hence, each oxygen molecule must contain at least two atoms and must consist of an even number of atoms.

The reaction of hydrogen and chlorine to form hydrogen chloride provides evidence for molecules with the formulas H_2 and Cl_2. Each molecule of hydrogen chloride must have at least one atom of hydrogen and one atom of chlorine. The hydrogen atoms (combined with chlorine atoms) that go into 2 L of hydrogen chloride are contained in just 1 L of hydrogen gas. This requires hydrogen molecules to be *diatomic.* A similar argument indicates that chlorine must also be diatomic. The assumption that the smallest even integer 2 is correct in the formulas for molecules of oxygen, hydrogen, and chlorine, has been proven to be correct.

Avogadro. *(NBS Archives)*

The law of Gay-Lussac together with Avogadro's hypothesis can be applied to determine the formula of water. It is possible to surmount the practical difficulties present in the experiment. (Mixtures of oxygen and hydrogen may explode. To measure the volume of water as a gas requires a temperature above 100°C, the boiling point of water.) Experimentally, it is found that 2 L of hydrogen reacts with 1 L of oxygen to form 2 L of water vapor. The 2 L of hydrogen gas has the same number of hydrogen atoms as 2 L of water vapor. Each hydrogen molecule contributes its two hydrogen atoms to one molecule of water. Because 2 L of water vapor are formed from just 1 L of oxygen gas, one molecule of water has one-half the number of oxygen atoms as one molecule of oxygen gas. Since oxygen gas is diatomic, water must have the formula H_2O.

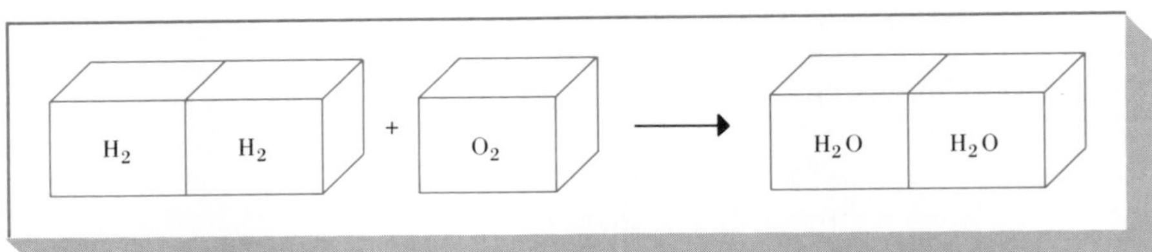

FIGURE 2.5
Since 2.0 L of H_2 combine with 1.0 L of O_2 to form 2.0 L of water vapor, water must have the formula H_2O. Each molecule of water has the same number of atoms as one molecule of hydrogen gas, and it has one-half the number of oxygen atoms as one molecule of oxygen gas.

SAMPLE CALCULATION

When 1.00 L of the gas methane, composed of the elements carbon and hydrogen, is burned, 1.00 L of carbon dioxide, CO_2, and 2.00 L of water vapor are formed. What is the formula for methane?

Because one volume of methane contains the same number of carbon atoms as are found in one volume of carbon dioxide CO_2, each molecule of methane contains one carbon atom. And inasmuch as each volume of methane contains the same number of hydrogen atoms as are found in two volumes of water, each methane molecule must contain four atoms of hydrogen. The formula of methane is CH_4.

CHEMICAL EQUATIONS DESCRIBE REACTIONS

Chemists write *chemical equations* to summarize chemical reactions. Chemical equations can be interpreted on a macroscopic or observational level. In a reaction of gases, coefficients appearing before formulas may represent relative volumes. (If a coefficient in a balanced equation is not given, it is assumed to be 1.) For example, 2 L of nitric oxide gas combines with 1 L of oxygen gas to form 2 L of nitrogen dioxide gas.

$$2\,NO + O_2 \rightarrow 2\,NO_2$$

On the microscopic level, equations represent molecular descriptions of reactions. For example, two molecules of nitric oxide react with one molecule of oxygen to form two molecules of nitrogen dioxide. Coefficients appearing before formulas represent relative numbers of molecules on a microscopic scale. Material is balanced, for there are equal numbers of nitrogen atoms and oxygen atoms on each side of the arrow in this equation.

The reactions presented in Figures 2.2 and 2.3 provide additional examples of chemical equations.

$$H_2 + Cl_2 \rightarrow 2\,HCl$$
$$2\,H_2 + O_2 \rightarrow 2\,H_2O$$

The use of chemical equations to calculate the masses of reactants and products involved in chemical reactions is elaborated in an Appendix.

RELATIVE ATOMIC WEIGHTS

Using combining weights of elements and the formulas of compounds, *relative atomic weights* can be calculated. With the correct formula H_2O and the weight ratio of oxygen to hydrogen in water, 8 : 1, it is possible to calculate the atomic weight of oxygen relative to hydrogen. Since one atom of oxygen weighs eight times as much as two atoms of hydrogen, an oxygen atom must be sixteen times as heavy as a hydrogen atom.

Chemists found it useful to have an agreed standard for expressing relative atomic weights. It would be possible to express atomic weights relative to bromine, or to sulfur, or to still another element. A standard element had to be selected and a value had to be assigned to that element.

Hydrogen was one possible candidate for that standard. It is the lightest element. If the relative atomic weight of hydrogen were exactly 1 then all other elements would have atomic weights greater than one. However, hydrogen proved to be a poor candidate in other ways. Few compounds containing only hydrogen and a metal were known. Of those known, some were very reactive and difficult to analyze. Since weighing chemical samples on a balance was the principal procedure used in the determination of chemical atomic weights, these shortcomings were major obstacles.

Chemists initially adopted a scale that compared the atomic weight of other elements to that of oxygen. Most elements form compounds with oxygen. The relative atomic weight of oxygen was defined to be 16. The value of 16 for oxygen had the advantage of assigning a value near 1 to the element hydrogen.

Using charge-to-mass measurements similar to those described in Chapter 3, physicists generated a different scale of relative weights. In 1961, a new, unified scale was chosen. The standard for the new scale is presented in Chapter 3. On this scale, the relative atomic weight of oxygen is 15.9994 and that of hydrogen is 1.0079.

DETERMINING ATOMIC WEIGHTS FOR METALLIC ELEMENTS

Chemists applied the law of Gay-Lussac together with Avogadro's hypothesis to determine atomic weights of gaseous elements, elements such as bromine and iodine that could readily be vaporized, and elements such as carbon that form a number of gaseous compounds. However these methods could not be applied to the determination of atomic weights for many metals.

Measurements of specific heats were used to help determine atomic weights of metallic elements. The *specific heat* of a substance is the quantity of heat required to increase the temperature of 1 g of the material 1°C. Specific heats can be measured using an ice calorimeter. When a piece of metal of known weight is taken from a boiling-water bath and dropped into a container of cracked ice, the metal cools to the temperature of the ice water and some ice melts. Water from the melted ice can be separated and weighed, and the specific heat of water is known. [The specific heat of water measured at 14°C is exactly 1 cal/(g · °C).] For a fixed weight of a metal, the larger its specific heat the greater the amount of ice that is melted.

Two French scientists Pierre DuLong (1785–1838) and Alexis Petit (1791–1820) observed that for many metallic elements, the product of their specific heat times their atomic weight is a constant. The *law of DuLong and Petit*, formulated in 1819, states that for a metal the specific heat of an element times its atomic weight (in grams) is approximately 6 cal/°C.

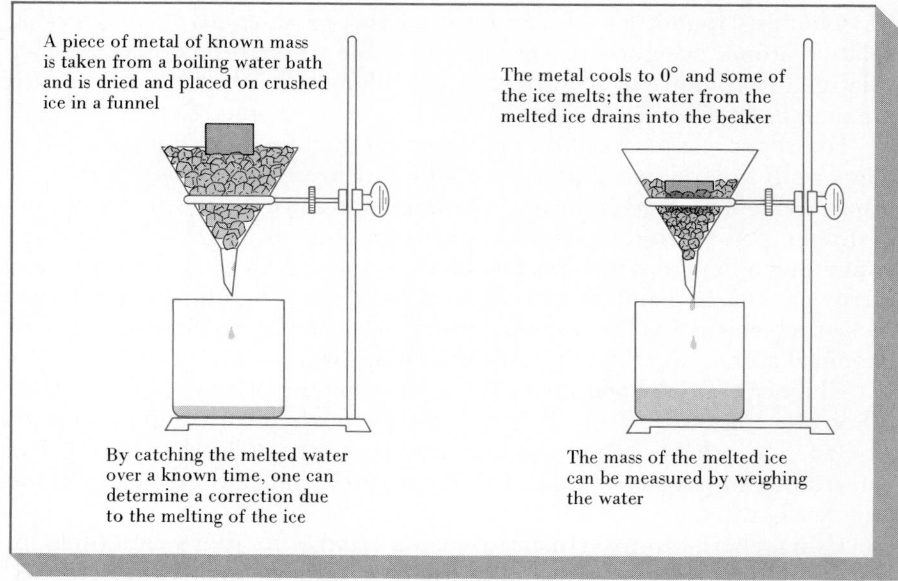

A piece of metal of known mass is taken from a boiling water bath and is dried and placed on crushed ice in a funnel
The metal cools to 0° and some of the ice melts; the water from the melted ice drains into the beaker
By catching the melted water over a known time, one can determine a correction due to the melting of the ice
The mass of the melted ice can be measured by weighing the water

FIGURE 2.6
An ice calorimeter can be used to determine the heat capacity of metals. To melt 1 g of ice, 80 cal are required.

The relation described by DuLong and Petit is only approximate and does not apply to all solid elements. However it did enable early chemists to correct atomic weights calculated for copper, zinc, nickel, and iron. Early atomic weights for these elements had been calculated from analyses of their oxides by assuming a formula MO_2 for each oxide. When the relative atomic weights were recalculated by assuming the correct formulas were MO, the calculated atomic weights of the metals were in accord with the law of DuLong and Petit.

SAMPLE CALCULATION

In a determination of the atomic weight for copper reported in a 1826 table of atomic weights, it was found that 63.31 g of copper combined with 16 g of oxygen in copper oxide. What was the atomic weight of copper?

TABLE 2.1

DATA ILLUSTRATING THE LAW OF DuLONG AND PETIT

Metal	Atomic Weight (AW)	Specific Heat (SH) cal/°C	AW × SH cal/°C
Aluminum	27.0	0.215	5.81
Iron	55.8	0.1075	6.00
Copper	63.5	0.0924	5.87
Silver	107.9	0.0562	6.06
Gold	197.0	0.0308	6.07
Lead	207.2	0.0305	6.32

Assumed Formula	Atomic Weight of Copper
CuO	63.31
CuO_2	126.62

The approximate atomic weight of copper was found by applying the law of DuLong and Petit.

$$AW \times SH = 6.0$$
$$AW = 6.0/SH = 6.0/0.0924$$
$$= 65$$

The approximate weight permits the formula CuO to be assigned. Therefore, the atomic weight of copper was determined to be 63.31. The accepted value today is 63.546.

MEASURING QUANTITIES IN CHEMICAL REACTIONS, THE MOLE CONCEPT

Chemists use a chemical counting unit, the mole, that can be measured with a balance. A mole of atoms of an element is the mass in grams equal to the relative atomic weight of the element. A mole of hydrogen atoms is 1.008 g; a mole of oxygen atoms is 16 g. A mole of atoms of every element contains the same very large number of atoms.

A mole of a particular compound is the quantity having a weight in grams equal to the formula mass. The formula weight of a compound is numerically equal to the sum of relative atomic weights of the atoms in the formula unit. (If the compound is known to consist of molecules, the formula weight is usually referred to as the molecular weight.)

The weight of 1 mol of hydrogen gas, H_2, is

$$1 \text{ mol } H_2 \frac{2 \text{ mol H}}{1 \text{ mol } H_2} \frac{1.008 \text{ g}}{1 \text{ mol H}} = 2.016 \text{ g}$$

The molecular weight of water, H_2O, is

$$\frac{2 \text{ mol H}}{1 \text{ mol } H_2O} \frac{1.008 \text{ g}}{1 \text{ mol H}} + \frac{1 \text{ mol O}}{1 \text{ mol } H_2O} \frac{16 \text{ g}}{1 \text{ mol O}} = 18.02 \text{ g}$$

SAMPLE CALCULATION

Calculate the mass of 1 mol of calcium carbonate, $CaCO_3$.

$$1 \text{ mol Ca} \frac{40 \text{ g}}{1 \text{ mol Ca}} + 1 \text{ mol C} \frac{12 \text{ g}}{1 \text{ mol C}} + 3 \text{ mol O} \frac{16 \text{ g}}{1 \text{ mol O}}$$
$$= 40 + 12 + 48 \text{ g}$$
$$1 \text{ mol of } CaCO_3 = 100 \text{ g}$$

A balanced chemical equation can be interpreted either as a statement about atoms and molecules or as a statement about moles in chemical reactions. In the balanced equation for the formation of carbon dioxide from carbon monoxide and oxygen, 2 mol of carbon monoxide react with 1 mol of oxygen to give 2 mol of carbon dioxide.

$$2\ CO + O_2 \rightarrow 2\ CO_2$$

GROUPS OF ELEMENTS WITH SIMILAR CHEMISTRY

Chemists classify most elements as either *metals* or *nonmetals.* Metals are shiny solids that conduct heat and electricity well. Mercury is an exception; it is the only metal that is liquid at room temperature. When combined with one another, metals form alloys that may vary widely in composition. Most nonmetals, including all the gaseous elements, do not conduct electricity. Nonmetals form compounds, having fixed composition, with both metallic and nonmetallic elements.

As observations accumulated from various laboratories, a body of descriptive chemistry of elements and their compounds grew. Strong chemical and physical similarities were noted for small groups of elements. Members of these groups or *chemical families* have very different atomic weights. (In contrast, elements with atomic weights close to one another often have very different chemical properties.)

The earliest chemical family to be recognized may be the *coinage metals,* copper, silver, and gold. The coinage metals are among the very few metals to be found uncombined in nature. Artisans have shaped these lustrous metals into jewelry and vessels for religious celebration. From the time of the pharaohs, people have valued these metals.

In contrast, the *alkali metals,* lithium, sodium, and potassium, are soft, low-melting solids that react violently with water to form hydrogen gas and basic (alkaline) solutions. Compounds of these elements are crystalline solids that dissolve readily in water. The chlorides of these elements have the formulas LiCl, NaCl, and KCl.

The *halogens* are nonmetals with strikingly similar chemistry. All are diatomic molecules in the elemental state. Fluorine is a pale yellow gas, and chlorine is a pale green gas. Bromine is a red-brown liquid that has a low boiling point. Iodine, an almost black crystalline solid, readily sublimes to a violet vapor. The heavier the halogen, the deeper its color. The lightest halogen is a gas; the heaviest, a solid. The hydrogen halides, HF, HCl, HBr, and HI, are gases that dissolve in water to form acidic solutions.

The elements carbon, silicon, germanium, tin, and lead form another chemical family. This group of elements shows greater differences and fewer similarities than the coinage metals, the alkali metals, or the halogens. The lower atomic weight elements, carbon and silicon, are nonmetals. The higher atomic weight elements, tin and lead, are metals. There are similarities, however. The two oxides of carbon, carbon monoxide, CO, and carbon dioxide, CO_2, and the two oxides of lead, PbO and PbO_2, have similar formulas. In contrast with the high-melting halides of the alkali metals, the halides of elements in this family are volatile liquids. Carbon tetrachloride, CCl_4, boils at 77°C; tin tetrachloride, $SnCl_4$, boils at 114°C.

MENDELEEV'S PERIODIC TABLE

During the period 1869–1871, Dmitri Mendeleev (1834–1907) formulated a *periodic table* to account for similarities in the chemistry of elements. Known *elements were placed in rows in order of increasing atomic weight.* Mendeleev selected the length of rows so that *elements in a chemical family were placed in the same column.* The elements of two rows of Mendeleev's table are shown below with the relative atomic weights given to the nearest integer value.

Li = 7	Be = 9	B = 11	C = 12	N = 14	O = 16	F = 19
Na = 23	Mg = 24	Al = 27	Si = 28	P = 31	S = 32	Cl = 35

This arrangement of elements is repeated below showing the formulas of compounds of these elements with chlorine and with hydrogen. Note

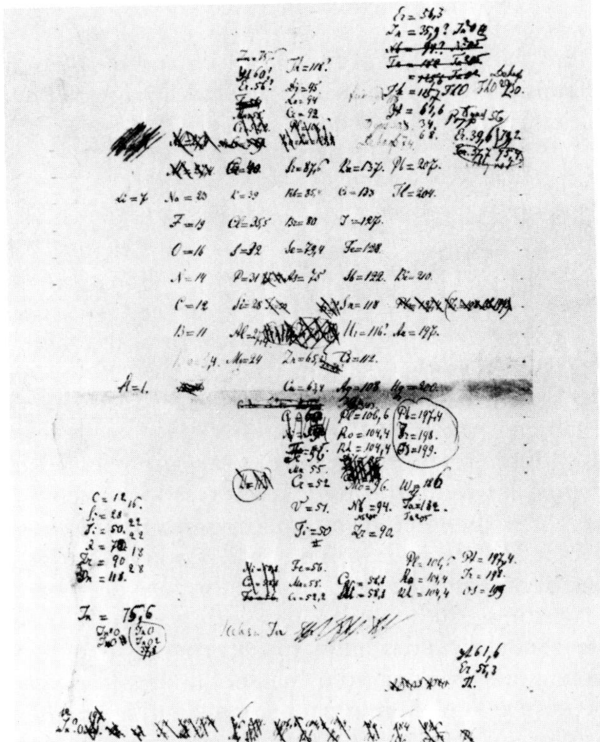

Mendeleev's periodic table. *(Burndy Library.)*

Tabelle II.

Reihen	Gruppe I. — R^2O	Gruppe II. — RO	Gruppe III. — R^2O^3	Gruppe IV. RH^4 RO^2	Gruppe V. RH^3 R^2O^5	Gruppe VI. RH^2 RO^3	Gruppe VII. RH R^2O^7	Gruppe VIII. — RO^4
1	H = 1							
2	Li = 7	Be = 9,4	B = 11	C = 12	N = 14	O = 16	F = 19	
3	Na = 23	Mg = 24	Al = 27,3	Si = 28	P = 31	S = 32	Cl = 35,5	
4	K = 39	Ca = 40	— = 44	Ti = 48	V = 51	Cr = 52	Mn = 55	Fe = 56, Co = 59, Ni = 59, Cu = 63.
5	(Cu = 63)	Zn = 65	— = 68	— = 72	As = 75	Se = 78	Br = 80	
6	Rb = 85	Sr = 87	?Yt = 88	Zr = 90	Nb = 94	Mo = 96	— = 100	Ru = 104, Rh = 104, Pd = 106, Ag = 108.
7	(Ag = 108)	Cd = 112	In = 113	Sn = 118	Sb = 122	Te = 125	J = 127	
8	Cs = 133	Ba = 137	?Di = 138	?Ce = 140	—	—	—	— — — —
9	(—)	—	—	—	—	—	—	
10	—	—	?Er = 178	?La = 180	Ta = 182	W = 184	—	Os = 195, Ir = 197, Pt = 198, Au = 199.
11	(Au = 199)	Hg = 200	Tl = 204	Pb = 207	Bi = 208	—	—	
12	—	—	—	Tb = 231	—	U = 240	—	— — — —

(From Annalen der Chemie, supplemental vol. 8, 1872.)

FIGURE 2.7
Mendeleev's horizontal periodic table of 1871. (From Annalen der Chemie, supplemental vol. 8, 1872.)

that the formulas of the chlorides and of the hydrides are characteristic of families. Note also that nonmetals are found on the right in these rows and metals are on the left.

LiCl	$BeCl_2$	BCl_3	CCl_4			
NaCl	$MgCl_2$	$AlCl_3$	$SiCl_4$			
			CH_4	NH_3	H_2O	HF
			SiH_4	PH_3	H_2S	HCl

One version of Mendeleev's periodic table is shown. Of the families discussed in this chapter, the coinage metals and the alkali metals are in column I, the family containing carbon and lead is in column IV, and the halogens are found in column VII.

If the relationships were as apparent as those shown thus far, one might wonder why the periodic law was so long in coming. Proceeding by trial and error with elements ranked in order of increasing atomic weight and with their chemistry described on 3 × 5 index cards, one would soon discover a periodic arrangement. In part, the answer is that in rows starting with Li and Na, the periodic similarities within a family and differences between families seem most dramatic. In later rows, Mendeleev had to place three similar elements, for example the metals iron, cobalt, and nickel, in an eighth column. The chemistry of some elements in the same column also differed in important ways indicating, with hindsight, that two families may be in

Metals
Nonmetals
Li
Na
K
Rb
Cs
Fr
Alkali metals
Cu
Ag
Au
Coinage metals
C
Si
Ge
Sn
Pb
F
Cl
Br
I
At
Halogens

FIGURE 2.8
The outline of the modern long version of the periodic table is shown with the symbols of the alkali metals, the coinage metals, the halogens, and the family that runs from carbon to lead. The heavy jagged diagonal line separates the metals on the left from the nonmetals on the right. A full version of the periodic table is shown inside the front cover.

each column. Mendeleev indicated that the chemistry of the metal manganese was both similar and different from that of the halogens by placing Mn with a predicted heavier metal on the left side of column VII and placing Cl, Br, and I on the right.

Current versions of the periodic table avoid these difficulties by having longer horizontal rows, and rows of different lengths. In these versions of the periodic table, metals are found on the left and nonmetals are found on the right. Until very recently, the numbering of columns in the long table reflected that of Mendeleev. For example, group VII of Mendeleev's table was split into two columns: VIIA containing Mn, and VIIB containing the halogens.

A greater understanding of the achievement represented by the development of a periodic law can be gained by looking at the problems Mendeleev surmounted. Not all the naturally occurring elements had been discovered. For some elements, reported atomic weights were in disagreement or in error. Mendeleev applied the law of DuLong and Petit to correct the atomic weights of some elements reported in the literature.

Mendeleev. *(National Portrait Gallery, Smithsonian Institution, Washington, D.C.)*

Mendeleev did not always order elements according to increasing atomic weights. For some pairs of adjacent elements, he reversed their order, asserting that the reported atomic weights must be wrong. The pair of elements tellurium and iodine are an example of an inversion cited by Mendeleev. Although iodine was reported to have a lower atomic weight than tellurium, Mendeleev placed iodine in the halogen column, and he put tellurium, which forms H_2Te and Na_2Te, under oxygen. (Mendeleev's assignment of family relationships for iodine and tellurium was correct although the atomic weight of iodine is, indeed, less than that of tellurium.)

More dramatically, Mendeleev proposed the existence of new elements and predicted their properties. To obtain correct family placement of heav-

Property	Eka Silicon	Germanium
Atomic weight	72	72.3
Density, g/cm^3	5.5	5.47
Specific heat, cal/(g · °C)	0.073	0.076
Boiling point of $GeCl_4$ °C	Under 100	86

ier elements, Mendeleev left four gaps in his periodic tables. Three of these gaps were filled by elements discovered during the following 15 years. An illustration of the power of the periodic table to correlate and predict chemical behavior is the prediction and, then, observation for the element germanium. Mendeleev asserted the existence of an element, eka silicon, to lie below silicon and above tin in column IV of his periodic table. Its properties and those of its compounds would be intermediate between those of silicon and tin and their compounds. When this element was discovered, it was named germanium.

Mendeleev's periodic table was provocative for it correlated much chemical information, yet the underlying basis of the correlation was elusive. Why should the ranking of elements in order of increasing atomic weight lead to an arrangement in which elements of similar chemistry are listed in columns in a table? As scientists learned more about the nature of atoms, they developed a more satisfying rationale for the periodic table. An explanation for chemical periodicity based on atomic structure is developed in Chapter 4.

QUESTIONS

1. Dalton assumed that water had the formula HO. What formula would he have attributed to hydrogen peroxide, H_2O_2, that would be consistent with his assumption?
2. A 2 L amount of the gas ammonia is produced by the reaction of 1 L of nitrogen and 3 L of hydrogen, H_2. What is the simplest formula for nitrogen? For ammonia?
3. Isotopes of uranium have been separated by the diffusion of a gaseous uranium fluoride, UF_x. A 1 L amount of UF_x can be prepared by reacting 3 L of fluorine gas with solid uranium. What is the simplest formula for this uranium fluoride? If fluorine is diatomic, what is the formula for uranium fluoride?
4. A 1 L amount of a hydrocarbon C_xH_y burns completely with oxygen to form 2 L of CO_2 and 2 L of water vapor. What is the formula for the hydrocarbon?
5. Balance the following equations by supplying the missing coefficient.
   (a) $C_3H_8 + ?\ O_2 \rightarrow 3\ CO_2 + 4\ H_2O$
   (b) $C_6H_6 + \frac{15}{2}\ O_2 \rightarrow ?\ CO_2 + ?\ H_2O$
   (c) $H_2 + Br_2 \rightarrow ?\ HBr$

6. When 1.00 L of ammonia gas, NH_3, is mixed with 1.00 L of gaseous hydrogen chloride, HCl, all the gas reacts and a white solid is produced. What is the formula for the white solid?
7. When a direct electric current is passed through a dilute aqueous salt solution, the water reacts to form hydrogen gas at one electrode and oxygen gas at the other. How many milliliters of hydrogen form for every 25 mL of O_2?
8. A confused chemistry student asks you the difference between an atom and a molecule. How would you explain the difference? What experimental evidence would you cite for atoms? For molecules?
9. Why would the confusion of atoms and molecules make it difficult to determine accurate atomic weights?
10. The heat capacity of the element lead is one-fourth that of iron. What is the approximate atomic weight of lead if the atomic weight of iron is about 56?
11. A metal oxide with the formula MO contains 4.09 g of the metal for every gram of oxygen. Calculate the relative atomic weight of the metal on a scale with O = 16. Use the periodic table to identify the metal.
12. Use the information in Table 2.1 to estimate approximate values for the heat capacity of lithium (AW = 6.94) and tin (AW = 118.7).
13. Calculate the mass of one mole of each of the following compounds.
    (a) SO_2
    (b) SO_3
    (c) CaO
    (d) $CaSO_4$
14. How many moles of oxygen are required to burn 1 mol of each of the following hydrocarbons?
    (a) $CH_4 + ?\ O_2 \rightarrow CO_2 + 2\ H_2O$
    (b) $C_7H_8 + ?\ O_2 \rightarrow 7\ CO_2 + 4\ H_2O$
15. On what basis did Mendeleev group elements into families? On what basis did he order elements across the periodic table? Give some specific examples of periodic similarities and differences in your answer.
16. Use the periodic table to write formulas for the compounds formed between the following pairs of elements.
    (a) Potassium, K, and bromine, Br
    (b) Barium, Ba, and chlorine, Cl
    (c) Gallium, Ga, and fluorine, F
    (d) Silicon, Si, and iodine, I
17. Use the periodic table to predict the formulas of the compound formed between hydrogen and each of the following elements.
    (a) Arsenic, As
    (b) Germanium, Ge
    (c) Selenium, Se
    (d) Iodine, I
18. Selenium is in the same family of the periodic table as oxygen. Which of the following properties of selenium can be predicted on the basis of family relationships? If a property can be predicted, give a reason why.

(a) Small amounts of selenium are essential to a good diet.
(b) Hydrogen and selenium combine to form a compound H_2Se.
(c) Selenium is more dense than sulfur, S, and less dense than tellurium, Te.

19. Mendeleev predicted the existence of eka boron and eka aluminum, elements to fill empty positions below B and Al in the periodic table. What are the symbols and names of these elements?
20. Which of the following elements has chemistry most similar to antimony Sb? Briefly justify your answer.
    (a) Tin, Sn
    (b) Bismuth, Bi
    (c) Nitrogen, N
    (d) Tellurium, Te
21. What does the work of Mendeleev suggest about the role of scientists' belief about order in nature?

CHAPTER

THREE

RADIOACTIVITY AND THE STRUCTURE OF ATOMS

The pace of discovery in physics and chemistry quickened as the nineteenth century drew to a close. Growing knowledge of the electrical nature of matter together with new discoveries in physics provided a background for an inquiry into the structure of the atom.

In 1896, Henri Becquerel (1852–1908) discovered radioactivity. Following his discovery, scientists found that atoms of radioactive elements eject charged particles to become atoms of different elements. The exploration of the properties of charged particles and of the interaction of radiation with matter soon led to the discovery of subatomic particles and to the formulation of a nuclear model for atomic structure.

THE DISCOVERY OF RADIOACTIVITY

The discovery of radioactivity began with a false hypothesis. When fluorescent minerals are illuminated with ultraviolet light, they emit visible light. Since x-rays also cause some compounds to glow or fluoresce, Becquerel thought that fluorescence might be related to newly discovered x-rays; in particular, he believed the x-ray emission might accompany fluorescence.

To test this hypothesis, Becquerel used the observation that x-rays penetrate matter and cause film to become fogged. (Physicians routinely use x-ray pictures to examine broken bones, and dentists use x-rays to examine teeth.) Becquerel placed fluorescent materials over photographic plates wrapped in black paper. He exposed the materials to light so they fluoresced, and then he developed the plates to look for evidence of x-rays. Potassium uranyl sulfate, a salt of uranium, caused the plates to appear exposed, but most other fluorescent substances did not cause this appearance.

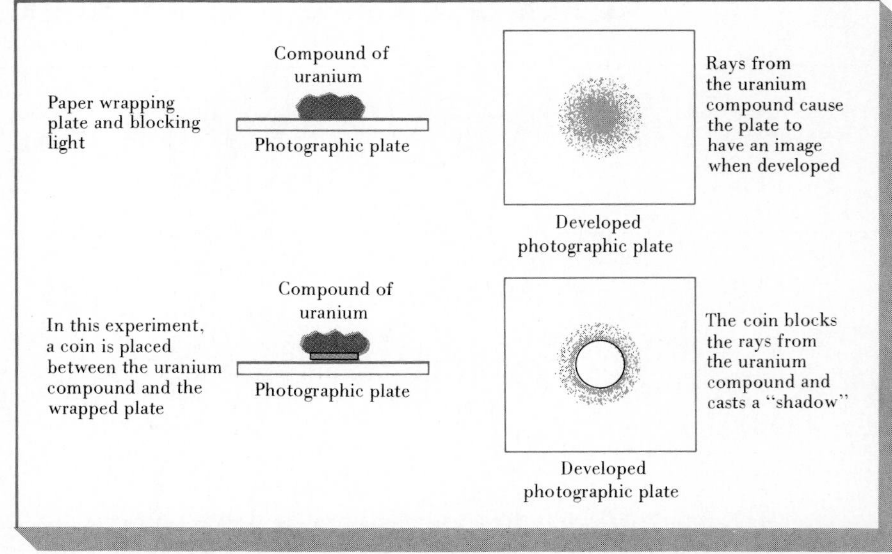

FIGURE 3.1
Henri Becquerel discovered that radiation from compounds of uranium could pass through paper and leave an image on photographic plates. These rays did not penetrate a coin so that a "shadow" could be cast by blocking some of the radiation from the uranium.

Marie Curie. *(New York Public Library.)*

A chance observation produced evidence that led to the discovery of radioactivity. During a period of cloudy weather, some wrapped photographic plates and the fluorescent salt of uranium lay together in a drawer. This time the compound had not been exposed to light, but Becquerel developed the plates just the same. He found a characteristic fogged halo below where the compound had been. When materials were placed between the compound and plates, they shielded the plates from exposure and left shadows. To account for these observations, Becquerel postulated that radiation from uranium caused the plates to fog.

Samples of pitchblende, an ore of uranium, emit more radiation than do pure uranium compounds. Marie Curie (1867 – 1934) attributed the radia-

FIGURE 3.2
Marie Curie separated traces of highly radioactive radium from pitchblende by precipitating it with a salt of barium, an element in the same family of the periodic table. From several tons of the ore, she obtained less than 0.1 g of radium chloride following repeated recrystallizations.

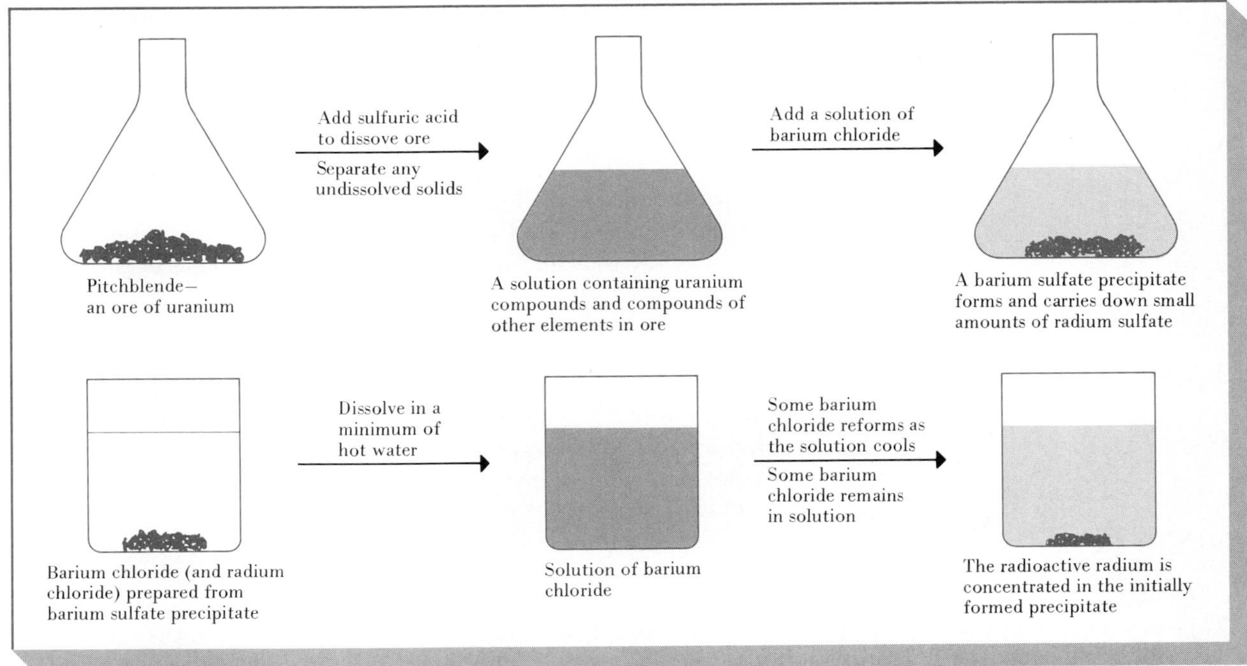

tion in pitchblende to the presence of one or more new radioactive elements, and, with her husband Pierre (1859–1906), she sought to isolate them. The amounts of these radioactive elements were too small to separate by themselves. Her plan was to precipitate insoluble compounds of known elements that might carry the new elements with them.

Marie Curie dissolved the pitchblende in acid, then added compounds of known elements to the solution. When she precipitated these "carrier" elements, some samples contained increased portions of radioactivity. (Radioactive elements in the pitchblende separated with carrier elements of the same family in the periodic table, for they have similar chemistry.) By redissolving radioactive samples and reisolating portions by precipitation, she obtained fractions richer in radioactivity. In 1898, the Curies isolated both the radioactive element polonium and a salt of radium, an element so radioactive that it glows in the dark. Becquerel and the Curies shared a Nobel prize in 1903.

THE ELECTRICAL NATURE OF MATTER

Solutions of salt in water conduct electricity. This can be demonstrated using a light bulb in an electric circuit that has two leads immersed in water. When the circuit is connected to a power source, no current flows if the water is pure, for water acts as a large resistance to the flow of electricity. But when common table salt, sodium chloride, is dissolved in the water, current flows and the bulb lights up.

In solutions that conduct electricity, there are carriers of electric charge called *ions.* Positive ions are called *cations,* and negative ions are called *anions.* In a sodium chloride solution the carriers of electric charge are the sodium cations and the chloride anions. When an external voltage is applied, cations move toward the negative electrode, and anions move toward the positive electrode.

When a direct current is passed through molten sodium chloride, sodium metal is produced at one electrode, and chlorine gas is formed at the other. Positive sodium ions gain negative charges to form metal at one electrode, and chloride ions lose negative charges to form gas at the other electrode. (In 1807, Sir Humphrey Davy had first isolated the alkali metals, sodium and potassium, by passing an electric current through fused samples of compounds of these metals.)

These experiments demonstrate the electrical properties of some chemical compounds and the electrical nature of some chemical reactions. To further understand the charged particles found in some chemical compounds and the electrical nature of some chemical reactions, it is first necessary to consider the structure of the atom.

SUBATOMIC PARTICLES AND THE NUCLEAR ATOM

Experiments in physics at the beginning of the twentieth century dramatically changed scientists' model of the atom. From the concept of an indivisible atom postulated by Dalton, scientists moved to a picture of the atom

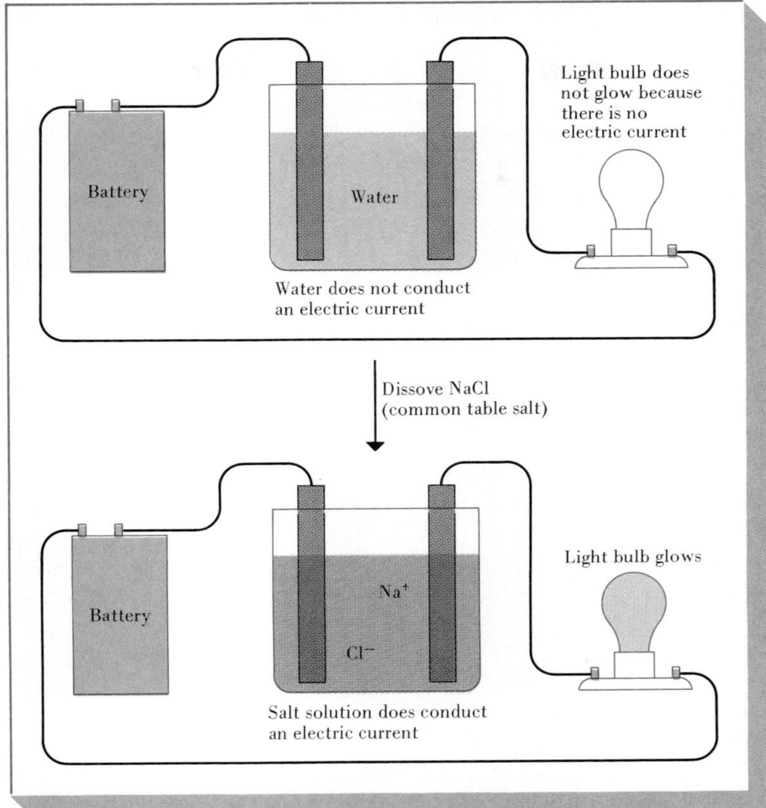

FIGURE 3.3
When common salt is dissolved in water, the resulting solution can conduct an electric current. Positive sodium ions and negative chloride ions are the carriers of electric charge in solution.

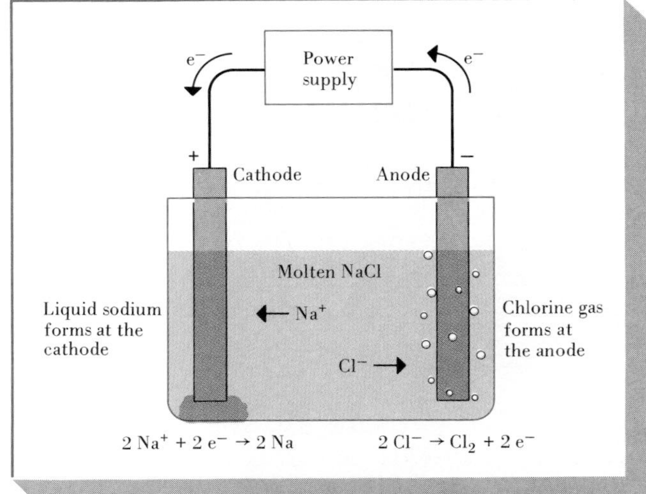

FIGURE 3.4
When a direct current is passed through molten sodium chloride, sodium metal is formed at one electrode and chlorine gas is formed at the other. Positive sodium ions gain negative charges to form metal, and negative chloride ions lose negative charges to form chlorine gas.

consisting of charged subatomic particles. The resultant model of the atom is so important for an understanding of bonding and reactions in chemistry that it is desirable to understand its experimental basis.

Consider the design of three basic experiments. In each of these experiments, scientists interpreted observations to infer properties of subatomic particles. These experiments are simple in concept but rely on a knowledge of physics that is beyond the background of most students taking an introductory chemistry course. One is able to connect each concept to its experimental basis without going into the physics used in calculations. In a qualitative way, we can understand the model of the atom developed by physicists near the turn of the century.

ELECTRONS

In 1897, J. J. Thomson (1856–1940) discovered the electron in the course of his investigation of cathode rays. (Cathode ray tubes are ancestors and cousins of the picture tubes used in television sets.) When a high voltage is imposed across a vacuum tube containing electrodes, the end of the tube farthest away from the cathode glows. If this end is then coated with a phosphorescent material, as it is in picture tubes, the glow is greatly en-

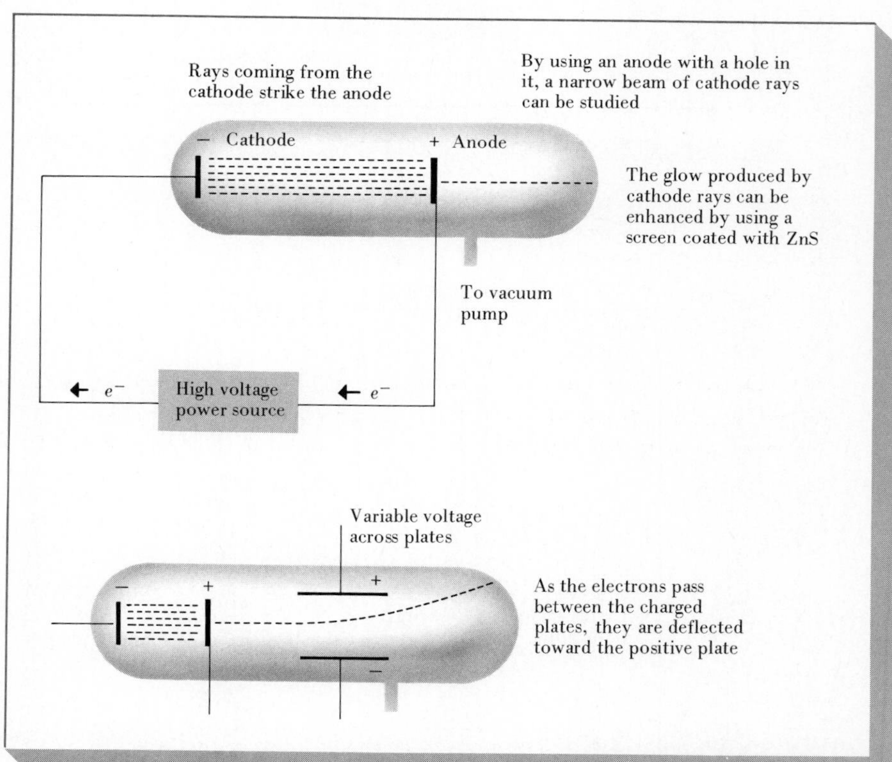

FIGURE 3.5
When a high voltage is applied to an evacuated tube, cathode rays are produced. Scientists showed that cathode rays were composed of negative particles called electrons.

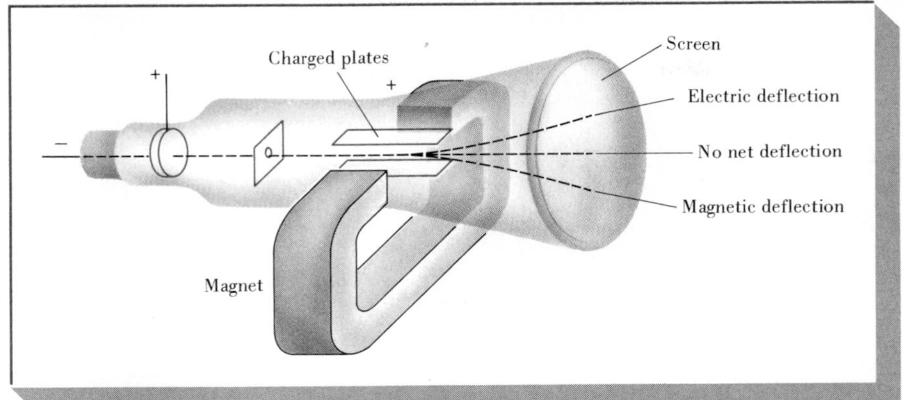

FIGURE 3.6
J. J. Thomson measured the charge-to-mass ratio, e/m, for the electron. A moving charge is deflected in a magnetic field. Thomson adjusted the voltage across the plates so that the electric force and the magnetic force on the electrons were equal and there was no net deflection. He also measured the deflection in the electric field alone. From these two measurements, he could calculate the value of e/m.

hanced. Objects placed between the cathode and the screen cast shadows indicating that the glow is caused by rays emanating from the cathode.

Thomson studied the nature of cathode rays. By using an anode with a slit in it, he could block most of the cathode rays and observe the position of a glowing spot at the end of the tube. When the narrow beam was passed through an electric field imposed between two plates, the spot was deflected. This indicated that the rays were charged particles rather than a form of light. Because the direction of the deflection was toward the positive plate, Thomson knew he was studying negatively charged particles. These negative particles are called *electrons.*

Thomson determined the charge-to-mass ratio e/m of the electron. He measured the deflection of a beam of electrons by an electric field of known strength. The amount of deflection depended on the charge of the electron, the mass of the electron, and the velocity of the electron since, for a given force an object is deflected less if it is heavier or if it is moving faster. Thomson then used a magnet to deflect the electrons in the direction opposite to that caused by the electric field. When the electric and magnetic forces acting on the electrons were made equal in strength, no deflection of the beam occurred. With this additional measurement, Thomson could calculate the value of e/m for the electron.

DETERMINING THE CHARGE OF THE ELECTRON

In 1908, Robert Millikan (1868–1953) measured the charge on the electron by studying the movement of charged oil droplets in an electric field. In the absence of a field, the tiny droplets fell at a constant speed. The downward gravitational force was offset by a frictional force, as is the case for a falling man suspended from a parachute. Millikan used a radioactive source to generate ions, and put negative charges on some of the oil drops. In an electric field, the charged drops rose at a constant speed with an electric force, an altered frictional force, and a gravitational force acting on them.

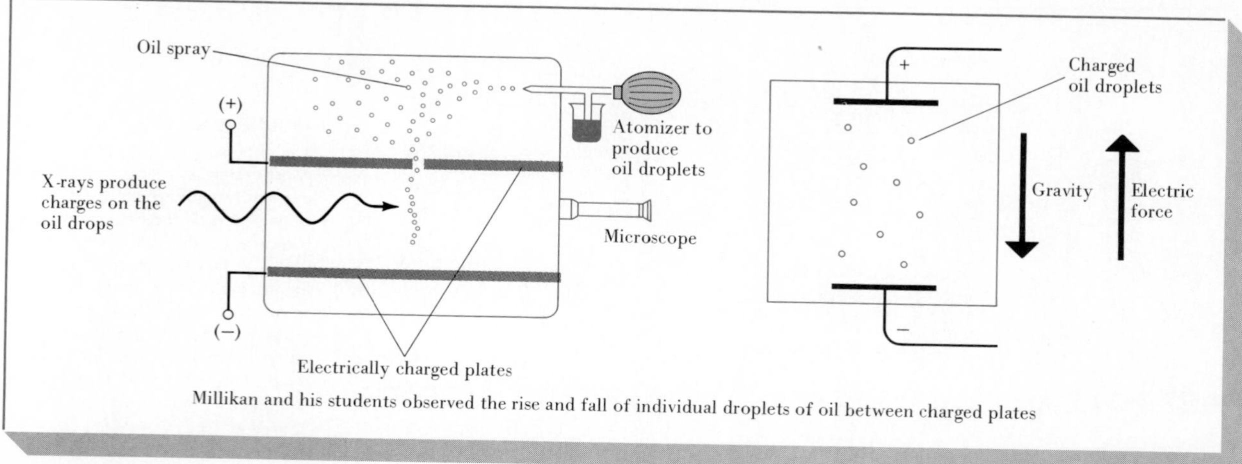

FIGURE 3.7
Robert Millikan determined the charge on a single electron by studying the motion of droplets of oil in an electric field. From the rate of fall in the absence of the field, he could calculate the size of an individual drop. From the rate of its rise in an electric field, he could then calculate the charge on the drop. All of the measured charges on individual drops could be expressed as an integer times the smallest unit of charge. That smallest unit of charge is the charge on a single electron.

By turning the voltage on and off repeatedly, Millikan and his students could observe individual drops rise and fall. The size of any drop could be calculated from the rate at which it fell. The charge on any drop could be calculated from its size and the rate at which it rose. By determining the charges on many drops, Millikan accumulated a table of charges.

Millikan found that the measured charges could all be expressed as multiples of a single charge. This largest common divisor of the charges was the smallest unit of charge, the charge on a single electron. That charge e is 1.602×10^{-19} C. Since the ratio of charge-to-mass, e/m, is known from Thomson's experiment, it is also possible to calculate the mass of an electron. The mass of an electron is $\frac{1}{1840}$ times the mass of the lightest atom, hydrogen.

THE NUCLEAR ATOM

Are electrons found in atoms? Scientists explored the consequence of this assumption. If atoms do contain negative electrons, they must also contain positive particles, for atoms are neutral. Because an atom is much heavier than its electrons, these positive particles must be much heavier than electrons.

One can ask, what is the distribution of these particles in space? Are negative and positive particles distributed randomly, are they spread out evenly throughout an atom, or are they clustered? Ernest Rutherford (1871–1937) provided the answer to this question by studying the deflection of *alpha particles,* products of radioactive decay, by a very thin gold foil. In 1911 he proposed the nuclear atom to account for his observations.

Rutherford studied the interactions of the most massive known radiation with gold foil, the thinnest possible solid. *Alpha particles are high-speed*

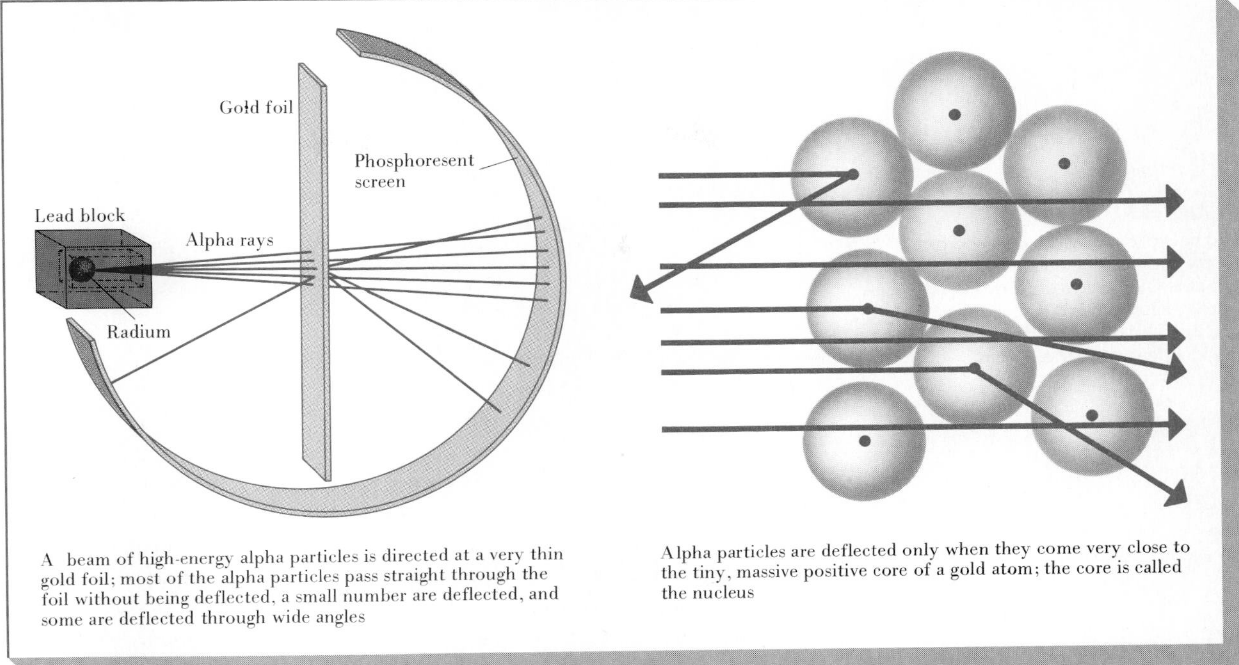

Gold foil
Phosphoresent screen
Lead block
Alpha rays
Radium
A beam of high-energy alpha particles is directed at a very thin gold foil; most of the alpha particles pass straight through the foil without being deflected, a small number are deflected, and some are deflected through wide angles
Alpha particles are deflected only when they come very close to the tiny, massive positive core of a gold atom; the core is called the nucleus

FIGURE 3.8
Ernest Rutherford proposed a nuclear model of atomic structure to account for the deflection of alpha particles by a thin gold foil.

positive particles. Each is about four times as heavy as a hydrogen atom and about 7000 times as heavy as an electron. Since a collision of an alpha particle with an electron would resemble that of a high-speed bowling ball with a ping pong ball, the path of an alpha particle would scarcely be perturbed by interactions with electrons in the foil. Alpha particles serve as a probe to measure the distribution of positive charge and mass.

Rutherford measured the number of alpha particles deflected by the foil as a function of the angle of deflection. Most alpha particles passed through the foil undeflected. A small fraction were deflected, some through large angles. Some came back toward the alpha emitter source. Rutherford later commented, "It was about as credible as if you had fired a 15-in shell at a piece of tissue paper and it came back and hit you."

To account for the large deflections, Rutherford proposed a model for the atom which consists of a tiny, massive, highly charged positive core called the *nucleus* that is surrounded by electrons. Since most alpha particles were not deflected, Rutherford concluded that most of the gold foil consisted of empty space or space occupied only by electrons. When infrequent collisions did occur, rebounding alpha particles were deflected through large angles both because there were large forces in the collisions and because massive gold nuclei recoil far less than lighter alpha particles. The large forces were due to the repulsion of the positive alpha particles by the large positive charge concentrated in the very small nucleus.

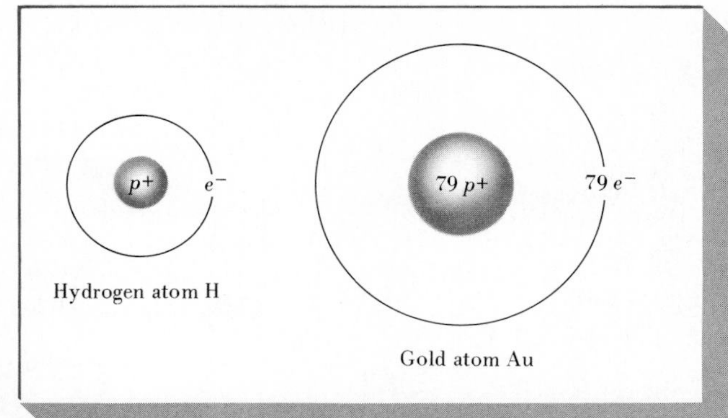

FIGURE 3.9
In a neutral atom the positive charge on the nucleus is balanced by the negative charge of electrons. A hydrogen atom has a +1 nucleus and one electron. A gold atom has a +79 nucleus and 79 electrons.

ATOMIC STRUCTURE

After Rutherford's formulation of a nuclear model for the structure of atoms, scientists made additional discoveries that enabled them to relate differences between elements to their different nuclear structures. Nuclei are composed of subatomic building blocks called *protons* and *neutrons. A proton is a positive particle with a charge equal in magnitude to the negative charge on an electron. A neutron is an uncharged particle with a mass close to that of a proton.* Both the proton and the neutron are about 2000 times as heavy as an electron. (Both protons and neutrons are called *nucleons.*)

All atoms of a given element have the same number of protons in their nuclei. The number of protons in an atom of an element, known as the *atomic number* of that element, is identical to the ranking of that element in the periodic table. (The atomic weight of elements, the quantity used by Mendeleev to rank elements, usually increases in the order of increasing numbers of protons in the nucleus.)

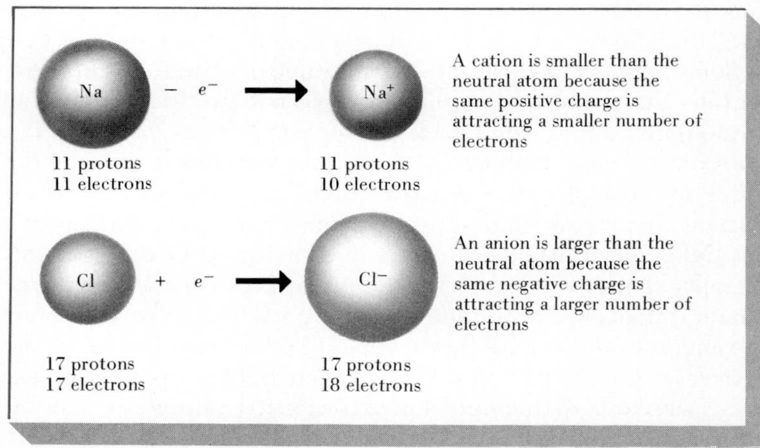

FIGURE 3.10
Cations are formed from atoms by the loss of one or more electrons. Anions are formed from atoms by the gain of one or more electrons.

In a neutral atom, the positive charge on the nucleus is balanced by the negative charge of electrons about the nucleus. An atom of hydrogen, the lightest element, has a single proton and a single electron. An atom of gold has 79 electrons encircling a nucleus that contains 79 protons.

The description of an atom consisting of a positive nucleus surrounded by electrons helps explain the formation of ions. *Cations are formed from atoms by the loss of electrons* while *anions are formed by the gain of electrons.* An atom of sodium can lose an electron to form a cation. The resulting ion Na^+ has a nucleus with 11 protons surrounded by 10 electrons. An atom of chlorine can gain an electron to form an anion. The Cl^- anion has 18 electrons surrounding a nucleus with 17 protons.

ASIDE

COUNTING THE ATOMS IN A MOLE

How many atoms are in a mole? The determination of the charge on the electron enabled chemists to answer this question. When a direct current of electricity passes through a dilute solution of a silver compound, silver metal is deposited at one electrode. Scientists measured the quantity of electric charge (current × time) required to deposit 1 mol of silver (107.9 g). This quantity of charge, 9.649×10^4 C, is called a faraday, F. Because it takes one electron to convert each Ag^+ to a silver atom, the charge required to yield 1 mol of silver metal is the charge on 1 mol of electrons. By dividing 1 F of electric charge by the charge on one electron, scientists calculated Avogadro's number, the number of electrons in a faraday and the number of silver atoms in a mole of silver metal. Avogadro's number N has the value 6.022×10^{23}.

$$N = \frac{F}{e} = \frac{9.649 \times 10^4}{1.602 \times 10^{-19}} = 6.022 \times 10^{23}$$

ISOTOPES

Isotopes are atoms of an element containing the same number of protons but different numbers of neutrons in their nuclei. The most abundant isotope of hydrogen has a nucleus that consists of a single proton. A small fraction of the atoms of hydrogen are those of a heavier isotope called deuterium that has both a proton and a neutron in its nucleus. The least abundant isotope of hydrogen, tritium, has one proton and two neutrons in its nucleus. Isotopes of an element have the same chemical properties, but they differ in their mass and nuclear chemistry. For example, tritium is the only radioactive isotope of hydrogen.

Naturally occurring fluorine consists of a single isotope with 9 protons and 10 neutrons in its nucleus. The *mass number* for an isotope is the sum of the number of protons and the number of neutrons in the nucleus. Fluorine has a mass number of 19. *In the symbol of an isotope, the mass number is*

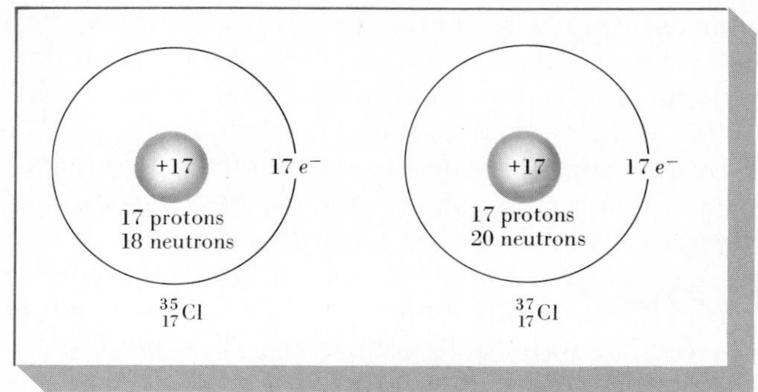

FIGURE 3.11
Isotopes are atoms of the same element that differ in the number of neutrons in their nuclei. These two isotopes of chlorine both have 17 protons and 17 electrons, but one has 18 neutrons and the other 20 neutrons.

given in the superscript and the atomic number is given in the subscript. The symbol of the naturally occurring isotope of fluorine is $^{19}_{9}F$.

An isotope of carbon, $^{12}_{6}C$, was chosen as the standard for measuring isotopic masses and atomic weights. The atomic mass of the $^{12}_{6}C$ isotope is defined to be exactly 12 atomic mass units (amu). For every isotope other than $^{12}_{6}C$, the mass is close but not identical to the mass number. For example, the atomic mass of naturally occurring fluorine is 18.998403 amu.

Many elements are found in nature as mixtures of isotopes. For example, there are two naturally occurring isotopes of chlorine, $^{35}_{17}Cl$ and $^{37}_{17}Cl$.

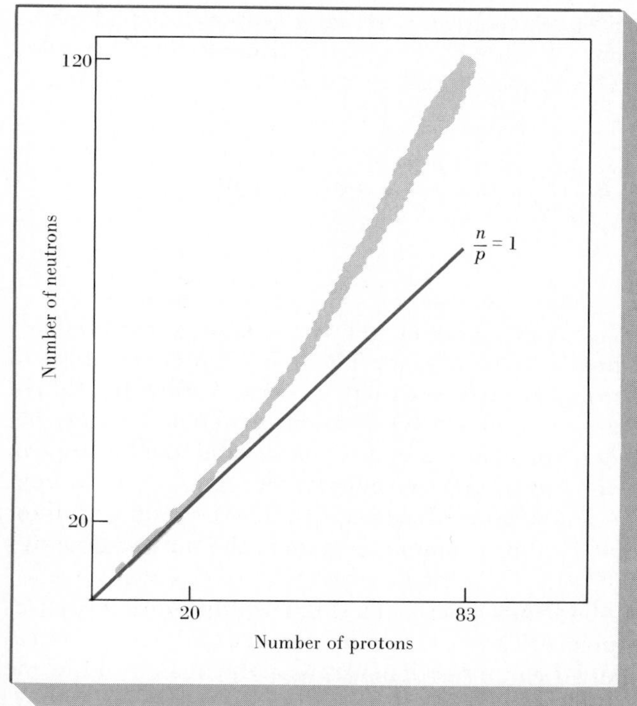

FIGURE 3.12
Stable nuclei are found in the band shown in the graph of the number of neutrons vs. the number of protons in a stable isotope. For elements with an atomic number greater than 20, there are more neutrons than protons in their stable nuclei. No stable isotope has more than 83 protons in its nucleus.

Each isotope of chlorine has 17 protons in its nucleus. The more abundant isotope of chlorine has 18 neutrons in its nucleus. The mass number of this isotope is 35 (17 protons + 18 neutrons = 35 nucleons). The isotope with mass number 37 has 37 nucleons − 17 protons = 20 neutrons. Atoms of each isotope of chlorine have 17 electrons to balance the +17 charge of the nucleus. The atomic weight of chlorine, 35.453, is the weighted average of the masses of the two isotopes.

A graph showing the trend in the number of protons vs. the number of neutrons in stable nuclei is shown in Figure 3.12. For light elements other than hydrogen, nearly equal numbers of protons and neutrons are found in their most abundant nuclei. For heavier elements, the number of neutrons in the nucleus exceeds the number of protons. Heavy elements have relative atomic weights that are more than double their atomic numbers since their nuclei contain more neutrons than protons.

NUCLEAR REACTIONS

Scientists write equations for nuclear reactions using symbols that identify both the atomic number and the mass number of each reactant and product. In nuclear reactions, both electrical charge and the number of nucleons are conserved. The sum of the charges on the products of a nuclear reaction must equal the sum of the charges on the reacting nuclei, and the total number of nucleons in the products must equal the total number of nucleons in the reactants. An equation for a nuclear reaction is balanced when the charges and the mass numbers are balanced.

The following equations illustrate characteristic modes of nuclear decay. (High-energy electromagnetic radiations called *gamma rays* are produced in many nuclear reactions, but they do not appear in the balance of charge or mass number.) All nuclei containing more than 83 protons are radioactive. Many of them undergo alpha decay.

$$^{238}_{92}\mathrm{U} \rightarrow {}^{234}_{90}\mathrm{Th} + {}^{4}_{2}\mathrm{He} \qquad \text{(alpha decay)}$$

An alpha particle is a helium nucleus emitted from a radioactive nucleus. Nuclei with a high neutron-to-proton ratios often undergo beta decay.

$$^{14}_{6}\mathrm{C} \rightarrow {}^{14}_{7}\mathrm{N} + {}^{0}_{-1}e \qquad \text{(beta decay)}$$

A beta particle is an electron emitted from a radioactive nucleus. In the case of beta decay, a neutron in the reacting nucleus emits an electron to become a proton in the product nucleus. Some nuclei with a low neutron-to-proton ratios decay by positron emission, and others decay by electron capture.

$$^{30}_{16}\mathrm{S} \rightarrow {}^{30}_{15}\mathrm{P} + {}^{0}_{+1}e \qquad \text{(positron emission)}$$
$$^{7}_{4}\mathrm{Be} + {}^{0}_{-1}e \rightarrow {}^{7}_{3}\mathrm{Li} \qquad \text{(electron capture)}$$

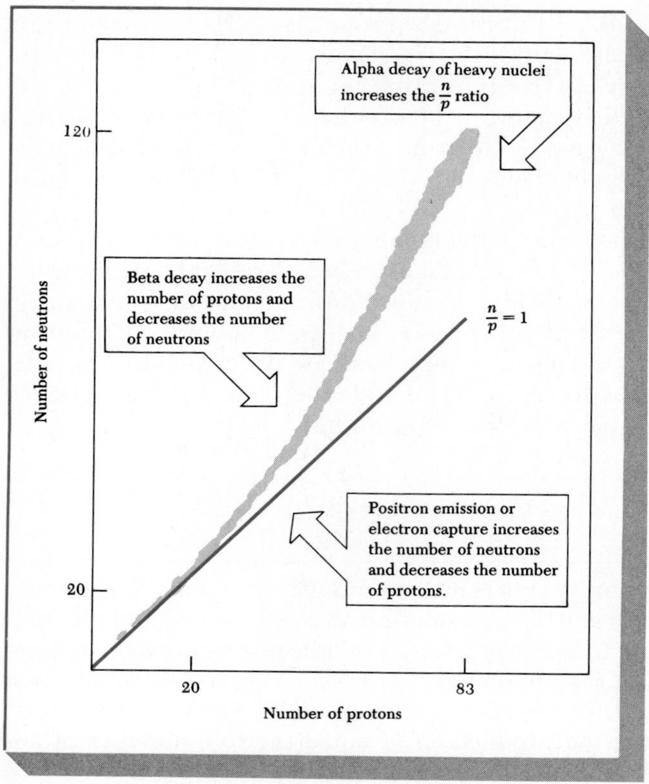

FIGURE 3.13
For radioactive elements, the mode of decay is in the direction of the zone of stability.

A *positron* has the same mass as an electron but it carries a +1 charge. When a positron is emitted, a proton is converted to a neutron. (Positrons are short-lived. Upon collision with an electron, both particles are annihilated and gamma radiation is produced.) In the case of *electron capture,* an electron is captured by the nucleus to convert a proton to a neutron. Both positron emission and electron capture raise the neutron-to-proton ratio.

The mass number and the atomic number of the products of nuclear reactions can be readily calculated. By balancing the charge numbers and the mass numbers, one can readily complete nuclear reactions. However, it is necessary to know either the starting isotope and one of the products or the two products of a nuclear decay reaction.

$$^{226}_{88}\text{Ra} \rightarrow {}^{4}_{2}\text{He} + ? \qquad \text{(alpha decay)}$$
$$\text{Mass number of product} = 226 - 4 = 222$$
$$\text{Atomic number of product} = 88 - 2 = 86$$

The symbol of element 86 is Rn so the product is $^{222}_{86}$Rn.

$$^{214}_{83}\text{Bi} \rightarrow {}^{0}_{-1}e + ? \qquad \text{(beta decay)}$$
$$\text{Mass number of product} = 214 - 0 = 214$$
$$\text{Atomic number of product} = 83 - (-1) = 84$$

The product of the beta decay reaction is $^{214}_{84}Po$.

Identify which of the following nuclear reactions of thorium isotopes produces an alpha particle and which produces a beta particle.

$$^{230}_{90}\text{Th} \rightarrow {}^{226}_{88}\text{Ra} + ?$$
$$^{234}_{90}\text{Th} \rightarrow {}^{234}_{91}\text{Pa} + ?$$

Products of nuclear reactions may, themselves, be radioactive. For example, a chain of nuclear decay reactions that starts with $^{238}_{92}U$ does not stop until it reaches the stable isotope $^{206}_{82}Pb$. Both polonium and radium, the radioactive elements first isolated by Madame Curie from a uranium ore, are "daughters" of uranium 238. Intermediate radioactive isotopes in this decay series undergo either alpha decay or beta decay. (Note that considerations about the safe use of a radioisotope need to take into account the possible production and safety of radioactive daughters.)

A S I D E

OBSERVING PROTONS AND NEUTRONS

It is worthwhile to step back and take a look at other evidence for the nuclear model of atomic structure. On the basis of his gold foil experiment, Rutherford formulated the model of an atom in which electrons surrounded a small, massive positive core. The text describes the nucleus as consisting of protons and neutrons. However, the gold foil experiment does not require that nuclei consist of protons and neutrons. In fact, the existence of protons was known before the Rutherford experiment, and neutrons were not observed directly until two decades later.

Positive particles, both protons and ions of elements other than hydrogen, were observed in experiments similar to that in which J. J. Thomson measured e/m for the electron. Positive particles were produced by electric discharges through evacuated glass tubes containing small residues of gases. These positive particles could be deflected by electric or magnetic fields. Scientists measured the charge-to-mass ratio, e/m, for these ions. The positive particle with the largest value of e/m was produced using tubes containing hydrogen gas. This was the proton $^{1}_{1}H$ produced by the loss of an electron from a hydrogen atom. (When the gas was oxygen, the positive particle produced was O^+ with 16 times the mass of H^+.)

There remained the task of demonstrating the existence of protons and neutrons in nuclei. Rutherford and his associates bombarded a wide variety of targets with alpha particles. Using nitrogen gas as the target, they produced new positive particles. These particles passed through foils that would stop alpha particles. By deflecting the new particles with a magnetic field, they showed them to be hydrogen nuclei.

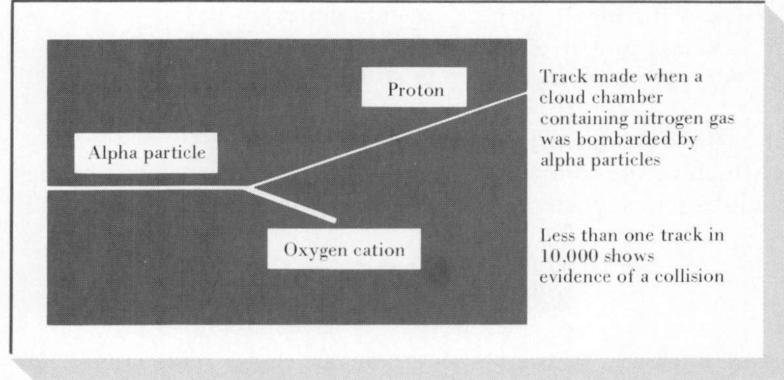

FIGURE 3.14
When an alpha particle strikes the nucleus of a nitrogen atom, a nuclear reaction produces a proton and an oxygen ion. In a cloud chamber, moving charges give rise to a trail of water droplets. These droplets are illuminated and photographed against a black background. The slower moving a charged particle, the more ions it produces and the thicker its track.

Measurements using a cloud chamber showed that these protons were produced by the reaction of an alpha particle with a nitrogen nucleus.

$$^{14}_{7}N + ^{4}_{2}He \rightarrow ^{17}_{8}O + ^{1}_{1}H$$

Charged particles moving through air knock electrons from some atoms to form ions. These ions initiate the formation of water droplets from water vapor. Scientists use the trail of water droplets to track the motion of charged particles. From thousands of cloud pictures, scientists found a small number in which the trail of an alpha particle ended abruptly and a new forked trail began. The alpha particle had been absorbed and a proton and an oxygen ion had been produced.

When a beryllium target was bombarded with alpha particles, a new intense radiation was produced. This radiation penetrated materials that stopped charged particles. It interacted with paraffin, a substance rich in hydrogen, to knock out protons. (Gamma radiation does not do this.) In 1932, James Chadwick showed that this penetrating radiation consisted of neutrons. These long-sought particles had been produced by the reaction of alpha particles with beryllium nuclei.

$$^{9}_{4}Be + ^{4}_{2}He \rightarrow ^{12}_{6}C + ^{1}_{0}n$$

NUCLEI VARY IN STABILITY

The mass of a nucleus is less than the sum of its separate parts. For example, the mass of a helium nucleus is slightly less than the sum of the masses of two protons and two neutrons.

$$\begin{array}{ccccc} 2\,^{1}_{1}H & + & 2\,^{1}_{0}n & \rightarrow & ^{4}_{2}He \\ 1.0078\text{ amu} & & 1.0087\text{ amu} & & 4.0026\text{ amu} \end{array}$$

$$\Delta m = 4.0026\text{ amu} - 2(1.0078\text{ amu}) - 2(1.0087\text{ amu}) = -0.0304\text{ amu}$$

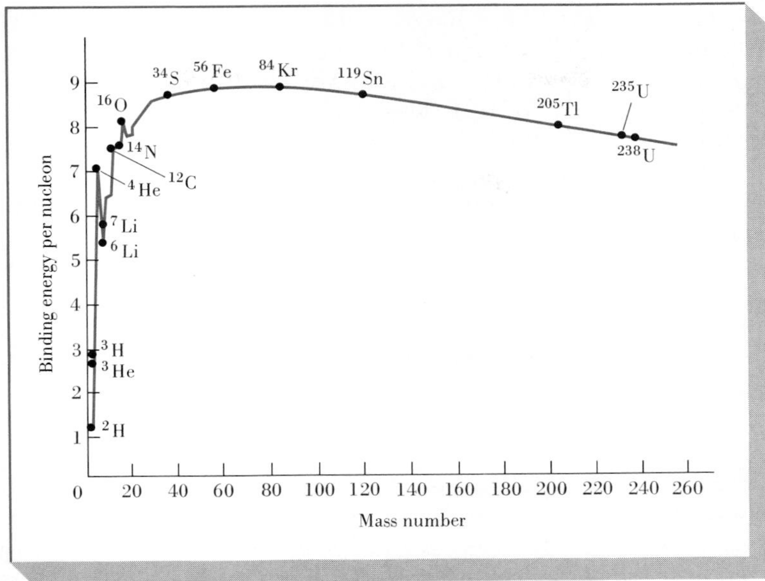

FIGURE 3.15
The binding energy per nucleon as a function of mass number. The most stable nuclei are at the top of the curve. The most stable nucleus is $^{56}_{26}Fe$.

Albert Einstein postulated that matter and energy are related. Einstein's equation is one of the more widely known equations of physics.

$$E = mc^2$$

Energy is equal to mass times the square of the speed of light. In many nuclear reactions, the mass of the products is measurably less than the mass of the reactants. So the m in the Einstein equation is the mass converted into energy—the mass of reactants minus the mass of the products.

The difference in mass between the mass of a nucleus and the sum of the masses of its nucleons is called the *mass defect.* The energy calculated from the mass defect is called the *binding energy,* because that much energy would be required to break the nucleus into its separate nucleons. The binding energy per nucleon first increases as the number of nucleons increases, and then it decreases. The most stable nuclei, those of iron and elements near iron in atomic number, have a greater binding energy per nucleon than the nuclei of lighter or heavier elements.

EFFECTS OF RADIATION

Particles ejected in nuclear reactions interact strongly with matter. Alpha and beta particles produce ions by knocking electrons from atoms along their path. Gamma radiation can also cause ionization. In addition, radiation can interact with molecules to break them into highly reactive fragments.

Although the more massive, higher-charged alpha particles cause greater damage along their paths, they are more easily stopped than beta particles. Alpha particles can be stopped by a thin piece of paper. Lighter beta particles penetrate further but are still readily contained by thin shields. Thick concrete or heavy lead shields are required to contain high-energy gamma radiation.

Radiation can cause mutations in living organisms. Both the ionization caused by radiation and the production of highly reactive fragments of molecules serve to bring about changes in DNA, the genetic material of cells. Because the body has some ability to repair radiation damage, the biological effects of long-term exposure to very low levels of radiation are controversial. Some people feel that any increase in radiation levels above background levels constitutes a health hazard; others do not.

ISOTOPES ARE TOOLS IN BIOLOGY, GEOLOGY AND MEDICINE

Scientists use both man-made and naturally occurring isotopes as research tools. Most elements occur as a mixture of isotopes. Nuclear reactions are used to produce radioactive isotopes. These isotopes can be incorporated into compounds and used to follow underwater drainage or the fate of a compound in metabolism. The following examples describe some of the range of applications of isotopes in research and medicine. Other examples will be presented later in the text when additional features of radioactive decay have been developed.

In 1952, Hersey and Chase used radioactive isotopic tracers to demonstrate that the DNA of viruses rather than their protein is responsible for

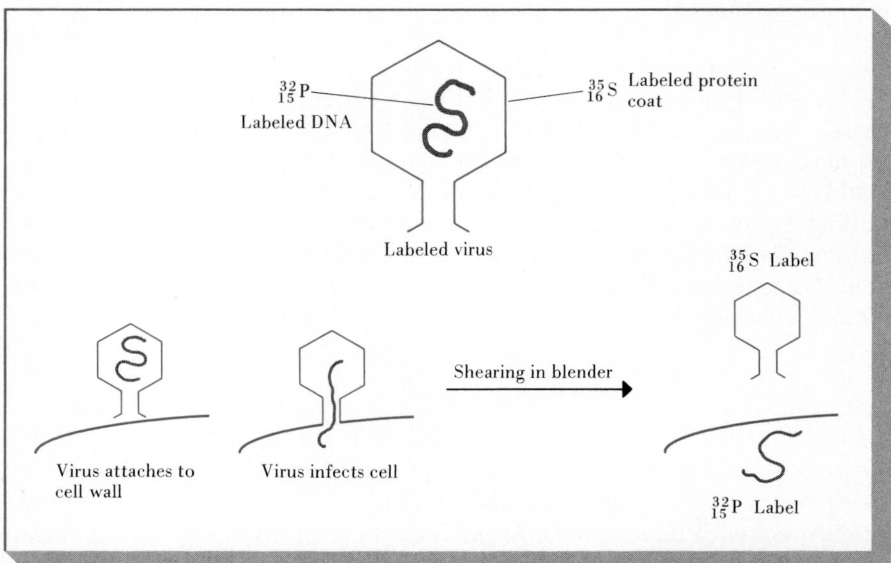

FIGURE 3.16
Hersey and Chase used isotopic markers to show that DNA rather than protein carried the genetic information of viruses.

infection. Many viruses that infect normal cells and cause them to make new viral particles consist only of protein and a small amount of DNA. Proteins contain a small amount of sulfur but no phosphorus. Phosphorus, and not sulfur, is present in DNA. Hersey and Chase grew viruses in a medium that contained $^{32}_{15}P$ and $^{35}_{16}S$, man-made radioactive isotopes. Then they used the isotopically labeled viruses to infect normal cells. After a short time the mixture was placed in a blender to shear away viral material adhering to the outside of cell surfaces and they tested the infected cells for the presence of $^{32}_{15}P$ and $^{35}_{16}S$. They found the phosphorus isotope but not the radioactive sulfur. Viral DNA had entered the cells while viral protein remained outside. They had conclusively demonstrated that DNA carried the instructions for the synthesis of new viral particles.

Radiologists use radioactive isotopes to help treat cancers. High levels of radiation are directed at tumors to kill cancerous cells. For example, needles of radioactive cobalt are placed about the site of a tumor to irradiate the tumor for a controlled time and dose. Since they multiply more rapidly, cancer cells are more vulnerable to radiation than normal cells. Because both normal and cancer cells suffer damage, radiation therapy can only be used intermittently. Doctors often use radiology in combination with chemotherapy to treat tumors inaccessible to surgery.

TRANSMUTATION AND NUCLEAR FISSION

When alpha particles strike some nuclei, transmutations occur. One element is converted into another by a nuclear reaction. For example, the nuclear reaction of an alpha particle with nitrogen that produced a proton, also produced an isotope of oxygen; and the nuclear reaction of an alpha particle with beryllium that produced a neutron, also produced an isotope of carbon. Scientists looked for other nuclear reactions by using protons and neutrons to bombard atomic targets.

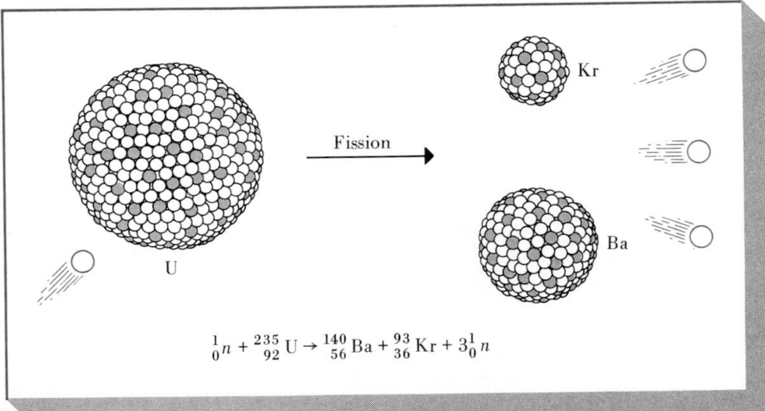

FIGURE 3.17
Fission occurs when a uranium 235 nucleus absorbs a neutron. The uranium nucleus splits into two product nuclei and produces additional neutrons. Large amounts of energy are released for the product nuclei have a greater binding energy per nucleon than the uranium nucleus.

Otto Hahn, a German chemist, separated the elements barium, lanthanum, and cerium with lower atomic numbers from a uranium source that had been bombarded by neutrons. He concluded that these lighter elements were produced by *fission.* When a nucleus of uranium 235 absorbs a neutron, it spontaneously splits to form two lighter nuclei and additional neutrons.

$$^{1}_{0}n + ^{235}_{92}U \rightarrow ^{140}_{56}Ba + ^{93}_{36}Kr + 3\ ^{1}_{0}n$$

Scientists immediately recognized that fission produces energy far in excess of that produced by any chemical reactions. In the fission of uranium 235, the mass of the products is less than the mass of the reactants. The missing mass is converted into energy. The energy appears as recoil energy of the particles produced by fission and as gamma rays.

Hahn communicated his discovery to Lisa Meitner, a former coworker who had fled Germany in 1938. Through scientists who were refugees from fascism in Europe, the news reached the scientific community in America. In a letter dated August 2, 1939, Albert Einstein, at the urging of fellow physicists, wrote to President Roosevelt calling attention to the use of fission as a source of energy and advising further investigation. This letter paved the way for the Manhattan project, an effort that culminated in the production of the first atomic bombs dropped on Japanese cities at the close of World War II. (The peaceful use of atomic energy for the production of power is discussed in Chapter 11.)

QUESTIONS

1. Sir Humphrey Davy was the first person to isolate the metals magnesium, calcium, strontium, and barium as well as sodium and potassium. Write an electrode reaction for the formation of barium from the Ba^{+2} ion found in compounds.
2. Molten calcium bromide conducts electricity. What can you infer about the particles present in the melt?
3. In his discovery of radioactivity, did Becquerel actually observe the alpha decay of $^{238}_{92}U$? (Consider the penetrating power of different types of radiation.) What radiation might have caused the wrapped photographic plates to fog?
4. Why are alpha and beta particles deflected in opposite directions by an electric field? Why are beta particles deflected more than alpha particles?
5. What modification of Dalton's atomic theory is required by the discovery of radioactivity? (Which of his postulates must be modified?)
6. When did the statement "Uranium is radioactive" become scientific fact? Would it have been meaningful to Dalton? Are scientific facts separable from theory?
7. How did the isolation of a radioactive element by the Curies contribute to the understanding of radioactivity? (Without this discovery, could it have been known that radioactive decay is a nuclear reaction?)

8. In the Millikan oil drop experiment, the drops occasionally changed charge. How would a change in charge affect the rate of fall? Of rise? What size charge might a drop gain or lose? Why?
9. Why would it be necessary to measure the charge on many drops to obtain the charge on an electron using Millikan's method?
10. The charge on the electron is 1.602×10^{-19} C. The ratio e/m for the electron is close to 1.76×10^{8} C/g. What is the mass of an electron?
11. If Rutherford had used a foil of a metal with a much lower atomic weight than gold, some of his observations would have been different. Would the number of alpha particles passing through the foil have increased greatly? Would the number of alpha particles deflected through wide angles have decreased significantly? Why or why not?
12. In Rutherford's gold foil experiment, what is the significance of each of the following observations?
    (a) Most alpha particles are not deflected when they pass through the foil.
    (b) Some alpha particles are deflected through wide angles.
13. Between 1800 and 1900, the educational background of people entering science changed greatly. In an encyclopedia, locate information on Priestley, Lavoisier, and Dalton, and compare it to the corresponding information on Madame Curie, J. J. Thomson, and Ernest Rutherford.
14. Mass spectrometers are instruments used to separate positive ions. Positive ions can be accelerated, passed through a slit, and deflected by either a magnetic field or an electric field. Using a mass spectrometer, ions of isotopes of an element differing in the number of neutrons in their nucleus can be separated. Why is this possible? (*Hint:* Consider the value of e/m for each isotope.)
15. Complete the following table.

Element	Protons	Neutrons	Mass Number	Isotopic Symbol
U	92		238	$^{238}_{92}U$
C			14	$^{14}_{6}C$
	17	18		

16. In Figure 4.9, the drawing indicates that the Na^+ ion is smaller than an Na atom. Consider the attraction of the nucleus for electrons. Why should the removal of one electron produce a smaller species?
17. Why is the chloride ion, Cl^-, larger than the chlorine atom?
18. Complete the following table.

Species	Protons	Electrons	Net Charge
S^{-2}		18	−2
Cl^-	17		−1
	18	18	
K^+		18	
	20		+2

19. Complete the following nuclear reactions found in the U-238 decay series.
    (a) $^{226}_{88}Ra \rightarrow ? + ^{4}_{2}He$
    (b) $^{222}_{86}Rn \rightarrow ^{218}_{84}Po + ?$
    (c) $? \rightarrow ^{214}_{82}Pb + ^{4}_{2}He$
    (d) $? \rightarrow ^{214}_{83}Bi + ^{0}_{-1}e$
20. Complete the following nuclear reactions. Identify the reactions as examples of alpha decay, beta decay, positron emission, or nuclear fission.
    (a) $^{14}_{6}C \rightarrow ^{14}_{7}N + ?$
    (b) $^{38}_{19}K \rightarrow ^{38}_{18}Ar + ?$
    (c) $^{210}_{84}Po \rightarrow ^{206}_{82}Pb + ?$
    (d) $^{1}_{0}n + ^{235}_{92}U \rightarrow ^{90}_{38}Sr + ^{144}_{54}Xe + ?$
21. Frederic Joliot-Curie and his wife, Irene, daughter of Madame Curie, showed that the bombardment of aluminum by alpha particles produced both neutrons and positrons. When the alpha particle source was removed, only the positron emission continued. They concluded that they had produced a radioactive isotope of phosphorus, and that the newly formed isotope of phosphorus was a positron emitter. For their demonstration of artificially induced radioactivity, they shared the Nobel prize in chemistry for 1935. Complete and balance the equations for the nuclear reactions discovered by the Joliot-Curies.

$$^{4}_{2}He + ^{27}_{13}Al \rightarrow ^{?}_{?}P + ^{1}_{0}n$$
$$^{?}_{?}P \rightarrow ^{?}_{?}? + ^{0}_{+1}e$$

22. What was the significance of Otto Hahn's discovery that the element strontium had been produced in a sample of uranium that had undergone neutron bombardment?

CHAPTER

FOUR

THE MODERN ATOM AND THE PERIODIC TABLE

A revolution in ways to view the nature of matter was part of a larger revolution that occurred in physics early in the twentieth century. The terms classical physics and modern physics are sometimes used to designate the prerevolutionary and postrevolutionary eras. In classical physics, light was thought to be electromagnetic radiation and wavelike in character. An atom was thought to consist of electrons circling the nucleus much as our planet circles the sun. After the revolution in physics, light was seen to have particlelike properties, and the properties of electrons in atoms were seen to resemble waves.

The revolution in physics led to a model of the atom that helps account for the chemical similarities and differences summarized in the periodic table. This model goes far to explain the bonding of elements to form compounds and the properties of the compounds formed by chemical combination.

LIGHT, WAVES, AND PARTICLES

Light can be considered to consist of electromagnetic waves. For each wave, the distance from crest to crest is called the *wavelength*, λ. Units used for reporting wavelengths are frequently meters, m, or centimeters, cm. The number of waves passing a stationary point in a unit of time is called the *frequency*, ν. Units of frequency are cycles per second or s^{-1}. Note that one cycle per second is also called one hertz (Hz). *The speed of light is the product of the wavelength times the frequency.*

$$c = \lambda\nu$$

In the visible spectrum, red light has the longest wavelength and lowest frequency, and violet light has the shortest wavelength and highest frequency. Visible light comprises only a small portion of the electromagnetic spectrum. Lower-frequency, longer-wavelength radiation includes infrared rays, microwaves, and radio waves. Shorter-wavelength, higher-frequency radiation includes ultraviolet light, x-rays, and gamma rays. All these forms of electromagnetic radiation travel at the same speed *c, the speed of light.*

$$c = 3.00 \times 10^8 \text{ m/s}$$

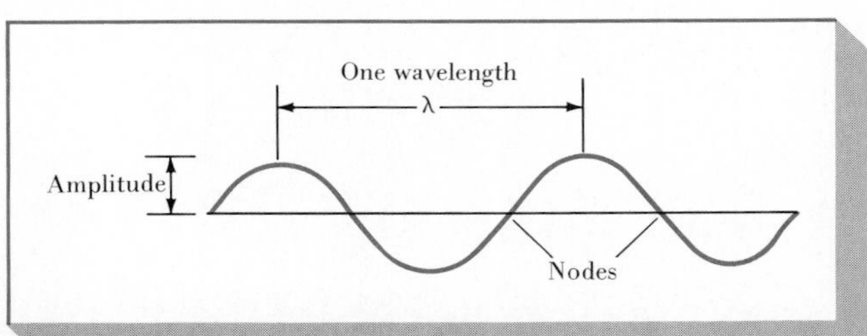

FIGURE 4.1
A wave is characterized by a wavelength, an amplitude, and a frequency. The frequency of a wave is the number of waves passing a fixed point per second.

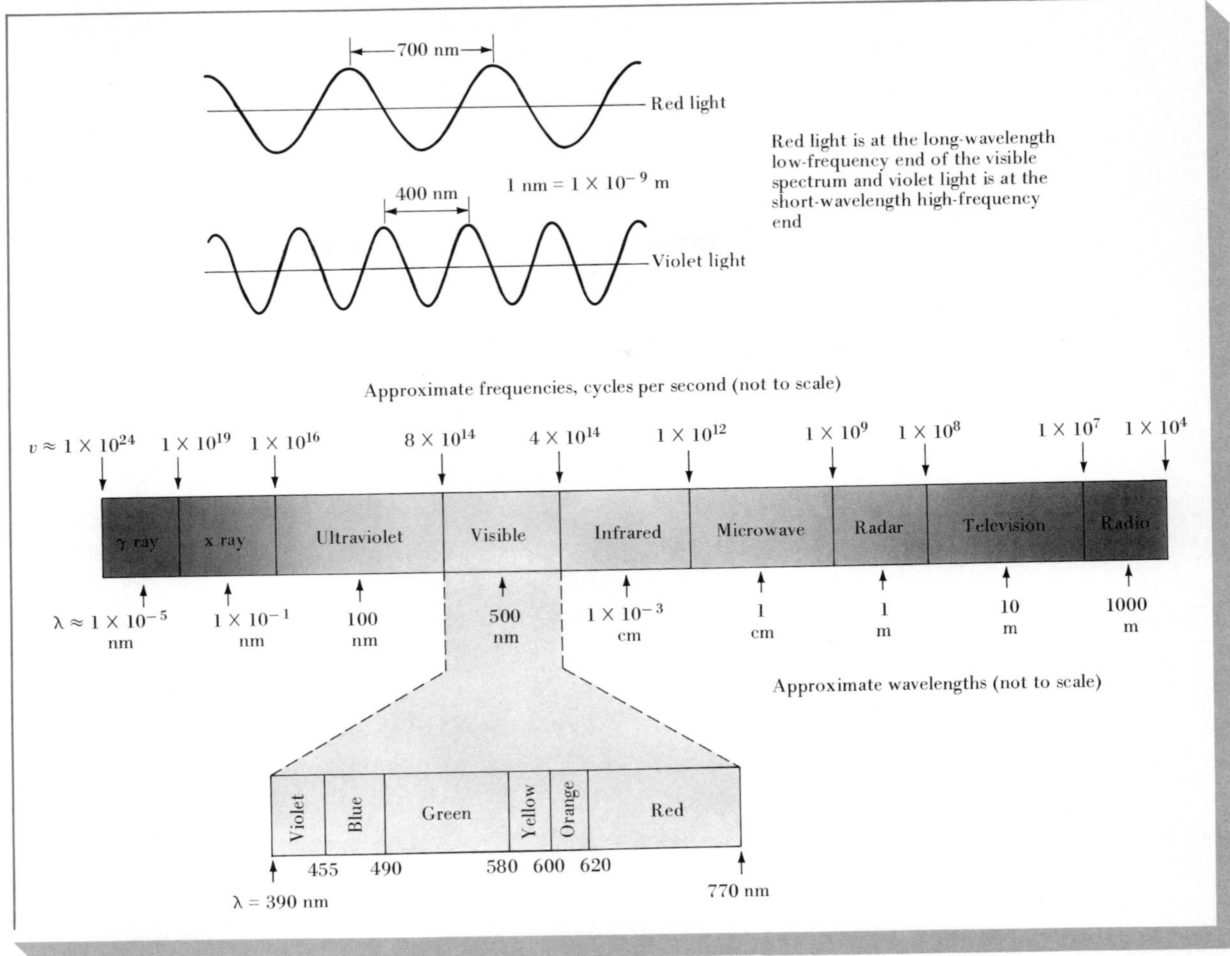

FIGURE 4.2
Visible light is a small portion of the electromagnetic spectrum.

A *wave model of light* is invoked to explain the bending of light by a prism or the diffraction of light by a grating. A prism separates a beam of white light into component colors because the bending of light at the surfaces of the prism depends on the wavelength. Longer-wavelength red light is bent less than shorter-wavelength blue. A diffraction grating is another device that separates white light into its colored components. The closely spaced rulings on a grating scatter light into its colored components because the direction for maximum reinforcement depends on the wavelength of light. Waves in phase reinforce one another. Waves out of phase cancel one another.

A *particle model of light* is required to interpret the *photoelectric effect.* Light with a frequency greater than or equal to a certain minimum value causes electrons to be emitted from some surfaces. Light of lower frequency, no matter how bright, has no effect. For example, yellow light

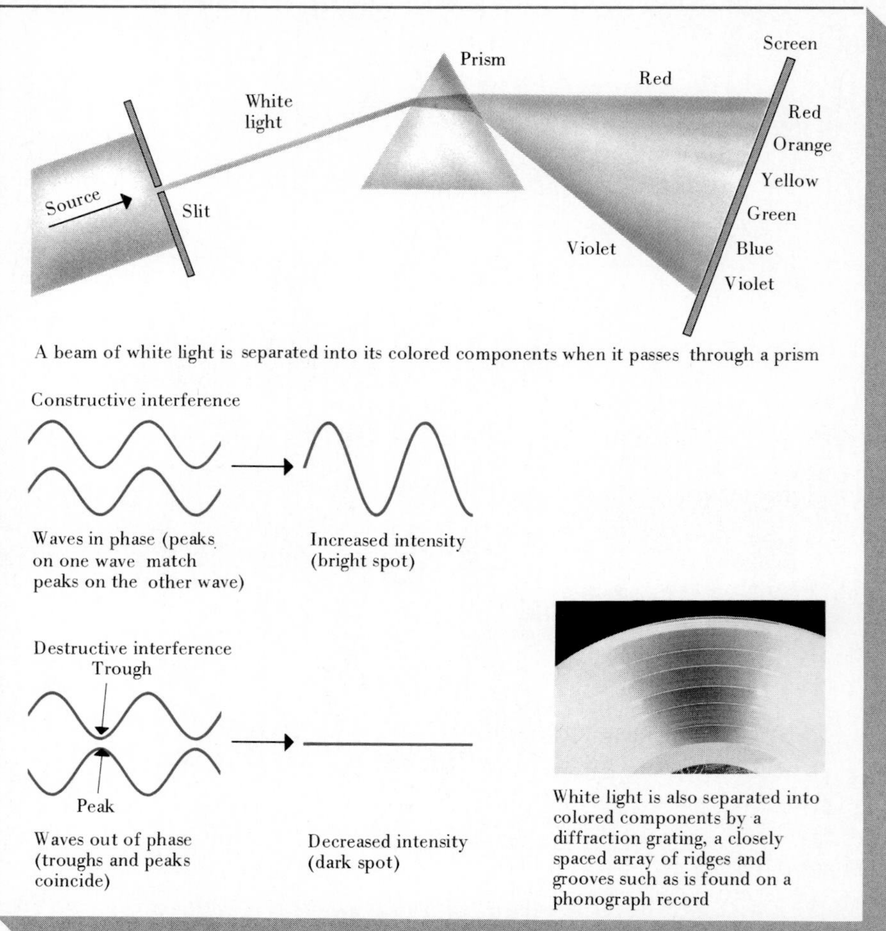

FIGURE 4.3
A wave model of light is used to explain the blending of light by a prism or the diffraction of light by a grating.

causes electrons to be emitted from potassium, but red light does not. This finding cannot be explained by the wave model of light. To explain the photoelectric effect, scientists postulate that *light consists of particles called photons,* and that a photon with sufficiently high energy can cause an electron to be ejected while one, or even many, lower-energy photons cannot.

In 1900, Max Planck had first proposed that light energy is emitted in discrete units called photons. According to Planck, the energy of a photon is proportional to its frequency. In *Planck's equation, E* is the energy of the photon, h is a constant known as Planck's constant, and ν is the frequency of the light.

$$E = h\nu$$

where h is Planck's constant and equals $6.63 \times 10^{-34}\ \text{J} \cdot \text{s}$.

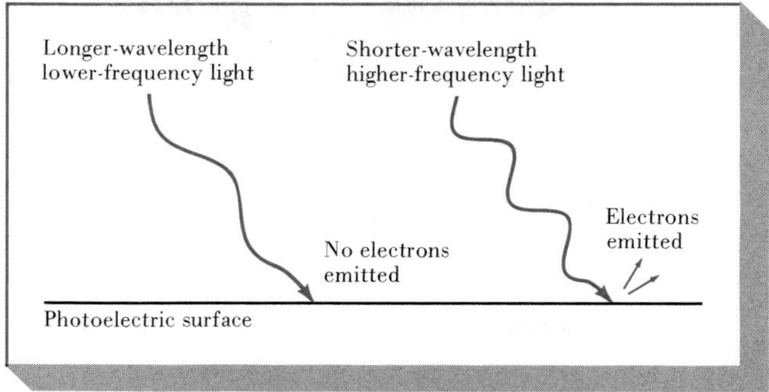

FIGURE 4.4
Light of high frequency causes electrons to be emitted from some surfaces. A particle model of light in which energy of the light depends on frequency, $E = h\nu$, is used to explain this phenomenon.

In some experiments light behaves like a wave, and in other experiments light acts as if it were a particle. The paradox that one experiment may be interpreted by a wave model of light and another experiment may be interpreted by a particle model is sometimes referred to as the *wave-particle duality of light.*

THE INTERACTION OF ATOMS AND LIGHT

A flame or an electric discharge can excite individual atoms to high-energy states. These excited atoms lose energy by emitting light. The brilliant reds, greens, and blues of fireworks are achieved by adding small amounts of selected salts to the powder used. Calcium chloride gives red; barium chloride, green; and copper chloride, blue. Colors observed in flame tests can be used to identify these and other elements. When a film of a chloride salt solution of an element is placed in a flame, the solution vaporizes, and a flash of characteristic color appears.

Unlike the continuous spectrum of light obtained from a glowing filament in a light bulb, the light from a flame test can be separated into a few lines with colors characteristic of the element being tested. When the light from the flame is passed first through a slit and then through a prism, the resulting spectrum consists of bright lines against a dark background. The red neon lights used in advertising, as well as the yellow sodium vapor lamps and greenish mercury vapor lamps used as streetlights, also produce discontinuous spectra with characteristic sharp lines.

A S I D E

SPECTROSCOPY LEADS TO THE DISCOVERY OF THE NOBLE GASES

Astronomers were the first to find evidence for a new element, helium, in the sun's corona during a solar eclipse in 1868. A dark line was present in the spectrum of the corona that could not be associated with the spectrum of any known element or

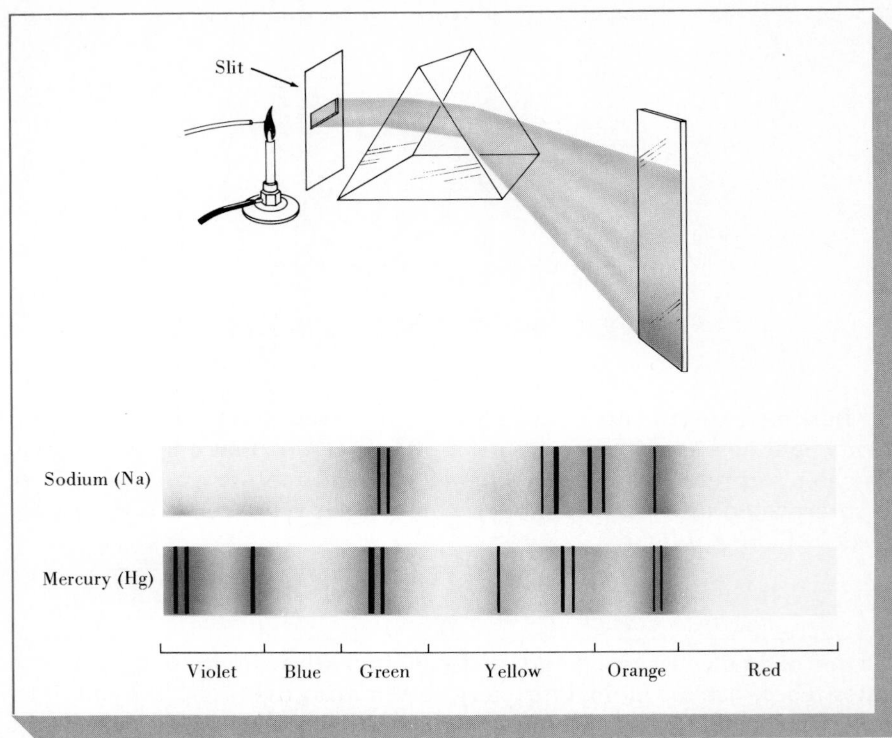

FIGURE 4.5
In a flame test, a drop of a metal halide solution is heated by an intense flame. If the light from the flame is passed first through a slit and then through a prism, the resulting spectrum consists of a series of bright lines that are characteristic for the metallic element. A similar line spectrum can be obtained if the vapor of the element is excited by an electric discharge.

compound. Light from the sun of the proper wavelength had been absorbed by atoms of an unknown element present in the sun. The element was named helium after the Greek "helios" for sun.

The dark line present in the spectrum of the sun was later shown to be an absorption line of a gaseous element that was first isolated from a uranium ore in 1888. The helium trapped in the ore was a product of radioactive decay. When alpha particles are slowed by their interactions with matter, they eventually pick up electrons to form helium atoms. When the ore was dissolved in acid, the helium gas was released.

A search that led to the discovery of other gaseous elements began when discrepancies were found in the density of nitrogen. Nitrogen gas prepared from compounds of nitrogen was less dense than nitrogen from air. William Ramsey and his associates sought to remove nitrogen and oxygen from air and examine the residue. In one set of experiments, oxygen-enriched air was repeatedly sparked to form nitrogen oxides which dissolved in sodium hydroxide solution. The remaining gas was more dense than nitrogen, and its spectrum contained new lines. Ramsey had isolated the element argon. In the following years, Ramsey also isolated neon, krypton, and xenon from liquefied air.

At various times these unreactive, monoatomic elements have been called noble gases, inert gases, and rare gases. Argon is 1 percent of the air, and helium is one of the most abundant elements in the universe. Not only are these elements abundant,

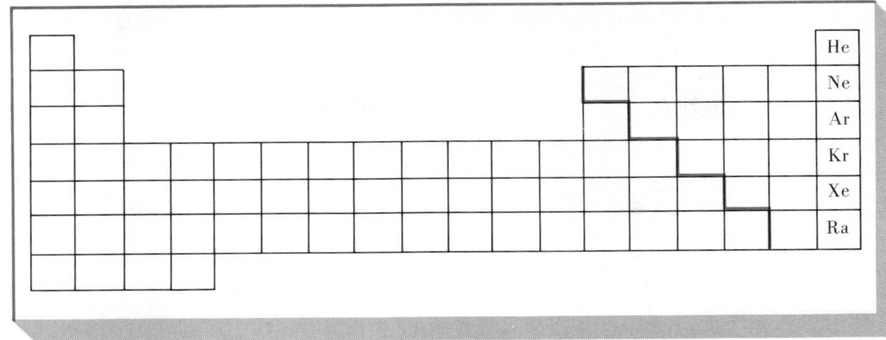

FIGURE 4.6
An entire family of elements, the noble gases, was discovered after Mendeleev had formulated the periodic table.

but some undergo chemical reaction. Although no compounds of any of these elements were known before 1956, a few compounds of the heavier elements of the family have been prepared since then. (Would the chemical reactions of xenon have been discovered earlier had the elements not been called the inert gases?)

A SIMPLE EQUATION DESCRIBES HYDROGEN SPECTRA

One of the simplest atomic spectra is that of hydrogen; its visible spectrum, known as the Balmer series, consists of a few lines that are increasingly close together toward the violet end of the spectrum. A simple empirical mathematic expression was developed by Rydberg, and showed a relationship between the wavelengths of lines in the spectrum of hydrogen.

$$\frac{1}{\lambda} = R_H\left(\frac{1}{2^2} - \frac{1}{n^2}\right) \qquad n = 3, 4, 5, \ldots$$

where R_H is Rydberg constant and equals 109,677.6 cm^{-1}. For the first three lines in the Balmer series, the expression in the parentheses has the values $\frac{1}{4} - \frac{1}{9}$, $\frac{1}{4} - \frac{1}{16}$, and $\frac{1}{4} - \frac{1}{25}$.

Scientists looked for and found additional series of emission lines at wavelengths predicted by substituting the integers 1 and then 3 for the

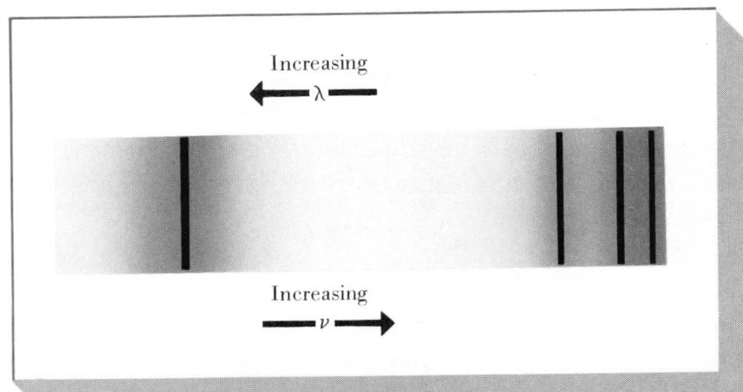

FIGURE 4.7
The emission spectrum of the hydrogen atom consists of a small number of lines increasingly close together at shorter wavelengths.

integer 2 in the Rydberg equation. Lines of the Lyman series have shorter wavelengths than visible light and are found in the ultraviolet portion of the spectrum. For the first three lines in the Lyman series, the expression in the parentheses has the values $1 - \frac{1}{4}$, $1 - \frac{1}{9}$, and $1 - \frac{1}{16}$. Lines of the Paschen series have longer wavelengths than visible light and are found in the infrared portion of the spectrum. For the first three lines in the Paschen series, the expression in the parentheses has the values $\frac{1}{9} - \frac{1}{16}$, $\frac{1}{9} - \frac{1}{25}$, and $\frac{1}{9} - \frac{1}{36}$.

THE BOHR ATOM

In 1913, Niels Bohr (1887–1951), a young Danish physicist, proposed a model of the atom to account for the line spectrum of the hydrogen atom. For Bohr, the fact that only discrete amounts of energy were emitted in the photons implied that only certain fixed energies were available to the electron in a hydrogen atom. Bohr proposed that in this simplest atom, an electron circles the nucleus as the earth circles the sun. *The circular path of the electron was called an orbit,* and the energy of an electron in an orbit was fixed.

Only a limited number of orbits were permitted in the *Bohr atom.* The radius of an orbit was required to be n^2r_0 where n is a positive integer and r_0 is the radius of the lowest-energy orbit.

$$r = n^2 r_0 \qquad n = 1, 2, 3, 4, \ldots$$

The integer n is called a *quantum number.* This quantum number also appears in the calculated value of the energy of each orbit in the Bohr atom.

Bohr expressed the energy of the electron in an orbit as the sum of an energy term that involved the attraction of the negative electron to the positive nucleus and an energy term due to the movement of the electron. He then solved the equation for the energy associated with each value of n. The energy of an electron in the Bohr atom depends upon the quantum number n, Planck's constant h, the mass m and charge e of the electron, and the speed of light c.

$$\text{Energy} = -\frac{2\pi^2 m e^4 c^4 \times 10^{-14}}{h^2}\,\frac{1}{n^2} \qquad n = 1, 2, 3, 4, \ldots$$

A totally separated electron and proton are considered as the starting point for measuring energy. The closer the electron is to the nucleus, the lower its energy. The lowest-energy state of the hydrogen atom is associated with the quantum number $n = 1$; higher values of n correspond to higher-energy states.

According to Bohr, an electron in a higher-energy orbit (farther from the nucleus) can fall to a vacant lower-energy orbit. To do so, the electron must lose its excess energy in the form of a photon of light. The energy of the emitted photon accompanying a change from one orbit to another can be calculated readily as the difference between the energies of the two orbits.

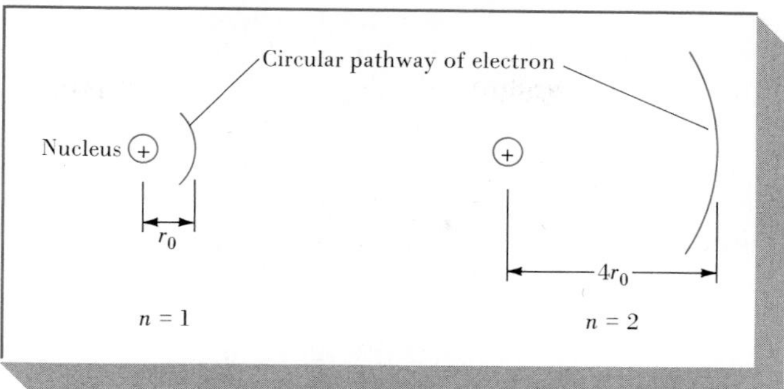

FIGURE 4.8
In the Bohr atom, an electron circles the nucleus in an orbit. The radius of the orbit depends on the value of the quantum number n.

$$E_{\text{upper}} - E_{\text{lower}} = \Delta E = h\nu = \frac{hc}{\lambda}$$

The energies calculated for changes in orbits in the Bohr description of the hydrogen atom corresponded to lines observed in the Lyman, the Balmer, and the Paschen series of emission lines for hydrogen. Using the values for energy calculated from the Bohr atom, a value for R, the empirical constant in the Balmer spectral series, was calculated. The value calculated, 109,677 cm^{-1}, is in striking agreement with the experimental value. These results implied a dramatic connection between line spectra and the electronic structure of atoms.

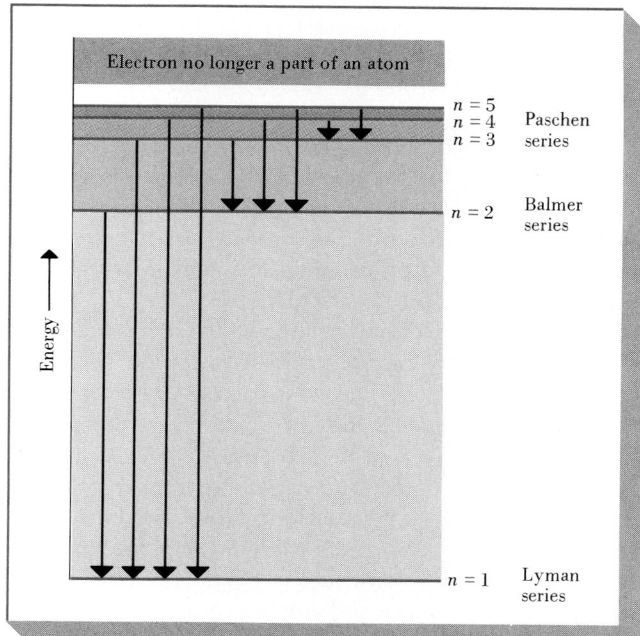

FIGURE 4.9
According to the Bohr model of the atom, an electron falls from a higher-energy orbit (larger value of n) to a lower-energy orbit (smaller value of n) by emitting a photon with energy equal to the energy difference between the two orbits.

PROBLEMS WITH THE BOHR ATOM

Although the Bohr atom accounted for the line spectrum of hydrogen, the appearance of integers or quantum numbers was arbitrary. This model had other deficiencies. According to classical physics, the orbits should not be distinct and stable. Instead, electrons should radiate energy and spiral into the nucleus. Because of the introduction of quantum numbers and the failure of physics to account for the path of the electron, the artificiality of this model for atomic structure was recognized even as it was proposed.

Why was the Bohr model of the atom important, even though it was recognized to be deficient? The Bohr model was the first to describe the electronic structure of atoms, and it successfully accounted for the emission spectrum of hydrogen. It was extended to account for some features of the periodic table. Although incorrect in some ways, the Bohr model of the atom pointed to new physical phenomena and guided scientists during a period of rapid change.

A S I D E

ATOMIC SPECTRA ARE IMPORTANT FOR ASTRONOMERS

Astronomers gather light from stars and galaxies to learn about these distant objects. Even in the continuous spectrum of a white star, a few lines may stand out. These lines are due to atomic absorptions of abundant elements in the stars and can be matched to absorption spectra of elements on earth. Astronomers use the relative intensities and shifts in the wavelengths of these lines to measure the temperature of stars, the rotation of galaxies, and the size of the universe.

Some stars absorb light with wavelengths found in the Balmer series of hydrogen. The temperature of these stars is between 7500 and 11000°C. In hotter stars, electrons are stripped from the hydrogen atoms and do not drop to such a low energy level. In cooler stars, electrons are not often excited to the higher energy levels needed for visible hydrogen absorptions to occur. Stars that do not show hydrogen lines but do show absorption lines from more easily excited calcium atoms are in the temperature range 5000–6000°C.

The wavelength of lines in the spectrum of a star depends on the motion of the star relative to the earth. Everyone is familiar with the Doppler shift heard in the roar of a passing motorcycle. For the rider, a constant roar is heard, but for the curbside observer, the pitch is elevated as the motorcycle approaches and drops as it goes by. The behavior of light is related, but more complex.

If a spectral source is moving away from us at a very high velocity, then the wavelengths of spectral lines are shifted toward the red end of the spectrum to a longer wavelength. The greater the red shift, the faster the object is receding. Because faint, distant galaxies are moving away from us faster than nearby galaxies, astronomers have concluded that the universe is expanding.

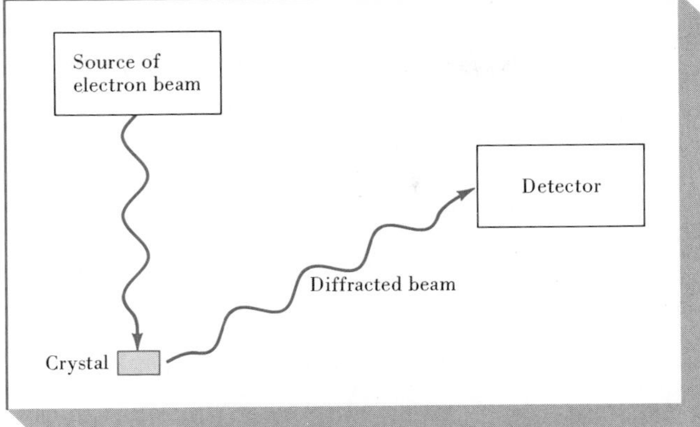

FIGURE 4.10
To demonstrate the wave nature of electrons, scientists studied their diffraction. In this experiment, electrons are diffracted by a crystal of nickel. The diffracted beam of electrons is observed at the same position as diffracted x-rays of the same wavelength.

ELECTRONS CAN BEHAVE AS WAVES

If light could behave as a particle, could an electron behave as a wave? The answer to this question is yes. In 1923, Louis de Broglie (1892–1987) postulated that a particle could behave as a wave with a characteristic wavelength, $\lambda = h/mv$, where λ is the wavelength, h is Planck's constant, m is the mass of the particle, and v is its velocity or speed.

The *de Broglie hypothesis* was tested experimentally using electrons. It was known that atoms in a crystal diffract x-rays. Electrons were accelerated to a velocity such that their wavelength, calculated using the de Broglie relationship, was equal to the wavelength of diffracted x-rays. When this beam of electrons was aimed at a nickel crystal, diffraction was observed. The de Broglie hypothesis was correct; *electrons can behave as waves.* (Scientists now use electron microscopy to visualize objects too small to be seen under light microscopes.)

THE WAVE MECHANICAL ATOM

In 1926, Erwin Schrodinger (1887–1961) proposed a *wave mechanical model of atomic structure.* In this model or theory, *an electron in an atom is considered to behave as a wave.* Familiar examples of waves are vibrating guitar strings and vibrating air columns in organ pipes. The physics of waves had been studied in detail, and the mathematics necessary to describe wave behavior was well-known. If electrons behave as waves in atoms, their behavior in some ways must be similar to that of vibrating strings, waves on water, and the sound of music.

The mathematical relationship of musical notes to atomic structure is important but difficult to visualize. The mathematics required to solve the equations for atomic structure is beyond the scope of an introductory chemistry course. We can, however, reason by analogy. Vibrating strings are familiar objects. If we consider the vibrations of a guitar string as a guide to the behavior of electrons in atoms, then we might better understand the description of the atom developed in the late 1920s.

Solutions of the wave equation for a vibrating string are quantized. Only certain wavelengths are allowed. Since a guitar string is anchored at each end, lasting vibrations or standing waves must consist of an integral number of half-waves. To put this relation in mathematical form, only wavelengths satisfying the following equation are allowed.

$$\frac{n\lambda}{2} = L \qquad n = 1, 2, 3, \ldots$$

where L is the length of the string. In the equation, the integer n is called a quantum number. Each solution to the wave equation has a different value of n.

For vibrating strings, the wavelength or the frequency of the vibration is related to pitch. The shorter the wavelength, the higher the frequency or pitch. When shortening a string by pressing on a fret, the guitarist creates a different string length, a new solution for the wave equation, and a higher note. In contrast, the energy of the wave is related to its amplitude. The greater the amplitude of the vibration, the louder the sound produced. The length and energy of a wave are independent. A musical note may be either high or low, loud or soft.

In what ways is the treatment of electrons in atoms similar to the treatment of a vibrating guitar string? Similarities are found in the mathematical form of the solutions. For the electron in the atom, there are boundary conditions analogous to the anchoring of a guitar string at its ends. Solutions to the wave equation for the atom are also characterized by integers called *quantum numbers*. For an electron in an atom, there are three quantum numbers, n, l, and m, corresponding to three dimensions in space, and a fourth quantum number, s, related to the magnetic properties of the electron.

Differences between wave descriptions of vibrating strings and of electrons in atoms lie in their physical interpretations. For the guitar string, amplitude is related to energy, and wavelength is related to pitch. For the

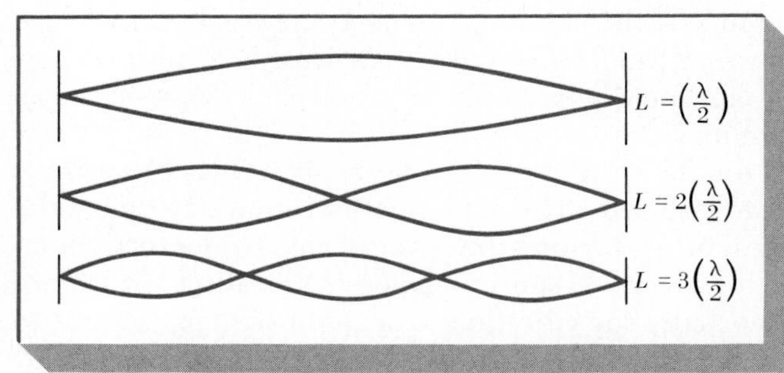

FIGURE 4.11
For a string anchored at both ends, standing waves are restricted to a small number of wavelengths. Each possible standing wave consists of an integral number of half-waves.

electron, amplitude is related to the preference of the electron for various positions in the atom, and wavelength is related to the energy of the electron.

ATOMIC ORBITALS AND EMISSION SPECTRA

Solutions to the wave equation for an electron in an atom yield orbitals. An *orbital* is a region in space where an electron can be found. Associated with each orbital is a unique set of quantum numbers, n, l, and m. *Each orbital can contain a maximum of two electrons,* for two electrons having different values of the fourth quantum number, s, can share an orbital.

Orbitals having the same value for n, the principal quantum number, belong to the same *shell.* The quantum number n can have the integer values 1, 2, 3, ... *The number of orbitals in a shell is equal to* n^2. The maximum number of electrons in a shell is $2n^2$. For $n = 1, 2, 3$, or 4, this limits the first four shells to 2, 8, 18, and 32 electrons, respectively.

Representations of three different orbitals are shown. One cannot speak of the pathway of an electron in an orbital but only of the probability of finding it in a region of space. The darker the shading in the diagram of an orbital, the greater the probability of finding an electron in a particular area. Note that orbitals labeled $2p$ and $3d$ have nodal planes. Nodes are a feature of the wave nature of electrons in atoms. An electron in a $2p$ orbital may be found above or below the nodal plane but never in the plane. The existence of nodal planes in the description of orbitals illustrates the dramatic difference between a wave description and a particle description of an electron in an atom.

Orbitals in the wave description of the hydrogen atom have the same energy as orbits in the Bohr atom. The explanation for the observation of line spectra is the same in both models of the atom. However, the appearance of quantum numbers is a natural outcome of the wave behavior of an electron in an atom rather than an arbitrary restriction as in the Bohr model.

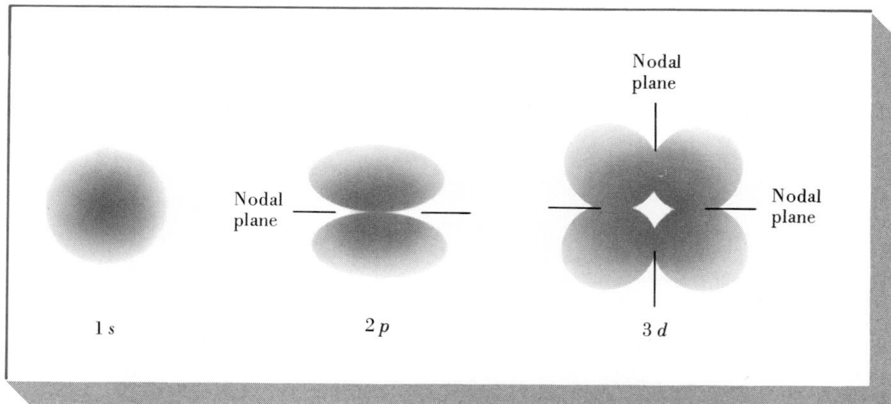

FIGURE 4.12
Atomic orbitals differ in their size and shape. The darker the shading in a diagram, the higher the probability of finding an electron in a small region of space. Note that some orbitals have nodal planes.

ASIDE

MODERN PHYSICS CHANGES OUR WORLD VIEW

Since the time of the Greeks, physics seemed to be an extension of common experience. Archimedes is said to have claimed that he could move the earth given a long enough lever and a fulcrum. Newton connected the movements of the planets to the fall of apples. After Newton, people knew that the same laws of physics apply on earth and in the heavens. Using newtonian mechanics, people calculated the motions of planets and of artillery shells.

In the twentieth century, physicists developed a radically different world view. A new generation of young physicists used new methods to change our view of nature. "Gedanken," or thought experiments, called into question the separation of mass and energy and of time and space.

Werner Heisenberg, who with de Broglie and Schrodinger developed a wave description of electrons in atoms, argued that there is a limit to which we can simultaneously measure position and movement. He said that we change a system when we measure it. We can measure the position of an electron by bouncing a photon of light off it. The light is collected by a lens and is returned to the observer. There is uncertainty to the path of the light through the lens; so, there is some uncertainty in the position of the electron. With shorter wavelength light, that uncertainty is reduced. However, since the light transfers some energy to the object being measured, the movement or momentum of the electron is changed. The shorter the wavelength of the light, the higher its energy; consequently, more energy will be transferred when a photon bounces off an electron. There is an absolute limit to which we can know both

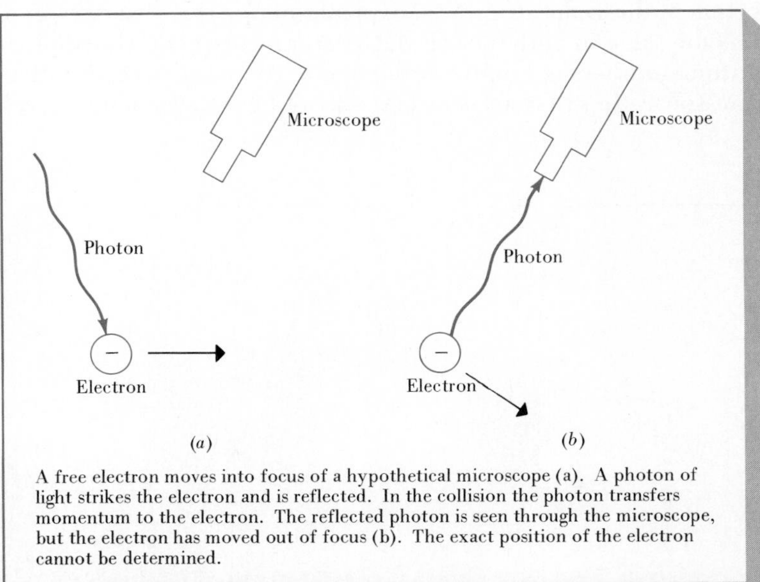

A free electron moves into focus of a hypothetical microscope (a). A photon of light strikes the electron and is reflected. In the collision the photon transfers momentum to the electron. The reflected photon is seen through the microscope, but the electron has moved out of focus (b). The exact position of the electron cannot be determined.

FIGURE 4.13
Werner Heisenberg used a thought experiment to demonstrate that there is a limit to which we can know both the position and path of an elementary particle at the same instant.

position and path at the same instant. The uncertainties are small, but they become important for the small particles found in atoms.

This change in world view from the certainties of Archimedes and Newton to the uncertainties symbolized by wave-particle dualities was wrenching even for the physicists involved. Heisenberg wrote:

> I remember discussions . . . which went through many hours till very late at night and ended almost in despair; and when at the end of the discussion I went alone for a walk in the neighboring park I repeated to myself again and again the question: Can nature possibly be as absurd as it seemed to us in these atomic experiments.

ATOMIC STRUCTURE AND THE PERIODIC TABLE

To extend the wave model of the atom to elements other than hydrogen, approximate methods are used. Orbitals are assumed to be like those of hydrogen. With each increase in atomic number, one proton is added to the nucleus and one additional electron is placed in an orbital. Orbitals of lower energy (closest to the nucleus) are filled before those of higher energy. In this way electron configurations are built up.

Electron shells contain *subshells* that differ in energy. Orbitals in a subshell share the same value of l, the second quantum number. Since the value of l goes from 0 to $(n - 1)$, the number of subshells in a shell is equal to n. Subshells with $l = 0$ are designated s subshells, those with $l = 1$ are p, and those with l values of 2 and 3 are labeled d and f. Within each subshell, there are $(2l + 1)$ orbitals. Each s subshell has one orbital, each p subshell has three orbitals, each d subshell has five orbitals and each f subshell has seven orbitals. Finally, each orbital can hold a maximum of two electrons. *Thus, an s subshell can hold 2 electrons; a p subshell, 6; a d subshell, 10; and an f subshell, 14.*

The design of a modern periodic table reflects the buildup of the electronic configuration of atoms. Rows in the periodic table correspond to the filling of subshells and shells. Atoms in the same column of the periodic

n	l	Designation of Subshell	Orbitals in Subshell	Orbitals in Shell
1	0	1s	1	1
2	0	2s	1	
2	1	2p	3	4
3	0	3s	1	
3	1	3p	3	
3	2	3d	5	9
4	0	4s	1	
4	1	4p	3	
4	2	4d	5	
4	3	4f	7	16

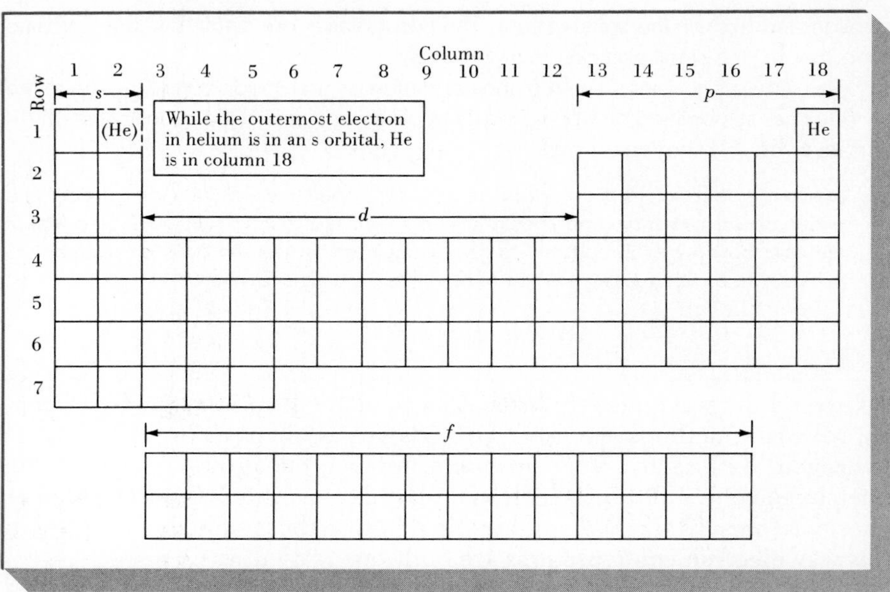

FIGURE 4.14
In modern versions of the periodic table, elements are grouped according to subshells being filled.

table have the same number of outer electrons in subshells of the same type. The lower the atom in a column, the more filled shells beneath the outer electrons.

Not all subshells within a shell are filled before the next shell is begun. Note, for example, that the 4*s* subshell is filled before the 3*d*. The filling of subshells can be recognized by the position of an element in the periodic table or by applying the following scheme. When the subshell notations are listed in indented fashion as shown, the columns then give the sequence of subshell filling.

1*s*						
	2*s*	2*p*				
		3*s*	3*p*	3*d*		
			4*s*	4*p*	4*d*	4*f*
				5*s*	5*p*	5*d* . . .
					6*s*	6*p* . . .
						. . .

ELEMENTS IN A CHEMICAL FAMILY HAVE SIMILAR ELECTRON STRUCTURES

The chemistry of an element is closely related to its outer electron configuration. Atoms of elements in the same family in the periodic table have the same number of outer electrons in the same type of subshell. Atoms of the alkali metals all have only one outer electron which is in an *s* orbital. The

Atomic Number	Element	Protons in Nucleus	Inner Electrons	Kernel Charge	Outer Electrons
3	Li	3	2	+1	1
11	Na	11	10	+1	1
19	K	19	18	+1	1
37	Rb	37	36	+1	1
55	Cs	55	54	+1	1
9	F	9	2	+7	7
17	Cl	17	10	+7	7
35	Br	35	28	+7	7
53	I	53	46	+7	7

halogens have seven outer electrons, two in an *s* subshell, and five in a *p* subshell.

Why do elements with the same outer electron configuration have similar chemistries? We can best see the reason if we develop one more concept, that of the *kernel. The kernel of an atom consists of the nucleus and the inner electrons.* Electrons are attracted to the nucleus and repelled by other electrons. Inner electrons, those with lower values of the principal quantum number, are closer to the nucleus than the outer electrons. The repulsion of an outer electron by each inner electron cancels the attraction due to one proton in the nucleus. The net charge attracting the outer electrons is the kernel charge, the charge on the nucleus minus the charge on the inner electrons. Members of a chemical family have the same kernel charge acting on the same number of outer electrons.

For most of the elements discussed in this text, the kernel charge and number of outer electrons can readily be calculated from the arabic number

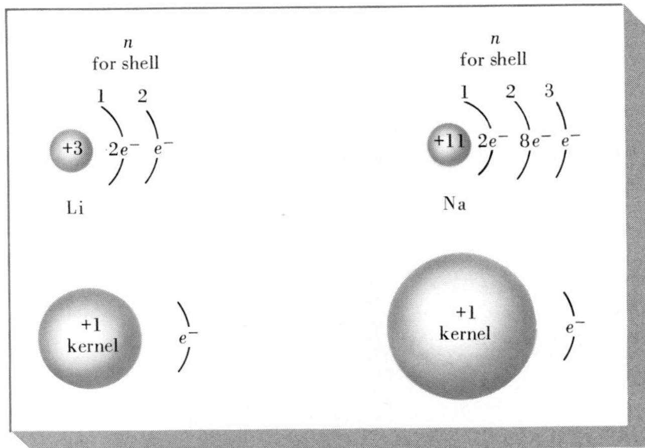

FIGURE 4.15
Within a family of the periodic table, atoms of different elements have the same kernel charge and the same number of outer electrons. For the alkali metals, lithium and sodium, one outer electron is attracted to a +1 kernel.

heading the column in the periodic table. For elements in columns 1 and 2, this number is simply the column number. For elements in columns 13 through 18, the number is equal to the column number minus 10. In the case of the halogens, this calculation would give the value $17 - 10 = 7$.

We have used the wave mechanical model of the electron to develop a picture of the atom in accord with the periodic table. Having developed this picture, we can now simplify our approach to chemistry. Atoms of elements in a family have similar chemistry because they have the same kernel charge and the same number of outer electrons. Those in different families have different kernel charges and different numbers of outer electrons. Chapter 5 will develop ideas about chemical bonding and properties of compounds by considering only the outer electrons of atoms and the positive kernels attracting those electrons.

QUESTIONS

1. Which of the following colors of visible light has the longest wavelength? The shortest?
   (a) Blue
   (b) Red
   (c) Yellow
2. In which of the following regions of the radiation spectrum are photons with the lowest energy found? With the highest?
   (a) Infrared
   (b) Visible
   (c) Ultraviolet
3. Which element in the second row of the periodic table is not known to form compounds? How many outer-shell electrons do atoms of this element have?
4. If scientists had not measured the spectral lines of all known elements, would the appearance of a dark line in the spectrum of the sun have been evidence for the existence of helium? (How did the routine measurement of the position of lines in spectra contribute to the discovery of new elements?)
5. Why were the noble gases discovered later than the elements in other families of the periodic table?
6. Why does the emission spectrum of hydrogen have fewer lines than those of other elements?
7. In the Balmer series for the hydrogen atom, the spacing of lines becomes closer toward the violet end of the spectrum. Why do they converge toward the high-energy side of the range?
8. The lines in the Balmer series of the hydrogen spectrum obey the following equation.

$$\frac{1}{\lambda} = R_H\left(\frac{1}{2^2} - \frac{1}{n^2}\right) \qquad n = 3, 4, 5, \ldots$$

Calculate the value of the fraction that is equal to $(1/2^2) - (1/n^2)$ for $n = 3, 4, 5$, and 6.

9. Justify the statement that for all lines in the Balmer series of the hydrogen spectrum $\frac{5}{36} R_H < 1/\lambda < \frac{1}{4} R_H$.
10. What is the value of n for the orbit in the Bohr atom with $r = 25r_0$?
11. What is the maximum number of electrons that can be accommodated in the shell with $n = 5$?
12. Sketch a wave for a vibrating string with ends fixed and a length L equal to $5\lambda/2$. How many nodes does this wave have?
13. Sketch a d orbital and identify its nodal planes.
14. Is an electron a wave or a particle in the Bohr model of the atom?
15. Does an electron behave as a wave or as a particle in an atom?
16. Henry Moseley found that the absorption of x-rays, high-energy electromagnetic radiation, provided an independent way to measure the number of protons in the nucleus. The frequency of x-rays emitted from target metals in x-ray tubes increases with increasing atomic number. This radiation arises when an electron in an outer orbital of an excited atom falls to a vacant position in an innermost orbital. Why should the energy emitted when an electron falls to an inner orbital increase with increasing atomic number? Why is the energy emitted not a periodic function of atomic number?
17. How many orbitals are there in each of the following subshells?
    (a) $3s$
    (b) $3p$
    (c) $3d$
    (d) $4f$
    (e) $5g$
18. The nonmetals, carbon and silicon, and the metals, tin and lead, are in column 14 of the periodic table. What is the kernel charge and the number of outer electrons for each of these elements? How does the size of atoms change going down this family of the periodic table? Why might lead, and not carbon, be a metal?
19. Complete the following table.

Atomic Number	Element	Protons	Inner Electrons	Kernel Charge	Outer Electrons
4		4	2		
	Mg	12		+2	
	Ca				2
	O	8		+6	
16			10		
34					6

20. Write a letter to Mendeleev telling him the modern basis for grouping elements into families in the periodic table.

21. For each of the following elements, list the subshell in which their outermost electrons are found.

Symbol	Element	Subshell Being Filled
O	Oxygen	
Ca	Calcium	
Fe	Iron	
Li	Lithium	
Ag	Silver	
Hg	Mercury	
U	Uranium	

CHAPTER

FIVE

BONDING IN ELEMENTS AND IN COMPOUNDS

hen a small piece of sodium metal is placed in a container of chlorine gas, a violent reaction occurs to form salt, sodium chloride. The two reactants and the product serve to illustrate metallic bonding, ionic bonding, and covalent bonding, the three major types of chemical bonding. Sodium is a metal; it conducts heat and electricity well. Sodium chloride is a rigid, high-melting solid that does not conduct electricity. However, when salt is dissolved in water or is molten, it does conduct an electric current. The electric current is carried by the movement of ions, which indicates that sodium chloride is a compound with ionic bonding. Chlorine is a diatomic gas. The two atoms comprising a chlorine molecule are joined by a covalent bond, but molecules of chlorine have only weak attractions for one another.

BONDING IN METALS

Metals conduct electricity and heat well. The shiny appearance of metals, and their rapid conduction of electricity and heat are associated with mobile electrons. Ductile metals can be drawn to form wires; malleable metals can be hammered into different shapes. Metals are ductile and malleable because bonding is retained even as atoms are pulled or pushed past one another, causing atomic neighbors to be changed.

In a simple model for *metallic bonding,* we can consider a metal to be a lattice of positive ions surrounded by a mobile electron gas. (A *lattice* is a regular, repeating array of particles in space.) In this model, the charged ions are atomic kernels, and the electron gas consists of the loosely held outer-shell electrons. Atoms of metallic elements have at most only a few outer electrons, and they all have empty orbitals in their outer shells. In a solid or liquid where atoms are close to one another, atomic orbitals on one atom overlap those on neighboring atoms. Electrons can flow from one atom to another along paths of overlapping atomic orbitals. The easy movement of electrons accounts for the conductivity of the metal.

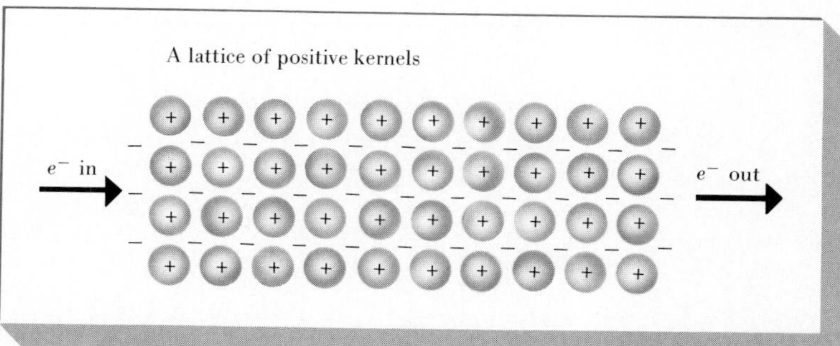

FIGURE 5.1
In one description of bonding in metals, a lattice of positive kernels is surrounded by a cloud of mobile outer electrons. The easy movement of electrons accounts for the conductivity of metals.

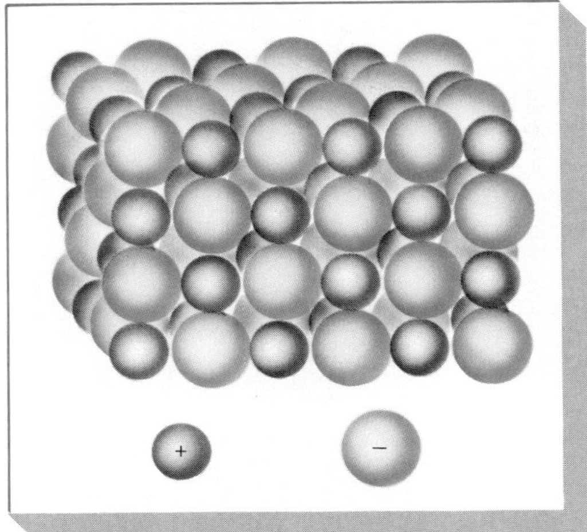

FIGURE 5.2
An ionic lattice is a regular array of positive and negative ions. Strong forces between nearby oppositely charged ions account for the high-melting point and hardness of ionic solids.

BONDING IN IONIC COMPOUNDS

Sodium chloride is a crystalline solid that melts at a high temperature. Its crystals are hard, yet they shatter when struck. Sodium chloride conducts electricity both when molten and when dissolved in water, but the solid itself does not conduct.

A *lattice of positive and negative ions* accounts for both the hardness and high melting point of sodium chloride as well as for its brittleness. Forces of attraction between positive and negative ions are great. Considerable energy needs to be supplied to melt the solid. The brittle nature of the solid results from the strong repulsive forces that result when the sliding of adjacent layers of ions brings positive ions near positive ions and negative ions near negative ions.

In the lattice, ions are not free to move. In solution or in the melt, oppositely charged ions can slide past one another; the movement of positive and negative ions in opposite directions conducts a current of electricity.

IONIC BONDING AND THE PERIODIC TABLE

Compounds formed between metals and nonmetals usually exhibit ionic bonding. Atoms of metals lose electrons to form cations, and atoms of nonmetals gain electrons to form anions. To understand why the formation of ionic compounds may be favorable, we need to consider the processes of cation formation and anion formation as well as the forces between cations and anions in a lattice.

The charges found on many cations in compounds are related to the group numbers of the elements in the periodic table. Sodium and other

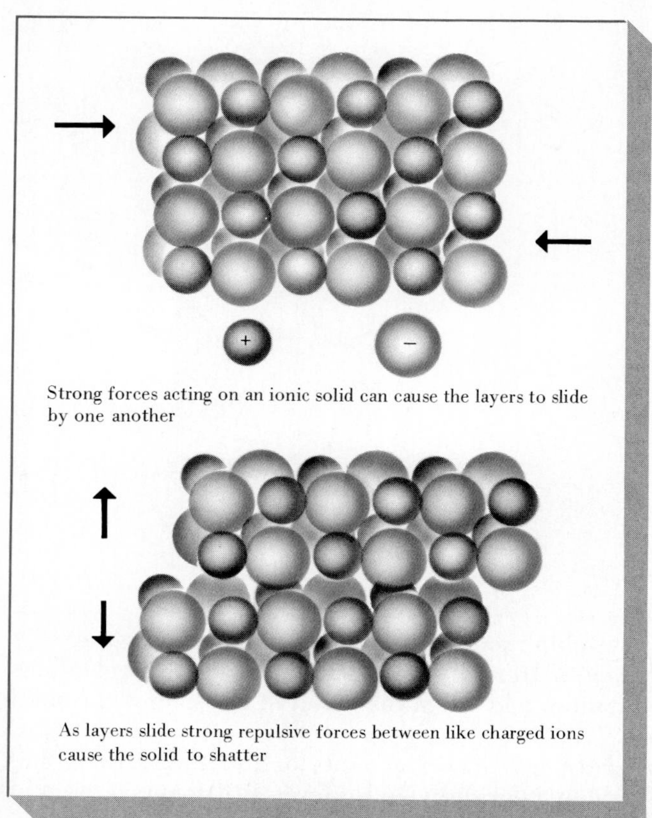

+
−
Strong forces acting on an ionic solid can cause the layers to slide by one another
As layers slide strong repulsive forces between like charged ions cause the solid to shatter

FIGURE 5.3
Although they are hard and high-melting, ionic solids also tend to be brittle. When like charges are brought near one another, there are strong forces of repulsion between them.

alkali metals in column 1 form +1 ions in compounds. Calcium and other alkaline earths in column 2 form +2 ions. Aluminum and other metals in column 13 form +3 ions ($13 - 10 =$ number of outer electrons on atoms of these elements). *Atoms of these metals tend to lose their outer electrons to form ions having octets of inner electrons.*

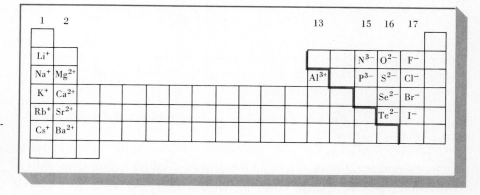

1	2	13	15	16	17
Li^+			N^{3-}	O^{2-}	F^-
Na^+	Mg^{2+}	Al^{3+}	P^{3-}	S^{2-}	Cl^-
K^+	Ca^{2+}			Se^{2-}	Br^-
Rb^+	Sr^{2+}			Te^{2-}	I^-
Cs^+	Ba^{2+}				

FIGURE 5.4
Ionic bonding is favored in compounds formed between the metals in columns 1 and 2 of the periodic table and the nonmetals in columns 16 and 17. The charge on these simple ions is directly related to their column number in the periodic table.

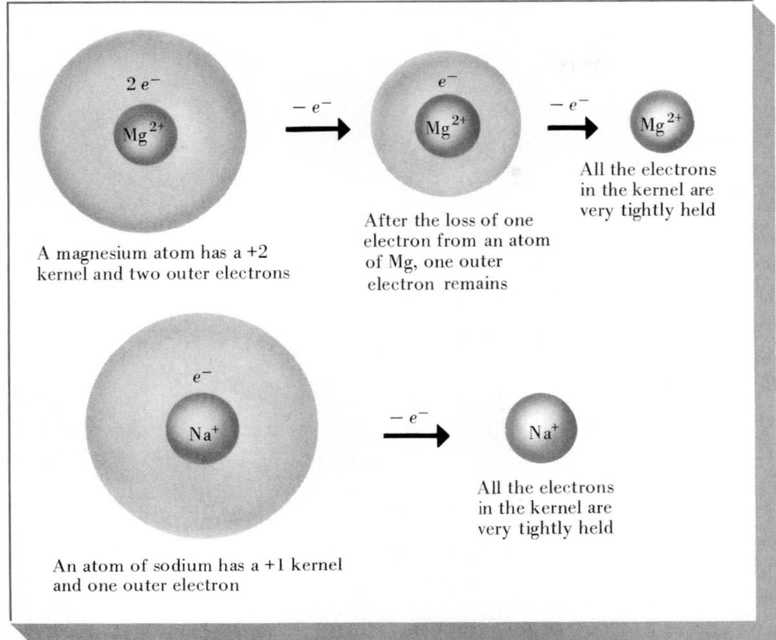

FIGURE 5.5
Less energy is required to form a +1 sodium ion than a +1 magnesium ion, but it is far easier to form a +2 magnesium ion than it is to form a +2 sodium ion.

The energy required to form a cation from an atom depends on both the charge of the ion and the position of the element in the periodic table. The energy required to remove a single electron from a sodium atom is relatively small. The outer electron is attracted only to a +1 kernel. It is energetically much more costly to remove a second electron to form +2 ions of sodium. Not only are the electrons in Na^+ closer to the nucleus than in the larger sodium atom, but they are in an inner shell that experiences a very strong attraction to the nucleus. It is energetically much less expensive to form Mg^{2+} than to form Na^{2+}, for both of the electrons removed from an atom of Mg come from the outer shell.

As one goes down a family in the periodic table, it becomes easier to form a cation. Although the kernel charge remains the same, the atoms are larger and the outer electrons are farther from the nucleus and more easily removed. For example, it is easier to form K^+ from an atom of potassium than Na^+ from an atom of sodium.

There is a corresponding relation for the charges on anions in compounds and the positions of nonmetals in the periodic table. Chlorine and the other halogens form −1 ions; members of the oxygen family form −2 ions. *These anions have an octet of outer electrons consisting of filled s and p subshells.*

The addition of one electron to a halogen atom, such as chlorine, is favored. The added electron is more attracted to the +7 kernel than it is repelled by the other seven outer electrons. The chloride anion is larger than the chlorine atom, for the +7 charge on the kernel cannot hold eight

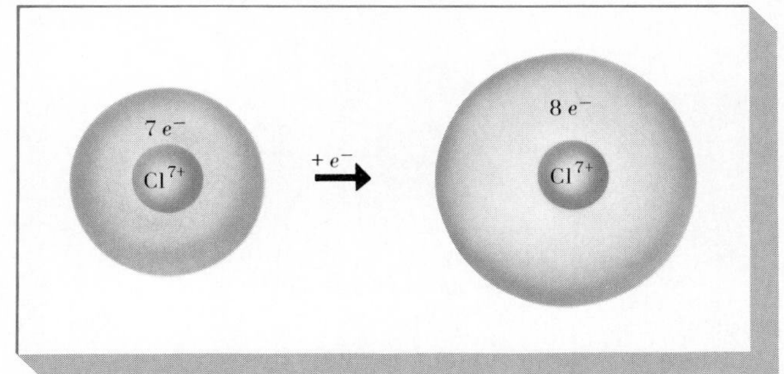

FIGURE 5.6
The strong attraction between the +7 kernel of a chlorine atom and its outer electrons permits the chlorine atom to gain an electron to form a −1 chloride ion.

electrons in the ion as closely as it holds seven electrons. Oxygen and other elements in the same family can gain two electrons to form anions with octets of outer electrons and −2 charges. The number of electrons an atom can gain equals 18 minus the column number of the element. Anions with more than eight outer electrons on a single atom are not found in nature.

The formation of compounds with ionic bonding can be considered to involve cost-benefit accounting at an atomic level. For example, the reaction of sodium with chlorine to form sodium chloride can be imagined as a series of steps. Many of the steps either are costly or pay small dividends. Sodium metal can be vaporized. Each atom of sodium can be stripped of an electron to form a sodium ion. Bonds holding the diatomic chlorine molecules together can be cleaved. Adding electrons to individual chlorine atoms forms chloride ions. When the energies in these steps are added, it is found that it is not favorable to form separated Na^+ and Cl^- ions from the elements.

The big payoff in this imagined pathway for the formation of NaCl is the formation of the ionic lattice. Since opposite electric charges strongly attract one another, bringing oppositely charged Na^+ and Cl^- ions together releases large amounts of energy. Because the benefits of lattice formation exceed the combined costs of the other steps, the formation of NaCl is favorable.

The charges on ions and the formulas of ionic compounds reflect the need for a favorable imbalance between the attractive forces of oppositely

Step		Energy cost or benefit
Vaporization	$Na(s) \rightarrow Na(g)$	Small cost
Cation formation	$Na(g) \rightarrow Na^+(g) + e^-(g)$	Small cost
Breaking Cl-Cl bond	$1/2\ Cl_2(g) \rightarrow Cl(g)$	Small cost
Anion formation	$Cl(g) + e^-(g) \rightarrow Cl^-(g)$	Small benefit
Lattice formation	$Na^+(g) + Cl^-(g) \rightarrow NaCl(s)$	Very large benefit

FIGURE 5.7
The formation of NaCl from sodium metal and chlorine gas can be imagined as a series of steps. The formation of a lattice with strong attractive forces between positive and negative ions provides the driving force for the reaction.

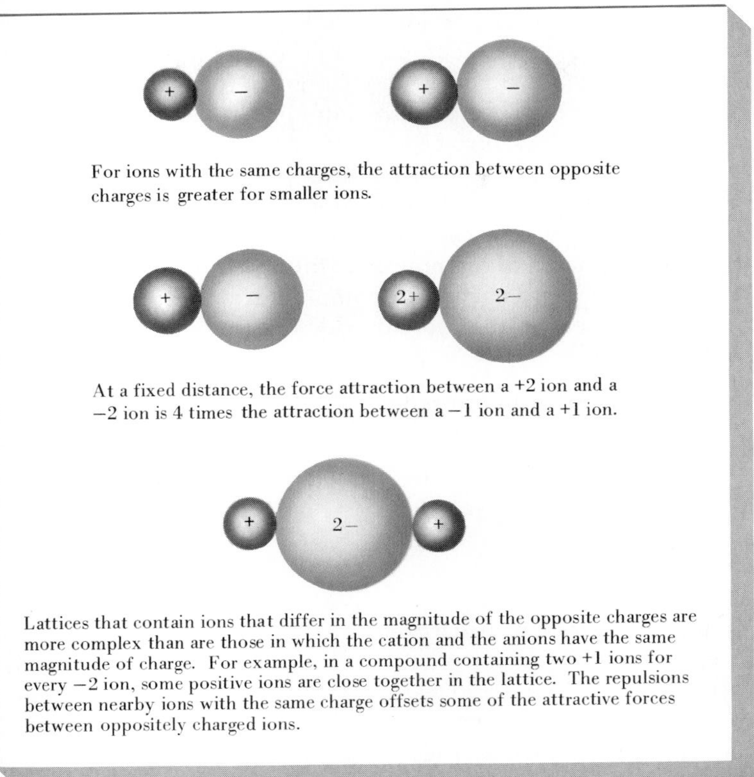

+
−
+
−
For ions with the same charges, the attraction between opposite charges is greater for smaller ions.
+
−
2+
2−
At a fixed distance, the force attraction between a +2 ion and a −2 ion is 4 times the attraction between a −1 ion and a +1 ion.
+
2−
+
Lattices that contain ions that differ in the magnitude of the opposite charges are more complex than are those in which the cation and the anions have the same magnitude of charge. For example, in a compound containing two +1 ions for every −2 ion, some positive ions are close together in the lattice. The repulsions between nearby ions with the same charge offsets some of the attractive forces between oppositely charged ions.

FIGURE 5.8
The forces between ions in an ionic crystal depend on distance, magnitude of charge, and the ratio of positive to negative ions.

charged ions and the cost of forming those ions. The higher the charges on ions, the greater the forces between them. For example, the force of attraction between +2 and −2 ions is four times as great as that between +1 and −1 ions at the same distance. However, the cost of forming cations increases rapidly as more electrons are being removed; indeed, ions with charges higher than +3 are not stable. It is also more costly to form multiply charged anions than singly charged anions. Although it is more difficult to form +2 cations and −2 anions than it is to form singly charged ions, the increased forces of attraction in the lattice can pay the added cost if it is not too great. For example, the lattice of CaO contains Ca^{2+} and O^{2-} ions, rather than Ca^+ and O^-.

FORMULAS AND NAMES OF COMPOUNDS

The formulas of ionic compounds give the ratio of cations to anions. In NaCl, there is one Na^+ for every Cl^-. If the charges on the cation and anion differ, the ions do not combine in a 1:1 ratio. This is illustrated by the formulas Na_2O and $CaCl_2$. It requires two +1 sodium ions to balance the charge on a −2 oxygen anion. Similarly, the charge on a +2 calcium cation is balanced by two −1 chloride anions.

The charges on ions in ionic compounds may not be apparent from the formulas. Consider the compound SrSe which contains strontium cations and selenium anions. To find the charge on these ions, one must look to the periodic table. Because Sr is in column 2, it has a +2 charge; selenium in column 16 has a −2 charge.

The naming of binary compounds (compounds formed between a pair of elements) is illustrated by the names of some common compounds. In the case of ionic elements, the name of the metallic element is given first, and the modified name of the nonmetal follows. The suffix *ide* usually indicates the compound as binary, or a compound formed between two elements. Two notable exceptions to this rule are sodium hydroxide, NaOH, and sodium cyanide, NaCN.

NaBr	Sodium bromide
K_2S	Potassium sulfide
CaO	Calcium oxide
HI	Hydrogen iodide

COVALENT BONDING

Nonmetallic elements vary greatly in their properties, and they combine to form compounds spanning a very wide range of properties including the gas carbon dioxide, the liquid water, and the mineral quartz. Atoms in diatomic elements such as hydrogen, as well as those in compounds such as water and carbon dioxide, are held together by *covalent bonds.* Covalent bonding can lead to the formation of discrete molecules with fixed formulas, or it can bind together very large numbers of atoms in a diamond or a grain of sand.

Atoms of nonmetallic elements can share electrons to form covalent bonds. By sharing electrons, atoms may combine to form lower-energy, more stable molecules. Consider the approach of two isolated hydrogen atoms to one another. As the atoms come close, the electron on each atom becomes attracted to both hydrogen nuclei. Because the electronic shell of each hydrogen atom is not filled, each electron is free to move about both nuclei. This sharing of electrons results in bonding. Because the hydrogen nuclei are attracted to the shared electrons, they are bonded to one another.

It is more precise to say that a bond is formed when there is a net increase in attraction between atoms, for there are opposing forces when atoms combine. There are attractive forces between the opposite charges of positive nuclei and negative electrons, and there are repulsive forces between like charges of positive nuclei and negative electrons.

Covalent bonding can only occur when low energy atomic orbitals are available to hold shared electrons. In a molecule of H_2, each of the hydrogen atoms can be considered to have a filled shell containing two electrons. No molecules having the formula H_3 are observed, because the electron from the third hydrogen atom would have to go into an outer orbital of the H_2 molecule where it has no attraction for the other atoms.

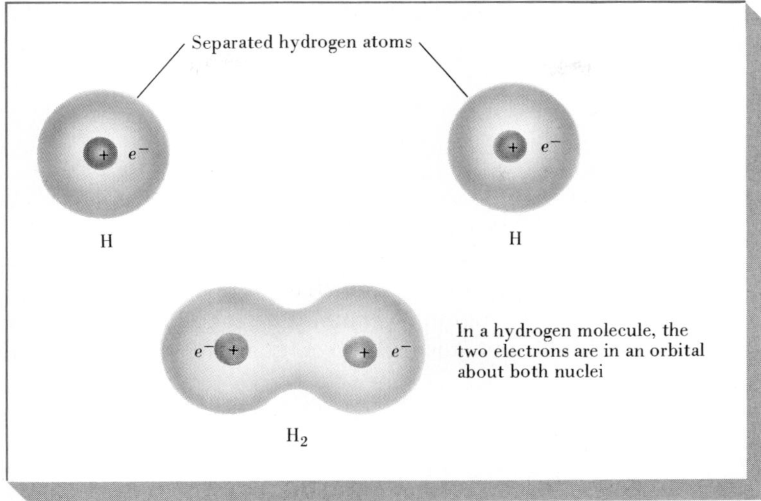

FIGURE 5.9
Atoms form covalent bonds by sharing electrons.

Note that the noble gases in column 18 do not readily form compounds. Atoms of these elements have an octet of electrons about them. (Helium, the exception, has a filled shell of two electrons.)

LEWIS STRUCTURES

Electron dot structures are used to represent bonding in covalent compounds. (Sometimes lines are used to represent shared pairs of electrons.) These structures are often referred to as *Lewis structures* after G. N. Lewis, a chemist at the University of California, who developed this model of covalent bonding. In a Lewis structure, all outer electrons are represented. Atoms share electrons to form filled shells. Each hydrogen atom can share two electrons, and each second-row element can share eight electrons.

The bonding in ammonia, water, and fluorine is shown in Fig. 5.10. In ammonia, there are covalent bonds between N and each H, and there is one pair of electrons on N that does not form a bond. It is called a *lone pair*. Water has two O—H bonds and two lone pairs of electrons on oxygen. In F_2, there is a covalent bond joining fluorine atoms, and there are three lone

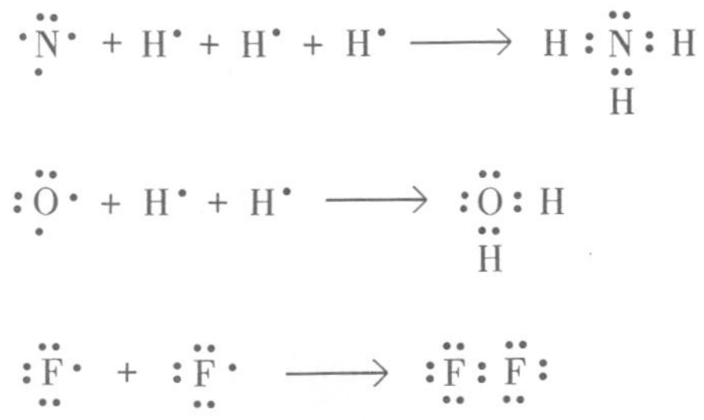

$$\cdot\ddot{N}\cdot + H\cdot + H\cdot + H\cdot \longrightarrow H:\ddot{N}:H$$
$$\qquad\qquad\qquad\qquad\qquad\qquad\quad H$$

$$:\ddot{O}\cdot + H\cdot + H\cdot \longrightarrow :\ddot{O}:H$$
$$\qquad\qquad\qquad\qquad\quad H$$

$$:\ddot{F}\cdot + :\ddot{F}\cdot \longrightarrow :\ddot{F}:\ddot{F}:$$

FIGURE 5.10
Bonding in covalent compounds is represented by electron dot structures. A shared electron pair represents a covalent bond. Pairs on a single atom are called lone pairs.

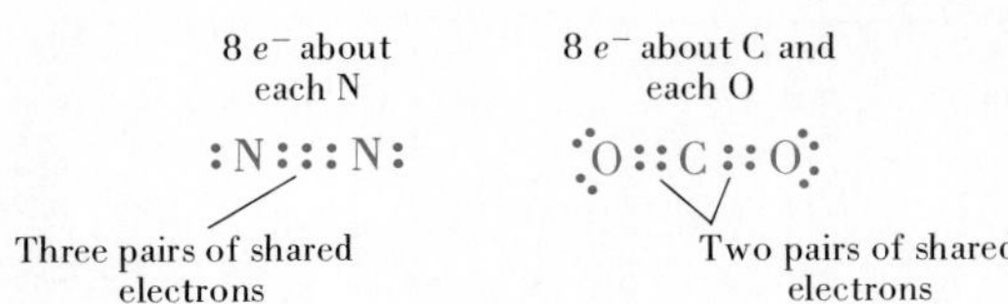

FIGURE 5.11
A molecule of nitrogen has a triple bond linking two nitrogen atoms. In carbon dioxide, double bonds link each oxygen atom to the carbon atom.

pairs on each fluorine atom. Note that atoms of the second-row elements N, O, and F have eight electrons about them. It is important to represent all outer electrons, both shared and unshared.

The bonding in nitrogen, N_2, cannot be accounted for by the sharing of a single pair of electrons. To attain an octet about each nitrogen atom, three

Draw a Lewis structure for carbon monoxide, CO.

Start with the correct atoms and the correct number of electrons

:C̈ :Ö: A correct structure must have 10 outer electrons

Join the atoms with a bond

:C̈:Ö:

Count the electrons about C and O

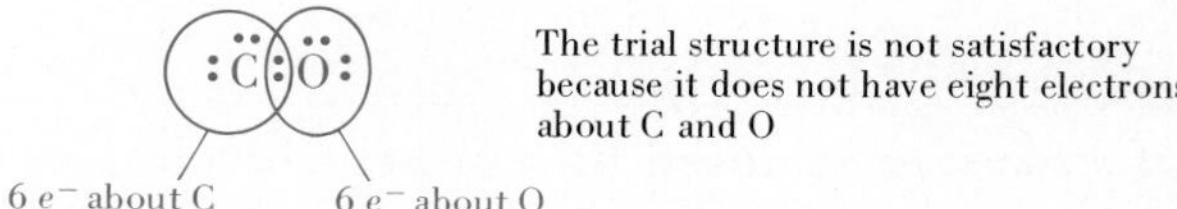

Since only the bonding electrons belong to both C and O, the only way to increase the number of electrons about one of the atoms is to move a lone pair into a bonding region

:C̈::O: :C::Ö: New trial structures

Count electrons about C and O

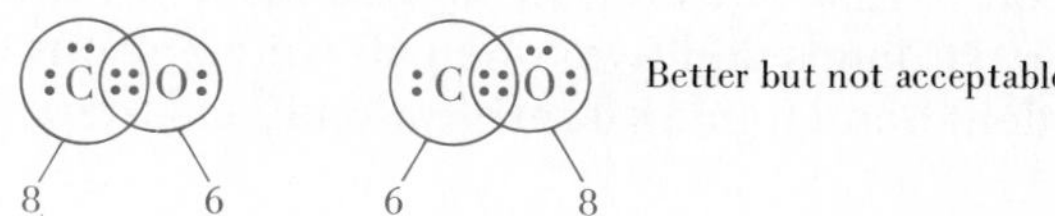

Move two electrons into the bonding region (from the atom that has an octet)

:C:::O:

Check that both C and O have an octet of outer electrons

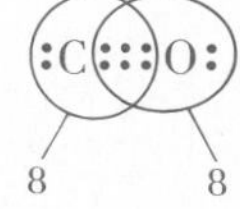

The structure is correct

FIGURE 5.12
A student should be able to write correct Lewis structures for simple molecules. One must begin with the correct number of electrons, but can proceed by trial-and-error by checking proposed structures and seeking to correct their deficiencies until a satisfactory structure is achieved.

pairs of electrons must be shared. Since three pairs of electrons hold the two nitrogen atoms together, chemists say that the nitrogen molecule has a *triple bond.*

A structure in which two pairs of electrons are shared between two atoms has a *double bond.* The Lewis structure of CO_2 has a double bond between the carbon atom and each of the two oxygen atoms.

The formulas of many covalent compounds can be determined by constructing Lewis structures. For example, what is the formula of the compound formed between the elements phosphorus and chlorine? Phosphorus is in column 15 of the periodic table, so an atom of phosphorus has five outer electrons. An atom of phosphorus can complete an octet by sharing three electrons from other atoms. Atoms of chlorine have seven outer electrons. Each chlorine atom can complete an octet by sharing one additional electron with another atom to form a covalent bond. Because three chlorine atoms can bond to each phosphorus atom, the compound formed between phosphorus and chlorine has the formula PCl_3.

Often incorrect Lewis structures can be recognized, for they fail to provide maximum bonding. When there are fewer than eight electrons on adjacent atoms, then a more stable arrangement of electrons can be reached by forming an additional bond. Note that in Lewis structures of compounds with multiple bonds, there is an octet of electrons about each carbon, nitrogen, or oxygen atom.

POLYATOMIC ANIONS

Covalent bonding is not only found in neutral molecules; it also occurs in polyatomic ions. For example, the sulfate ion has the formula SO_4^{2-} and a structure in which a central sulfur atom is surrounded by four oxygen atoms. In the nitrate ion, NO_3^-, three oxygen atoms surround a central nitrogen atom. The negative charges on sulfate and nitrate ions are not localized on single atoms; they are shared by atoms in the anions. Although the bonding within a polyatomic ion is covalent, ionic bonds hold a polyatomic ion and an ion of opposite charge in a compound.

The names and formulas of some compounds containing polyatomic anions are given. Note that parentheses are used when there is more than one polyatomic ion in a formula. In these cases, the formula of the ion is written within the parenthesis, and the subscript outside the parenthesis

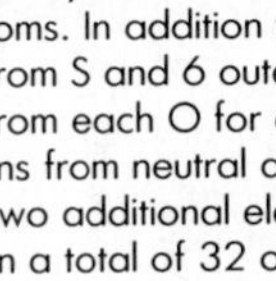

SO_4^{2-}

FIGURE 5.13
The sulfate ion has a central sulfur atom covalently bonded to four oxygen atoms. In addition to 6 outer electrons from S and 6 outer electrons from each O for a total of 30 electrons from neutral atoms, there are two additional electrons to give the ion a total of 32 outer electrons and a net charge of −2.

Name	Formula	Polyatomic Ion
Ammonium chloride	NH_4Cl	NH_4^+
Calcium nitrate	$Ca(NO_3)_2$	NO_3^-
Potassium sulfate	K_2SO_4	SO_4^{2-}
Sodium carbonate	Na_2CO_3	CO_3^{2-}

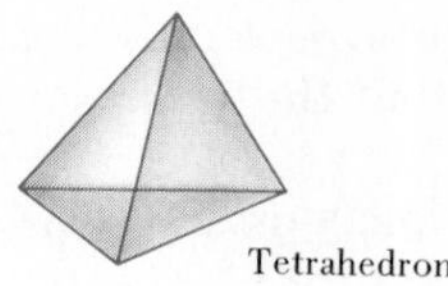

Tetrahedron | Trigonal pyramid | Bent

FIGURE 5.14
In the methane, ammonia, or water molecule, four electron pairs are located at the corners of a tetrahedron about a central C, N, or O. The shape of a molecule describes the location of nuclei.

indicates the number of anion units present. For example, $Ca(NO_3)_2$ has two nitrate ions per each calcium ion or two nitrogens and six oxygens per calcium.

SHAPES OF COVALENT MOLECULES

A simple theory that gives good predictions for the shapes of most molecules is called the valence shell electron-pair repulsion (VSEPR) model. The shapes of molecules are predicted on the basis of the number of outer-shell electron pairs about each atom. According to this theory, electron pairs repel one another; they stay as far apart as possible.

Methane, ammonia, and water are alike in that each molecule has four pairs of electrons about the central C, N, or O atom. To a first approximation, the four electron pairs in molecules of these compounds are located at the corners of a tetrahedron. (A tetrahedron has four identical faces; each face is an equilateral triangle. The four corners of a tetrahedron are equally far apart and represent the most favorable separation for four objects.) Lone pairs help determine the shape of a molecule, but they are ignored when the resulting shape is described. Methane is a tetrahedral molecule. Ammonia is a triangular pyramid, and water is bent. (Measurements of the shape of a molecule give positions of the nuclei rather than the electrons.)

$H_2C{=}CH_2$ $\qquad$ $H{-}C{\equiv}C{-}H$

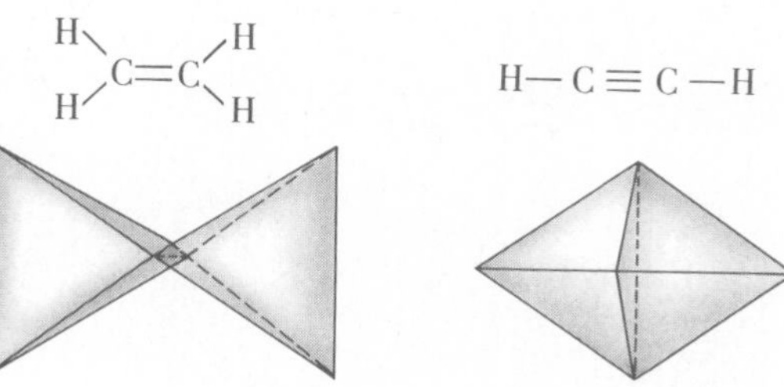

FIGURE 5.15
The atoms attached to the carbon atoms in a double bond lie in a plane. A double bond can be considered to be two tetrahedra sharing two corners or an edge. In the case of a triple bond, the atoms lie in a line. A triple bond can be considered as two tetrahedra sharing three corners or a face.

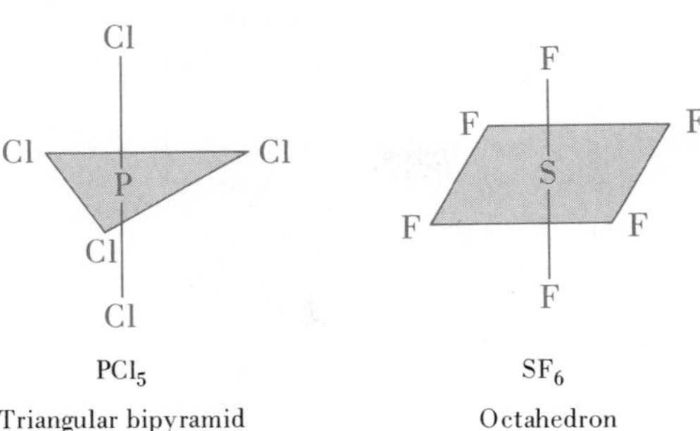

FIGURE 5.16
Elements in the third row of the periodic table can hold more than eight electrons in their outer shells. Valence shell electron-pair repulsion theory helps to predict the shape of these compounds.

This simple theory also describes the shapes of molecules of compounds with multiple bonds. In a double bond, two pairs of electrons or an edge of a tetrahedron are shared. In C_2H_4, all six nuclei lie in a single plane. In a triple bond, three pairs of electrons or a face of a tetrahedron are shared. The nuclei in C_2H_2 lie in a straight line.

Single, double, and triple bonds differ in their length. For example, when additional bonding pairs of electrons are shared between carbon atoms, the atoms are closer. Note that a carbon-carbon double bond is not half as long as a carbon-carbon single bond.

Some compounds are formed that do not have octets of electrons about central atoms. VSEPR theory also correctly predicts the shapes of these compounds. Elements in the third and higher-numbered rows of the periodic table can accommodate more than eight electrons in their outer shells. Phosphorus pentachloride, PCl_5, with 10 electrons about phosphorus is a triangular bipyramid. Sulfur hexafluoride, SF_6, with 12 electrons about sulfur is octahedral.

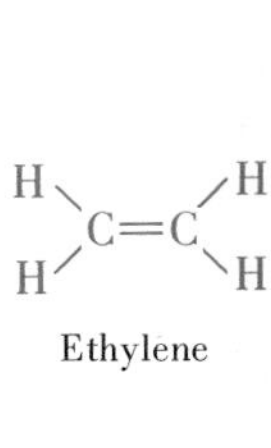

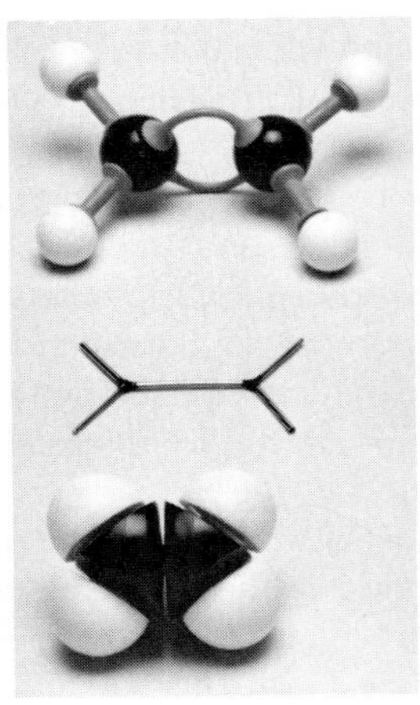

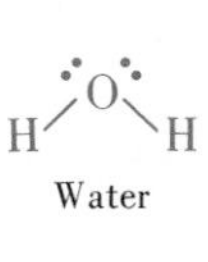

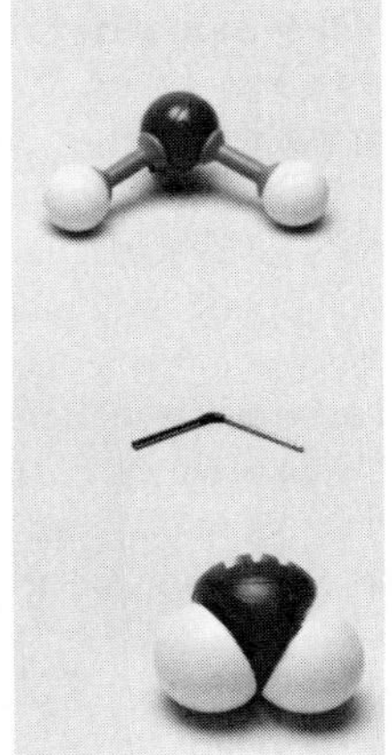

FIGURE 5.17
Three different physical models for the compounds ethylene and water are shown. The top model emphasizes connectedness by showing bonds. The middle model emphasized bond lengths and angles. The bottom model emphasizes the space occupied by electrons. (Randy Matusow.)

PHYSICAL MODELS OF COVALENT COMPOUNDS

A variety of molecular models are used by chemists to represent the three dimensional structures of compounds. Some resemble tinker toys; some are made of steel rods; others are made of a variety of interlocking molded plastic pieces.

It is instructive to compare different models of molecules. Models of two compounds are used to illustrate some of the variety of models. Ethylene, C_2H_4, has a carbon-carbon double bond. Water is a compound that has both shared and lone pairs of electrons about oxygen.

In a simple model, atoms of carbon are represented by balls with four holes directed at the corners of a tetrahedron. Oxygen is represented by a ball with two holes, and hydrogen, by a ball with one hole. Pegs connecting balls represent bonds. Chemical bonds are emphasized. In the model shown, the shape of each molecule is indicated. However, no provision is made to represent the lone pairs of electrons in water.

Dreiding models emphasize internuclear distances and angles. Rigid steel pieces are scaled to molecular dimensions. That ethylene is planar is apparent; that the planarity results from it having a double bond is not. Dreiding models of large molecules can be assembled and used to predict important spatial relationships between atoms.

A different scale model is made of interlocking molded plastic pieces. In this model, interatomic distance and angles are represented accurately. Unlike the Dreiding models, these are space-filling models. The size of a unit is proportional to the size of the electron clouds. Electron pairs are not represented distinctly, but their contribution to the shape and size of a molecule is.

A S I D E

REFLECTIONS ON MODELS

Which model is correct? Each model represents some features of a molecule closely and is wrong on other features. The proper question might be, "Which model is more useful for a given purpose?"

Scientists often choose to emphasize the utility of a model rather than its accuracy. Molecular models are designed to select and emphasize features of real compounds. By emphasizing some features and excluding others, a model may be simple or it may be complex. A model may or may not be useful for a given purpose. For example, either Dreiding or space-filling models would be preferred for estimating the dimensions of a complicated molecule. However, for illustrating Lewis structures, the simple ball and stick models prove superior.

It is apparent that a plastic ball is not an electron pair, or that a bent metal tube is not water. Molecular models are clearly artificial. They are also very useful, and the use of these models can often lead to a recognition of new relationships in actual compounds.

A model is a representation of a part of reality. It simplifies. It can be used to gain insight, and it can be manipulated to predict new relationships. A good model or theory may have high predictive power, or it may serve to relate many seemingly different phenomena in an intellectually satisfying way. The ultimate test of any model is always the degree to which it agrees with experimental observations of the properties of real substances. The greater the agreement (although not perfect), the more useful the model.

Scientists increasingly use the words "model" and "theory" interchangeably. Scientific theories are models of reality. Consider Lewis structures. Are electrons particles that can be represented by dots? Lewis structures ignore the wave character of electrons in atoms; however, they are very useful to an understanding of covalent bonding. Is Dalton's atomic theory an abstraction? It is formulated to account for combining weights and ratios. Properties of a compound such as bonding, taste, or smell are totally excluded.

QUESTIONS

1. What is the charge of the underlined ion in each of the following compounds?
   $K_3\underline{PO_4}$ $\quad$ $Na\underline{HCO_3}$ $\quad$ $\underline{Fe}SO_4$ $\quad$ $\underline{Fe}_2(SO_4)_3$
2. Use the periodic table to predict which of the following compounds are ionically bonded.
   NaI, CCl_4, BaO, P_4O_6, HCl, SiH_4
3. Using the periodic table, write formulas for the following ionic compounds: lithium iodide, potassium oxide, calcium sulfide, barium chloride, aluminum fluoride.
4. Name the following compounds.
   $AgCl$ $\quad$ $CaBr_2$ $\quad$ Al_2O_3 $\quad$ Na_2S $\quad$ Mg_3N_2
5. Sketch an ionic lattice for the compound MgO.
6. Compare the ionic compounds NaF and MgO. In NaF, the cation is (smaller, larger), the anion is (smaller, larger), and the lattice forces are (smaller, larger).
7. Rank the following cations of Group IIA metals in order of increasing size.
   Ba^{2+} $\quad$ Ca^{2+} $\quad$ Mg^{2+} $\quad$ Sr^{2+}
8. The following species have the same number of electrons. List them in order of increasing size.
   Ar $\quad$ Ca^{2+} $\quad$ Cl^- $\quad$ K^+ $\quad$ S^{2-}
9. Which of these ionic compounds of alkali metals with halogens should have the lowest melting point? Why?
   LiF $\quad$ NaCl $\quad$ KBr $\quad$ CsI
10. Name the following compounds.
    NH_4Cl $\quad$ $BaSO_4$ $\quad$ $CaCO_3$ $\quad$ $AgNO_3$
11. Write formulas for the following compounds: potassium nitrate, lithium carbonate, magnesium sulfate, ammonium bromide.
12. Which of the following compounds are covalently bonded?
    KI $\quad$ Pl_3 $\quad$ BaS $\quad$ AlF_3 $\quad$ $GeCl_4$

13. Using the periodic table, write formulas for the following covalent compounds: silicon fluoride, carbon sulfide, nitrogen chloride, phosphorus oxide.
14. Write Lewis structures for the following molecules or ions.
    (a) CH_4 and H_2O
    (b) CH_3CH_3 and CH_3OH
    (c) C_2H_4 and CH_2O
    (d) N_2 and CO
    (e) CO_2
    (f) H_3O^+ and OH^-
15. Draw Lewis structures of the following compounds. How many lone pairs of electrons are present in each?
    SiH_4 PH_3 H_2S HCl
16. Why are there neither ionic nor covalent compounds of the element neon, Ne?
17. What is the shape of each of the following molecules?
    PCl_3 H_2S CF_4 HF
18. What is the shape of each of the following ions?
    OH^- H_3O^+ NH_4^+ NH_2^-

CHAPTER

SIX

CHEMICAL STRUCTURE AND PHYSICAL PROPERTIES

Chemists seek to relate the physical and chemical properties of substances to their molecular structure. This approach was used to introduce bonding. Metals were described as conductors, and salts as rigid high-melting solids. The mobility of electrons in metals and the rigidity and strength of ionic solids were presented as consequences of types of bonding. In this chapter, we look briefly at reasons for metals differing in properties and for ionic substances not being alike.

The relationship between structure and properties is further explored for covalent compounds. The gases in the air and the grains of quartz on a sandy beach have covalent bonding, yet their properties seem to have nothing in common. How do chemists account for the properties of covalent compounds?

Much of our everyday chemical experience involves substances in solution. How can we explain the fact that ionic table salt dissolves in water but not in oil? And why is it that oil does not mix with water even though the bonding in both is covalent? In this chapter, the very special properties of water as a compound and as a solvent for other chemicals are emphasized.

METALLIC ELEMENTS DIFFER WIDELY IN PROPERTIES

Metals differ greatly in density. Some of these changes reflect trends in the periodic table. For example, the densities of metals increase dramatically going across a row of the periodic table from columns 1 and 2 and then 13. These densities of metals in the same row reflect atomic size, and the size of metal atoms decreases from left to right across the periodic table. Size differences are greatest in moving from the family containing sodium to that containing aluminum. Because atoms of aluminum are much smaller than atoms of sodium, aluminum has nearly three times the density of sodium. In contrast, the atomic sizes of the *transition metals* titanium through zinc, having partially filled or filled *d* subshells, show much more gradual changes.

The temperatures at which metals melt vary greatly. Mercury (column 12) is a liquid, while the column 1 elements are soft and have low-melting

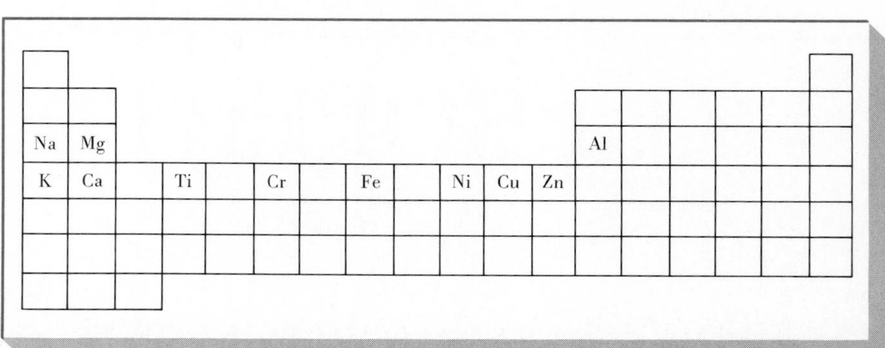

FIGURE 6.1
The metals whose symbols are given are used to illustrate changes in properties occurring from family to family across the periodic table and those occurring within families going down columns.

The Density of Elements, g/cm^3

	Sodium 0.97	Magnesium 1.74	Aluminum 2.7		
	Potassium 0.86	Calcium 1.55			
Titanium 4.5	Chromium 7.2	Iron 7.9	Nickel 8.9	Copper 9.0	Zinc 7.1

points; other metals have high strength even at elevated temperatures. The stronger the bonding in a metal, the higher the temperature required to melt it. The strength of bonding in a metal depends in part on the size of atoms and the way they pack in crystals, as well as the kernel charge and number of outer electrons of the particular element. In melting points, as in densities, many of the transition metals exhibit similarities.

For some people, the coinage metals, copper, silver, and gold, may seem the most "metallic" of metals. They are excellent conductors, are very malleable, and are quite unreactive. Atoms of copper, silver, and gold have one outer electron and a filled *d* subshell. They are considerably smaller and more tightly packed in the metal than their cousins, the alkali metals.

In contrast, a chemist usually considers the alkali metals, lithium, sodium and potassium, to be the most metallic of elements. They, too, are excellent conductors, but they are the most reactive of metals. In the alkali metals, the outer electrons are, on average, much further from the kernel and less tightly held. One way that a chemist looks at metals is as potential electron donors in chemical reactions.

IONIC VS. COVALENT BONDING

Both ionic and covalent compounds of some elements are known. For example, oxygen gains electrons to form oxide ions, O^{2-}, in compounds with sodium or calcium, but it shares electrons with hydrogen to form the compound water. In each case, oxygen atoms are surrounded by eight outer electrons. What determines whether bonding between oxygen and another element is ionic or covalent?

The Melting Point of Elements, °C

	Sodium 98	Magnesium 651	Aluminum 660		
	Potassium 64	Calcium 846			
Titanium 1675	Chromium 1890	Iron 1535	Nickel 1453	Copper 1083	Zinc 419

Oxygen forms ionic compounds with metal atoms that have low kernel charges and loosely held outer electrons. Ionic bonding is favored when an element on the far right side of the periodic table combines with an element on the left side. Atoms of these elements differ greatly in kernel charge, and hence in their attraction for electrons.

Oxygen forms covalent compounds with nonmetal atoms that have higher kernel charges and more closely held outer electrons. In general, covalent bonding is found in compounds formed between elements on the right side of the periodic table where atoms of each element have high kernel charges and strong attractions for electrons.

The term *electronegativity* has been introduced to describe the attraction that atoms of an element have for electrons. The higher the electronegativity of an element, the greater the pull its atoms exert on electrons shared with a second element. Except for the noble gases, elements in the upper right of the periodic table have high electronegativities while those in the lower left have very low electronegativities. Fluorine with its small, highly positive kernel is the most electronegative element. Cesium, the largest naturally occurring alkali metal, has the lowest electronegativity of all elements. Within a family, electronegativities decrease with increasing atomic number.

Rules for bonding in binary compounds can be stated in terms of electronegativity differences. When the difference in electronegativity between two elements is large, the bonding is ionic. When the difference in electronegativity is small and one or both elements are nonmetals, the

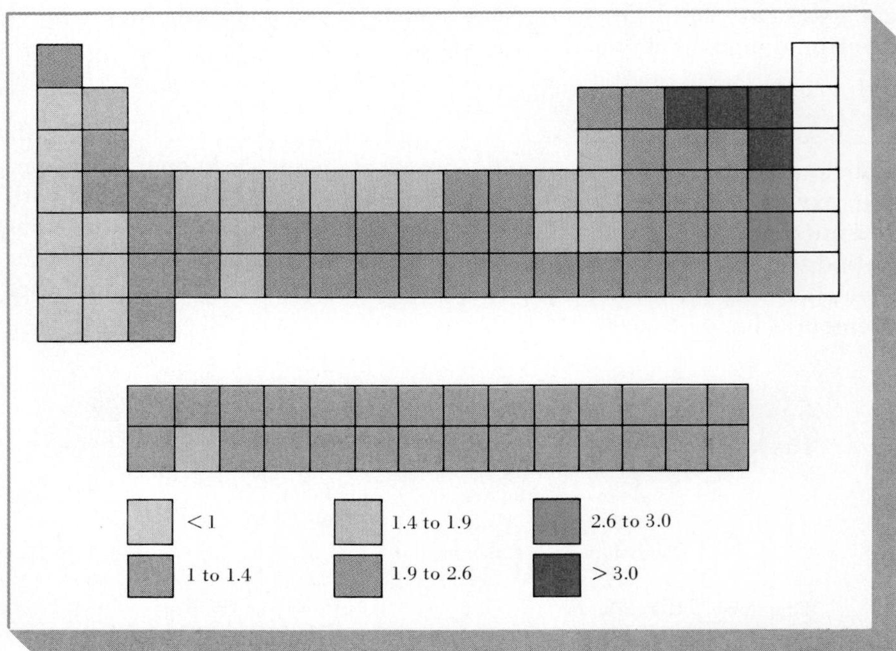

FIGURE 6.2
This periodic table indicates electronegativities of elements on a scale devised by Linus Pauling. On this scale, electronegativities range from a low of 0.7 for cesium to a high of 4.0 for fluorine.

bonding is covalent. And if the difference in electronegativity is small and both elements are metals, the bonding is metallic.

FORCES BETWEEN COVALENT MOLECULES

The boiling points of compounds vary greatly because the strength of intermolecular forces differ greatly. The boiling point of a compound provides a measure of the forces of attraction between molecules. When a liquid boils, molecules go from the liquid state in which they are close to one another into the vapor state in which they are further apart and more independent in movement. Energy needs to be supplied to overcome the forces of attraction between molecules in the liquid. The greater the forces of attraction in the liquid, the more energy must be supplied and the higher the boiling point.

Weak forces between covalent molecules depend in part on the total number of electrons present, which is given by the sum of the atomic numbers. For example, the total number of electrons is 18 in F_2, 34 in Cl_2, 70 in Br_2, and 106 in I_2. The boiling points in °C are -188 for F_2, -35 for Cl_2, 59 for Br_2, and 184 for I_2. At room temperature F_2 and Cl_2 are gases, Br_2 is a liquid, and I_2 is a solid.

Another way to increase the number of electrons in a molecule and the forces of attraction between molecules is to increase the number of atoms. We do not need to add the atomic numbers to see that CH_4, $C_{10}H_{22}$, and $C_{20}H_{42}$ contain increasing numbers of electrons. The forces between the molecules increase with the number of electrons, and at room temperature CH_4 is a gas, $C_{10}H_{22}$ is a liquid, and $C_{20}H_{42}$ is a waxy solid. Each carbon atom has about the same weak attraction for a neighboring molecule as does each hydrogen atom, but the attractions accumulate over all of the carbon atoms and hydrogen atoms in a molecule.

Atoms of different elements do not share electrons equally. A more positive kernel or a smaller kernel of the same charge will exert a greater pull on a shared pair of electrons. (This difference was introduced earlier with the concept of electronegativity.) The atom getting the greater portion of the shared electrons becomes slightly negative, and the atom losing some electron density becomes slightly positive. Chemists describe the covalent bond between such atoms as *polar*. It has a more positive end and a more negative end.

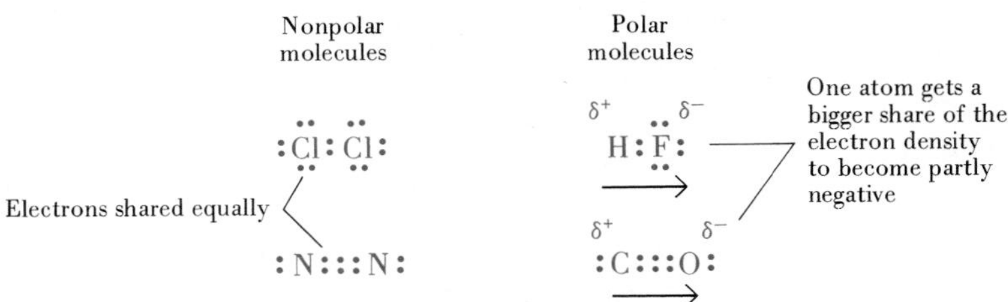

FIGURE 6.3 Covalent bonds may be either polar or nonpolar.

The *dipole moment* of a compound is a measurement of its net polarity. Both polar bonds and lone pairs of electrons can contribute to the dipole moment. Although bonds between different elements are always polar to some extent, not all compounds formed between different elements are polar. Molecules of a compound may have a symmetric shape that causes the bond dipoles to cancel. If that happens, the compound is nonpolar even though it has polar bonds.

Carbon monoxide has a dipole moment; carbon dioxide does not. Both compounds have polar carbon-oxygen bonds. However, in the linear carbon dioxide molecule, the carbon-oxygen bonds point in opposite directions and the two bond dipoles cancel.

Stronger intermolecular forces exist between compounds made up of polar molecules. Because the positive end of one dipole is attracted to the negative end of another molecule, polar molecules tend to have higher boiling points than nonpolar molecules of similar size and composition.

CO_2 AND SiO_2—INTRAMOLECULAR FORCES VS. INTERMOLECULAR FORCES

It is important to make a sharp distinction between those forces holding atoms together in a molecule and those forces that cause one molecule to be attracted to another. The properties of carbon dioxide and of silicon dioxide dramatically illustrate this distinction. Carbon dioxide is a gas at room temperature. Under high pressures, it liquefies. When liquid CO_2 comes out of a fire extinguisher, it forms a solid that passes directly from the solid state to the gas state on warming. In contrast, SiO_2 is sand or quartz. It is a solid at temperatures up to over 1600°C; it is both hard and strong.

There are similarities in the bonding of these very different compounds. Both CO_2 and SiO_2 are covalent compounds, and their elements carbon and silicon are in the same family of the periodic table. Atoms of both carbon and silicon each have four outer electrons and form four strong covalent bonds to oxygen. The intramolecular forces, the covalent bonds, are strong in both carbon dioxide and in sand.

However, carbon dioxide and silicon dioxide have very different structures. Silicon dioxide is a *giant molecule.* Each oxygen atom is a bridge

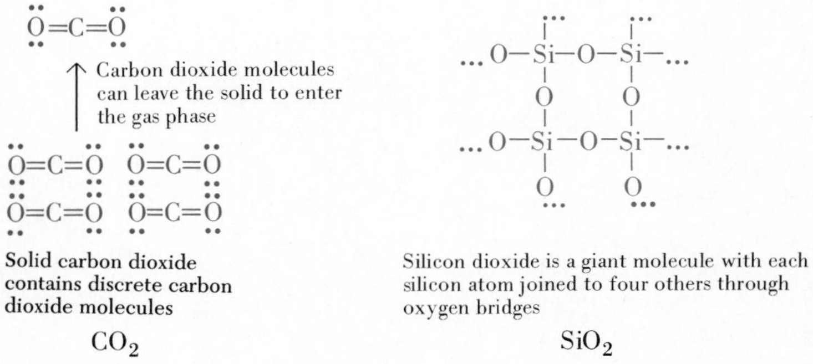

O=C=O

Carbon dioxide molecules can leave the solid to enter the gas phase

O=C=O O=C=O
O=C=O O=C=O

Solid carbon dioxide contains discrete carbon dioxide molecules

CO_2

...O—Si—O—Si—...
O O
...O—Si—O—Si—...
O O

Silicon dioxide is a giant molecule with each silicon atom joined to four others through oxygen bridges

SiO_2

FIGURE 6.4
Carbon dioxide and silicon dioxide have very different physical properties even though both compounds have covalent bonding. Carbon dioxide is a gas at room temperature, and silicon dioxide is a high-melting solid.

between two silicon atoms. A silicon dioxide crystal is one giant covalent molecule. In contrast, carbon dioxide consists of discrete CO_2 molecules. Although strong double bonds link each oxygen atom to a carbon atom, forces between different CO_2 molecules are weak. Little work against intermolecular forces is required to separate molecules, and thermal energy is sufficient to overcome intermolecular attractions even at low temperatures.

HYDROGEN BONDING AND THE PROPERTIES OF WATER

Water is a remarkable compound. Its boiling point is hundreds of degrees higher than those of the heavier diatomic gases O_2 and N_2, and it is higher than those of H_2S and H_2Se, the corresponding compounds of elements below oxygen in the periodic table. Water is a good solvent for ionic compounds and for many polar molecules, and it is the solvent for life processes occurring both inside and outside of cells. More energy is required to melt ice or vaporize water than is commonly required for the corresponding phase transitions of most other simple covalent compounds. Lakes and oceans buffer seasonal weather extremes, and the evaporation of moisture is an important part of the body's temperature control.

Water molecules are bent. Both the polar oxygen-hydrogen covalent bonds and the lone-pair electrons on oxygen atoms contribute to the dipole moment of water. The hydrogens on each water molecule are relatively positive, and the oxygen is relatively negative.

Strong intermolecular forces contribute to the special properties of water. There is a strong attraction between the positive hydrogen of one molecule and the lone pair of electrons on an adjacent molecule. These intermolecular attractions are called *hydrogen bonds*. Hydrogen bonds are the strongest intermolecular forces between covalent molecules. Attractive forces in water are especially large because each water molecule can be hydrogen-bonded to four neighbors.

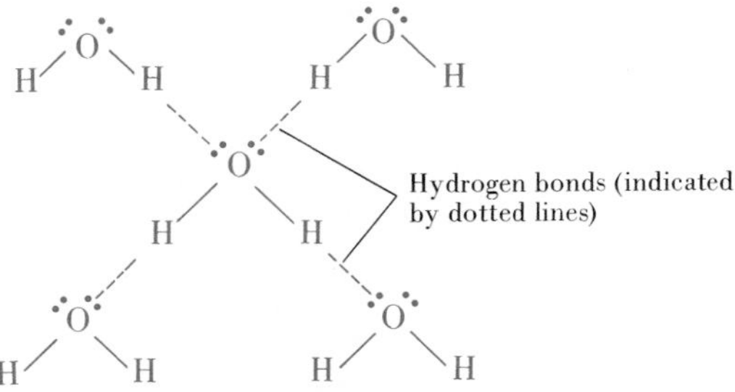

FIGURE 6.5
Hydrogen bonds are strong intermolecular forces which contribute to the special properties of water. In a hydrogen bond, the partially positive hydrogen atom of one water molecule is attracted to the lone pairs of electrons on an adjacent water molecule.

Water is not the only compound to exhibit hydrogen bonding. Both ammonia, NH_3, and hydrogen fluoride, HF, form strong hydrogen bonds. To form a hydrogen bond, both a relatively positive hydrogen (hydrogen attached to N, O, or F) and a lone pair of electrons on a small kernel (N, O, or F) are required. Hydrogen bonding contributes to the properties of many organic compounds and compounds of biological importance having OH or NH groups as part of their structure. The chemistry of some of these compounds is discussed in later chapters.

A S I D E

WHY ICE FLOATS

Few substances expand upon freezing. For most substances, the solid is more dense than the liquid. In a crystal lattice, molecules or ions are usually packed closely together and there is no room for particles to move. Upon melting, most substances expand, some holes are formed, and the presence of these holes permits the particles to move.

That water expands on freezing is essential to life in cold climates. If ice could sink and the water above it could continue to freeze, entire lakes would solidify during the winter. Less warming could occur during the summer because only the surface waters would warm initially and the accumulated ice at lower levels would not be exposed to the sun.

Why does ice float on water? Ice has a great number of hydrogen bonds and more regular hydrogen bonds than does water. Each molecule of H_2O can participate in up to four hydrogen bonds. In the lattice of ice, molecules are lined up to form strong hydrogen bonds rather than to be as close together as possible. This gives the ice lattice a cagelike structure in which there are small holes. Because water has a less regular structure and has fewer holes than does ice, water is the more dense substance.

WATER AS A SOLVENT

Although salt is held together by strong ionic lattice forces, it readily dissolves in water. In contrast, although forces between hydrocarbon mole-

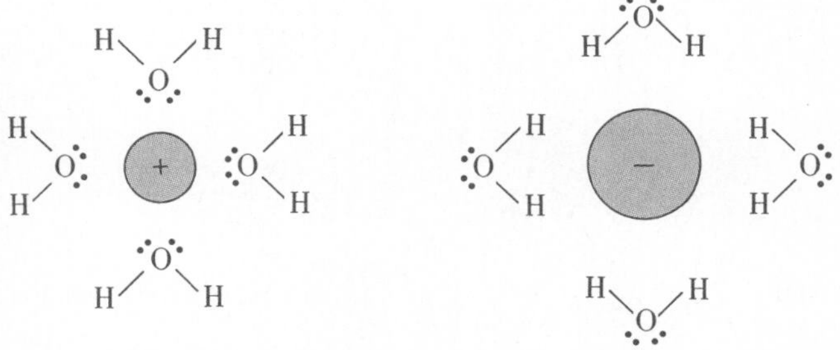

FIGURE 6.6
In a solution of a salt, the negative ends of water molecules are attracted to cations and the positive ends of water molecules are attracted to anions. The attractive forces between water molecules and ions help to overcome the lattice forces of an ionic compound causing it to dissolve.

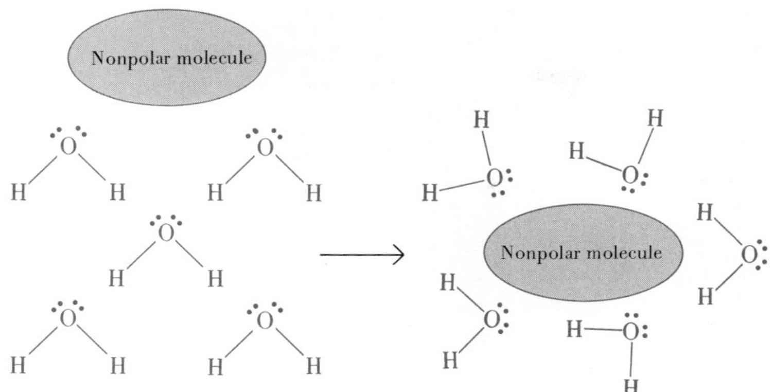

FIGURE 6.7
To dissolve a large nonpolar molecule in water, many hydrogen bonds between adjacent water molecules would be broken. There is no attraction between water molecules and nonpolar molecules to overcome the cost of breaking hydrogen bonds.

cules are weak, oil and water do not mix. The polar character of water and its strong network of hydrogen bonds contribute to these special properties of water as a solvent.

Many ionic compounds dissolve in water because there are strong attractive forces between water molecules and both positive and negative ions. Water molecules surround cations, with the negative end of the water dipole close to the positive ion. Around anions, the orientation of the water dipole is reversed; positive hydrogens are attracted to the negative ion. Chemists say that water solvates ions, because favorable forces between the water molecules and ions help bring ions into solution. In addition, the polar nature of water serves to shield the forces of attraction between separated, oppositely charged ions in solution.

Water is a poor solvent for most nonpolar compounds. Many hydrogen bonds between adjacent water molecules would have to be broken for a large molecule of a nonpolar compound to enter water. Because there are no strong forces between nonpolar molecules and water molecules to help pay this cost, oil-like compounds do not dissolve in water.

A S I D E

SOLUBILITY RULES FOR IONIC COMPOUNDS

Chemists have formed the following generalizations about the solubilities of ionic compounds in water:

1. All ammonium, potassium, and sodium compounds are soluble.
2. All nitrates are soluble.
3. Most chlorides, bromides, and iodides are soluble. (Exceptions are the halides of silver, lead, and mercury.)
4. Compounds of +2 and +3 cations with −2 or −3 anions generally have low solubility. For example, $CaCO_3$ and $BaSO_4$ are both insoluble.

Compounds dissolve when the solvation of ions by water can overcome the forces between ions in the lattice of the solid. The first three rules describe compounds in which the lattice forces are relatively weak. The ions named have only +1 or −1 charges, and many of these ions are relatively large. When the lattice forces are much greater, solvation of ions by water cannot overcome the lattice forces, and the solids tend to be insoluble.

QUESTIONS

1. Which of the following compounds are covalently bonded? Which are ionically bonded?
   KI PI_3 BaS AlF_3 $GeCl_4$
2. Use the periodic table to determine which element in each of the following pairs is more electronegative.
   (a) Br and Cl
   (b) F and N
   (c) Al and Na
   (d) O and P
3. Silver fluoride, alone of the silver halides, is soluble in water. Why is the bonding in AgF ionic, while that in AgCl, AgBr, and AgI is covalent?
4. The compound ICl is polar. Draw a Lewis structure of this compound. Would I or Cl attract elements more? Which end of the ICl molecule is more negative?
5. The compound BF_3 has a zero dipole moment even though each of the three B—F bonds is polar. What is the shape of BF_3?
6. Which of the following compounds has no net dipole moment? (In each compound, a carbon atom is at the center with bonds directed toward the corners of a tetrahedron.)
   CH_2Cl_2 $CHCl_3$ CCl_4
7. Although the bonding in elemental carbon (diamond) and in elemental nitrogen (N_2) is covalent, these elements differ greatly in their physical properties. Why?
8. Why is the boiling point of NH_3 much higher than that of CH_4?
9. Why is ammonia, NH_3, much more soluble in water than is methane, CH_4?
10. Illustrate the solvation of Ca^{2+} and of Cl^- ions that enables the ionic compound $CaCl_2$ to dissolve in water.
11. If the water molecule were linear rather than bent, would ocean water be as salty? Why or why not?
12. Methane (CH_4), the major component of natural gas, is much more soluble in water than is C_8H_{18}, a component of gasoline. Why should the smaller hydrocarbon molecule be more soluble in water?
13. Sodium chloride dissolves in liquid HF but not in C_8H_{18}. Why should this ionic solid be more soluble in HF than in C_8H_{18}?
14. Seawater contains salts of sodium, potassium, and calcium. Why does a clam use the less-abundant calcium ions to help build its shell rather than more-abundant sodium or potassium ions?

CHAPTER

SEVEN

ACIDS AND BASES

In chemistry as in other disciplines, more than one classification scheme is used. A structural approach to chemistry was presented in the previous two chapters. Compounds were grouped into two major classes, those with ionic bonding and those with covalent bonding. Approaches used to describe chemical changes are presented in this chapter and the next. Reactions are classified, and roles are assigned to chemical participants.

Biologists and political scientists also use both structural and behavioral classifications. When discussing a plant or animal, a biologist may emphasize either its taxonomy or its ecological niche. Similarly, a political scientist might discuss either constitutional checks and balances or voting behavior when talking about democratic government.

Classification schemes are used to show order in chemical change just as the periodic table is used to show order in chemical structure. Reactions are classified to emphasize similarities and differences.

ACIDS AND BASES

The reactions of acids with bases comprise a major category of chemical reactions. Earlier, acids were operationally defined by observations made when experiments were performed. *Acids* taste sour, fizz with soda, and cause some vegetable dyes to change color. *Bases* taste bitter and reverse any color changes that vegetable dyes undergo with acids.

What structural changes occur when an acid reacts with a base? Consider the following reaction. When fumes from a bottle of hydrochloric acid mix with fumes from a bottle containing a solution of ammonia in water, a white fog forms and a powder settles out. The reactants are hydrogen chloride and ammonia, and the product is ammonium chloride. In the reaction, the covalently bonded gases react to form an ionic solid.

$$\underset{\text{Hydrogen chloride}}{HCl(g)} + \underset{\text{Ammonia}}{NH_3(g)} \rightarrow \underset{\text{Ammonium chloride}}{NH_4Cl(s)}$$

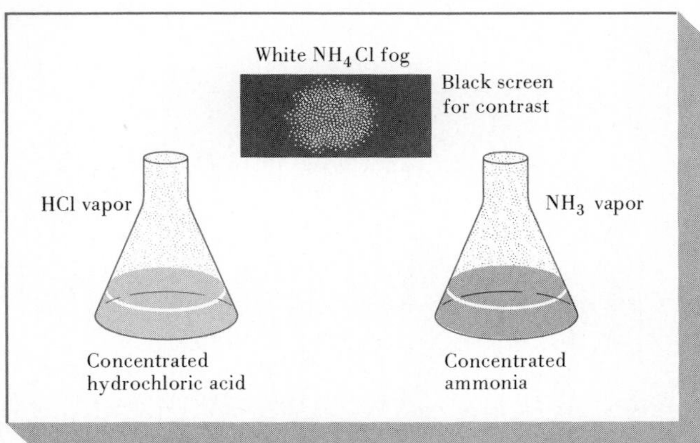

FIGURE 7.1
HCl vapor and NH_3 vapor react in an acid base reaction to form solid NH_4Cl.

$$\begin{array}{ccccccccc} \text{H} & & & & & \text{H}^{+} & & & \\ \text{H}:\ddot{\text{N}}: & + & \text{H}:\ddot{\underset{\cdot\cdot}{\text{Cl}}}: & \longrightarrow & & \text{H}:\ddot{\underset{\cdot\cdot}{\text{N}}}:\text{H} & + & :\ddot{\underset{\cdot\cdot}{\text{Cl}}}:^{-} \\ \text{H} & & & & & \text{H} & & & \\ \text{Base} & & \text{Acid} & & & & & & \end{array}$$

FIGURE 7.2
In the acid-base reaction of hydrogen chloride and ammonia, a bond between H^+ and Cl^- is broken, and a new bond is formed between H^+ and NH_3. The NH_4^+ and the Cl^- ions are combined in the ionic solid NH_4Cl.

In this acid-base reaction, the acid HCl donates a proton (the nucleus of a hydrogen atom) to the base NH_3, leaving the chloride ion, Cl^-, as the anion in ammonium chloride. The base NH_3 accepts a proton to form the ammonium ion, NH_4^+, which is the cation in ammonium chloride. In this acid-base reaction, a covalent bond between H and Cl is broken, and a new covalent bond is formed between H and N.

According to a definition formulated by Brønsted and Lowry, *an acid is a proton or hydrogen ion donor*, and *a base is a proton acceptor*. According to this definition, *all acids contain a hydrogen atom that is donated as* H^+, and *all bases have an unshared pair of electrons to bond to the incoming proton*. In every acid-base reaction, a new covalent bond is formed.

Brønsted and Lowry extended their definitions to the products formed in an acid-base reaction. The species formed from an acid by the loss of a proton is called the *conjugate base* of the acid. Similarly, the species formed when a base gains a proton is called the *conjugate acid* of the base. Consider the reaction of HCl and NH_3 to form NH_4Cl. In this example, Cl^- is the conjugate base of HCl, and NH_4^+ is the conjugate acid of NH_3.

Note that the products of an acid-base reaction have the potential to act as acids and bases. The conjugate base of every acid has an unshared pair of electrons with which it can accept a proton. The conjugate acid of every base contains a hydrogen atom which it can donate as a proton.

STRONG ACIDS AND BASES IN WATER

A solution of the gas HCl in water is called hydrochloric acid. Hydrochloric acid is a good conductor of electricity, for it contains ions formed by the reaction of HCl with water.

TABLE 7.1

COMMON STRONG ACIDS

Name of Acid	Formula	Conjugate Base	Name of Base
Hydrochloric acid	HCl	Cl^-	Chloride ion
Nitric acid	HNO_3	NO_3^-	Nitrate ion
Sulfuric acid	H_2SO_4	SO_4^{2-}*	Sulfate ion

* Sulfate ion is formed by the loss in step sequence of two acidic protons from sulfuric acid. Strictly speaking, HSO_4^- is the conjugate base of sulfuric acid. However, in dilute solutions of sulfuric acid, HSO_4^- reacts extensively with water to form hydronium ions and sulfate ions.

$$\underset{\text{Acid}}{HCl(g)} + \underset{\text{Base}}{H_2O(l)} \rightarrow \underset{\text{Hydronium ion}}{H_3O^+(aq)} + Cl^-(aq)$$

Hydrogen chloride reacts as an acid to donate a proton to water. Water is the base in this reaction, for it accepts a proton to form hydronium ion, H_3O^+. Only a few molecules of HCl remain in a dilute solution of hydrogen chloride. Acids that almost completely react with water to form hydronium ions are called *strong acids.* Common strong acids are hydrochloric acid, nitric acid, and sulfuric acid.

Hydronium ions are the most powerful proton donors that can exist in the presence of water. Acid-base reactions of aqueous strong acids are thus reactions of H_3O^+.

The best proton acceptor found in dilute aqueous solutions is the hydroxide ion, OH^-. Bases that dissolve in water or react almost completely with water to form hydroxide ions are called *strong bases.* Sodium hydroxide, NaOH, is the most common strong base. The bonding in NaOH is ionic, and a solution of sodium hydroxide contains sodium ions and hydroxide ions. Other strong bases are potassium hydroxide and barium hydroxide. If any base stronger than OH^- is added to water, it reacts with water to remove a proton and form hydroxide ion.

REACTIONS OF STRONG ACIDS WITH STRONG BASES

When solutions of hydrochloric acid and sodium hydroxide are mixed, an acid-base reaction occurs. Hydronium ions and hydroxide ions react to form water. The reaction of an acid with a base is called a *neutralization* reaction. In this example, one might say that the strong acid has neutralized the base or that the base has neutralized the acid.

$$H_3O^+ + Cl^- + Na^+ + OH^- \rightarrow Na^+ + Cl^- + 2H_2O$$

Chemists write *net ionic equations* to emphasize the species that are actually undergoing reaction. If equal molar quantities of hydrochloric acid and sodium hydroxide are mixed, the resulting solution is identical in composition to a solution prepared by dissolving sodium chloride in water. Sodium ions and chloride ions are present in solution both before and after

TABLE 7.2

COMMON STRONG BASES

Name	Formula
Sodium hydroxide	NaOH
Potassium hydroxide	KOH
Barium hydroxide	$Ba(OH)_2$

the acid-base reaction. Since they do not react, they may be omitted from the equation for the reaction. The resulting equation, a net ionic equation, focuses on the chemistry of the reacting species.

$$H_3O^+ + OH^- \rightarrow 2H_2O \qquad \text{net ionic equation}$$

WEAK ACIDS AND BASES

Strong acids and bases play important roles in industry, but for many people acetic acid in vinegar, citric acid in fruit juice, and the base ammonia in window washing fluid are far more familiar. Solutions of these compounds are poor conductors of electricity because relatively few ions are present. Acetic acid and citric acid are *weak acids*, acids that react only to a small extent with water to form hydronium ions and their conjugate bases. Ammonia reacts only to a small extent with water to form hydroxide ions and ammonium ions; it is considered to be a *weak base*. Most acidic and basic substances react only to a small extent with water. The majority of acids are weak acids, and most bases are weak bases.

To illustrate the reactions of weak acids and bases, consider acetic acid and ammonia as representative examples. Their reactions with water occur only to a small extent.

$$\underset{\text{Acetic acid}}{HCH_3CO_2} + H_2O \rightleftharpoons \underset{\text{Acetate ion}}{CH_3CO_2^-} + H_3O^+$$

$$\underset{\text{Ammonia}}{NH_3} + H_2O \rightleftharpoons \underset{\text{Ammonium ion}}{NH_4^+} + OH^-$$

In the reactions shown above, arrows of different lengths are used to indicate the extent of reaction. Solutions of ammonia in water are sold under the label "ammonium hydroxide" although they contain far fewer ammonium ions and hydroxide ions than ammonia molecules.

However, acetic acid will neutralize sodium hydroxide, and ammonia will react completely with hydrochloric or sulfuric acid. The reaction of a weak acid with a strong base and the reaction of a weak base with a strong acid are relatively complete.

$$HCH_3CO_2 + OH^- \rightarrow CH_3CO_2^- + H_2O \qquad \text{reaction with a strong base}$$

$$NH_3 + H_3O^+ \rightarrow NH_4^+ + H_2O \qquad \text{reaction with a strong acid}$$

The conjugate bases of weak acids are much stronger bases than are the conjugate bases of strong acids. The more weakly an acid donates a proton, the more strongly the conjugate base accepts it.

One test used to identify an acid shows that acids fizz with baking soda, $NaHCO_3$. Carbonic acid, H_2CO_3, is in equilibrium with carbon dioxide and water. Carbonic acid can donate one proton to water and leave bicarbonate ion, HCO_3^-, a very weak base. If a second proton is donated, carbonate ion,

TABLE 7.3

COMMON WEAK ACIDS AND WEAK BASES*

Name	Formula	Conjugate Base	Name
Phosphoric acid	H_3PO_4	$H_2PO_4^-$	Dihydrogen phosphate ion
Dihydrogen phosphate ion	$H_2PO_4^-$	HPO_4^{2-}	Hydrogen phosphate ion
Hydrogen phosphate ion	HPO_4^{2-}	PO_4^{3-}	Phosphate ion
Acetic acid	$HC_2H_3O_2$	$C_2H_3O_2^-$	Acetate ion
Carbonic acid	H_2CO_3	HCO_3^-	Bicarbonate ion
Bicarbonate ion	HCO_3^-	CO_3^{2-}	Carbonate ion
Ammonium ion	NH_4^+	NH_3	Ammonia

* The conjugate of a weak acid is a base. All of the species in the right column except dihydrogen phosphate ion form basic solutions in water.

CO_3^{2-}, a stronger base, is left. Adding acid to a bicarbonate or carbonate compound reverses these reactions, which results in the evolution of carbon dioxide gas.

Reaction of carbonic acid with water:

$$H_2CO_3 + H_2O \rightleftharpoons HCO_3^- + H_3O^+$$

Reactions of carbonic acid and bicarbonate ion with hydroxide:

$$H_2CO_3 + OH^- \rightarrow HCO_3^- + H_2O$$
$$HCO_3^- + OH^- \rightarrow CO_3^{2-} + H_2O^+$$

Reactions of carbonate ion and bicarbonate ion with acid:

$$CO_3^{2-} + H_3O^+ \rightarrow HCO_3^- + H_2O$$
$$HCO_3^- + H_3O^+ \rightarrow H_2CO_3 + H_2O$$
$$H_2CO_3 \rightleftharpoons CO_2 + H_2O$$

Carbonated water has a slightly acidic or sour taste. A small amount of sweetener can give a carbonated beverage an attractive sweet, yet tart, taste. When vinegar, a stronger acid, is used in sweet and sour sauces, more sweetener is required. Carbonated water has the added advantages of low cost and no calories. (To hold the tartness even when the bubbles are gone, soft drink producers add phosphoric acid to their syrups.)

Why do some dyes from plant sources change color in acids and bases? Dyes used as indicators are themselves weak acids or weak bases. The colors of an indicator acid and its conjugate base are different. When a base reacts with the acidic form of an indicator, it removes a proton to convert the indicator to its conjugate base and produce a color change. Similarly, when an acid donates a proton to the basic form of an indicator, it brings about a color change.

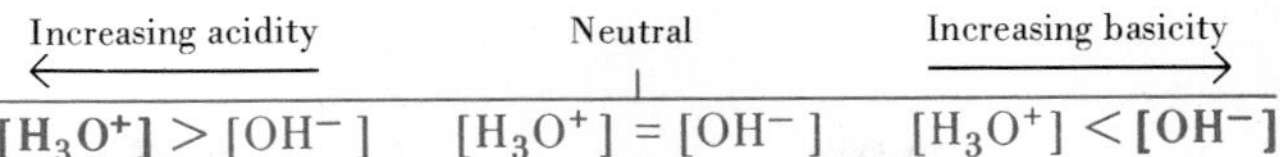

FIGURE 7.3
In a neutral solution, the concentration of hydronium ions equals the concentration of hydroxide ions. An acidic solution has an excess of hydronium ions, and a basic solution has an excess of hydroxide ions.

ACIDS AND BASES—A STRUCTURAL DEFINITION

Pure water or a neutral water solution, one that is neither acidic nor basic, contains equal concentrations of hydronium ions and hydroxide ions. The concentration of each of these ions in pure water is very small, 1×10^{-7} mol/L at 25°C, but these ions are present because the acid-base reaction of water with itself occurs to a small extent.

$$H_2O + H_2O \rightarrow H_3O^+ + OH^-$$

In a commonly used definition of acids and bases, a comparison to pure water is made when labeling a substance as an acid or a base. An *acidic solution* has an excess of hydronium ions over hydroxide ions; a *basic solution* has an excess of hydroxide ions over hydronium ions. There is a subtle shift of definition in this description, for the terms "acid" and "base" were introduced earlier to classify roles played in chemical reactions. The definition of acids and bases that involves comparisons to pure water makes no reference to reactions. This definition is based on chemical species present rather than chemical change.

pH

It is important to differentiate between the total acid present in a sample and the concentration of hydronium ions present. Many of the properties of

TABLE 7.4

pHs OF SOME REPRESENTATIVE SOLUTIONS

pH	$[H_3O^+]$	Solution
0	1.0	1.0 mol/L hydrochloric acid
1	1×10^{-1}	0.1 mol/L hydrochloric acid
2	1×10^{-2}	Lemon juice
3	1×10^{-3}	Vinegar
7	1×10^{-7}	Neutral solution (pure water)
7.4	4×10^{-8}	Blood
9	1×10^{-9}	Baking soda
11	1×10^{-11}	Household ammonia
13	1×10^{-13}	0.1 mol/L sodium hydroxide
14	1×10^{-14}	1.0 mol/L sodium hydroxide

acids and bases depend only on the concentrations of hydronium ions or hydroxide ions. A strong acid reacts almost completely with water to form hydronium ions and the conjugate base of the acid. In a solution of a weak acid, both hydronium ions and unreacted acid molecules are present. The concentration of an acid and the concentration of hydronium ions are not identical.

The pH scale is designed to express the concentration of hydronium ions present in a solution. Hydronium ion concentrations can vary over a very wide range. Concentrations of H_3O^+ from 1.0 mol/L to 1.0×10^{-14} mol/L are converted to a convenient scale with a 0 – 14 range by reporting the negative log of the hydronium ion concentration as *pH*. Each unit of pH corresponds to a tenfold change in concentration of H_3O^+.

A neutral solution has a pH of 7. The pH of acidic solutions is less than 7, and that of basic solutions is greater than 7. The lower the pH of a solution, the higher its acidity. The pHs of some representative solutions are given in Table 7.4.

TITRATIONS OF ACIDS AND BASES

The total acid in a solution can be measured by titration. A measured sample of an acid is titrated by adding a base of known concentration to the sample until the base has reacted with all of the acid present. A small amount of an indicator dye can be used to determine the endpoint. The indicator changes color when just enough base has been added to neutralize the acid. For every mole of protons donated by an acid, a mole of proton acceptors is required. While 1 mol of NaOH will react with 1 mol of either hydrochloric acid or acetic acid, 2 mol of NaOH react with 1 mol H_2SO_4 since both hydrogens of sulfuric acid are acidic.

For a solution, the concentration (in moles per liter) of a species multiplied by the volume (in liters) of solution is equal to the number of moles present. When a measured volume of acid is titrated with a base of known concentration, the concentration of the acid can be calculated by measuring the amount of base required to react with the acid.

$$\text{Moles}_{\text{acid}} = \text{moles}_{\text{base}}$$

The moles of the acid and moles of the base can be calculated by multiplying the concentration of each by the volume of each.

$$V_a C_a = V_b C_b$$

Dividing both sides of the equation by the volume of the acid gives an expression that can be solved for the concentration of the acid.

$$C_a = \frac{V_b C_b}{V_a}$$

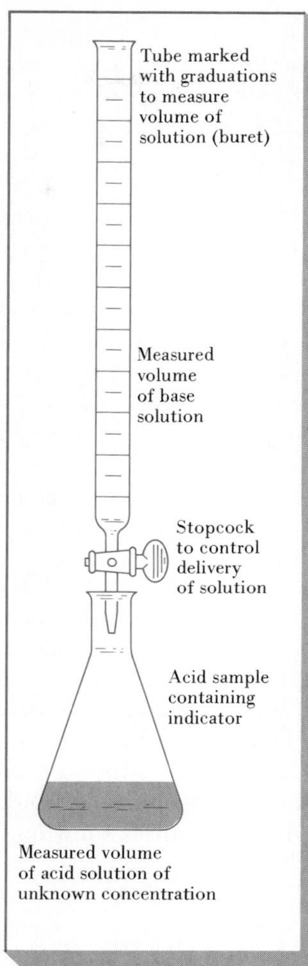

FIGURE 7.4
In a titration of an acid, an indicator is added to a carefully measured volume of a solution of an acid. To determine the volume of base required to neutralize the acid, a solution of a base is added until the indicator just changes color. If the concentration of the acid is known, the concentration of the base can be calculated. If the concentration of the acid is not known, it can be determined by using a base of known concentration.

SAMPLE CALCULATION

A 25.0-mL sample of a basic solution of unknown concentration is titrated with 0.100 mol/L hydrochloric acid. A total of 20.0 mL of acid is required to neutralize the base. What is the concentration of the base?

$$\text{Moles}_{\text{base}} = \text{moles}_{\text{acid}}$$

$$V_bC_b = V_aC_a$$

$$C_b = \frac{V_aC_a}{V_b}$$

$$C_b = \frac{20.0 \times 10^{-3}\ \text{L} \times 0.100\ \text{mol/L}}{25.0 \times 10^{-3}\ \text{L}}$$

$$C_b = 0.080\ \text{mol/L}$$

BUFFERS

A titration curve is a graph which shows the pH of a solution of an acid as a function of the volume of added base. The reaction of hydrochloric acid with sodium hydroxide illustrates the reaction of a strong acid and a strong base. When hydrochloric acid is titrated with sodium hydroxide, the pH changes only slightly until almost all of the strong acid has reacted. Near the endpoint, the pH rises steeply. At the endpoint, the indicator changes color to signal the end of the titration. If more base is added after all the acid has reacted, the curve begins to level and it approaches the pH of the solution of strong base being added.

The graph for the titration of a weak acid with a strong base has a significantly different shape. The slowest rate of increase of pH is at the halfway point of the titration. When half of the base required to neutralize the acid has been added, the pH of the solution has a value that is characteristic of the weak acid. At this midpoint pH, the concentration of the weak acid is equal to the concentration of its conjugate base. When the weak acid is acetic acid, this pH is near 4.7. If the acid is ammonium ion, this pH is about 9.3. The weaker the acid being titrated, the higher the value of the midpoint pH.

Note that the pH changes slowly on either side of the midpoint pH in the titration curve for acetic acid. In this middle region of the titration curve, the solution is said to be buffered. *Buffer solutions* change pH only slightly when a small amount of either strong acid or strong base is added. Buffer solutions are important to living organisms. Both blood and intercellular fluids are buffered to maintain a pH near neutrality.

How does a buffer work? A buffer solution contains about equal concentrations of a weak acid and its conjugate base. When a strong base is added to a buffer solution, the hydroxide ion reacts with the weak acid.

$$\underset{\text{Weak acid}}{HA} + OH^- \rightarrow \underset{\text{Conjugate base}}{A^-} + H_2O \qquad \text{addition of a strong base}$$

The concentration of the weak acid in the buffer solution is decreased and the concentration of its conjugate base is increased, but the change in the

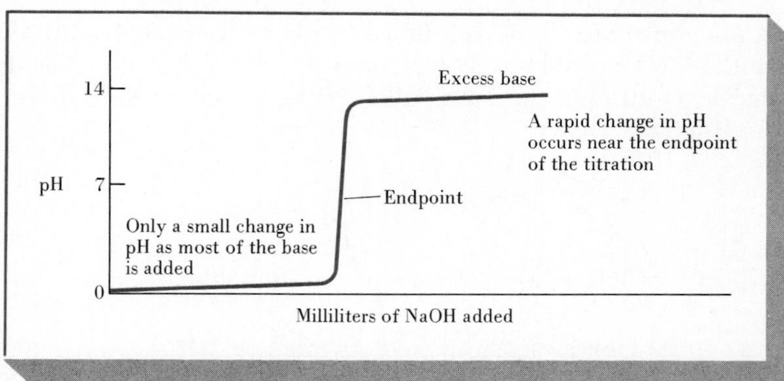

FIGURE 7.5
The curve obtained when 1.00 *M* hydrochloric acid is titrated with 1.00 *M* sodium hydroxide solution is an example of a titration curve for the reaction of a strong acid with a strong base.

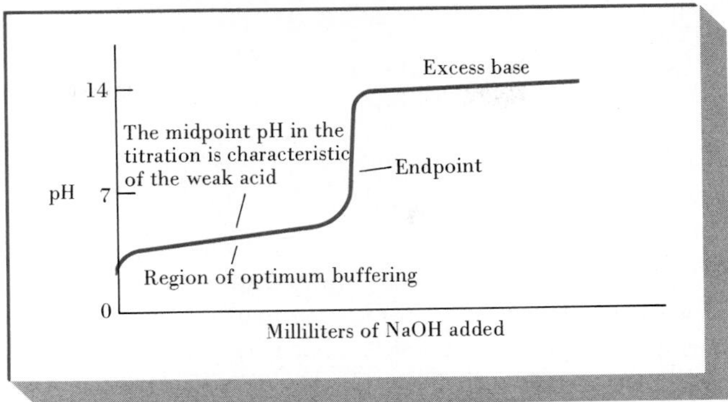

FIGURE 7.6
The curve for the titration of 1.00 *M* acetic acid with 1.00 *M* NaOH solution is an example of a titration curve for a weak acid with a strong base.

pH is small. When a strong acid is added to a buffer solution, the hydronium ions react with the weak base.

$$\underset{}{H_3O^+} + \underset{\text{Weak base}}{A^-} \rightarrow H_2O + \underset{\text{Conjugate acid}}{HA} \qquad \text{addition of a strong acid}$$

The concentration of the weak acid increases but the concentration of hydronium ions increases far less. As long as both the weak acid and its conjugate base are present, the pH of a buffer changes only slightly.

A phosphate buffer plays an important role in the body by maintaining a narrow pH range in living cells. It is the weak acid $H_2PO_4^-$ and the weak base HPO_4^{2-} that comprise the buffer that helps maintain a pH near 7 in cells.

$$H_2PO_4^- + OH^- \rightarrow HPO_4^{2-} + H_2O$$
$$HPO_4^{2-} + H_3O^+ \rightarrow H_2PO_4^- + H_2O$$

PERIODIC TRENDS IN ACID-BASE CHEMISTRY

A long standing generalization in chemistry concerns the oxides of metals and nonmetals. (Oxides are compounds of oxygen with a second element.) Oxides of metals, particularly those in columns 1 and 2, are bases in water, and the oxides of nonmetals are acids in water.

The purely ionic oxides of metals such as sodium and barium react with water to produce hydroxide ions. Oxide ion O^{2-} is a stronger base than OH^-. It accepts a proton from water to form hydroxide ion. This reaction is so favorable that these oxides rapidly react with water vapor if exposed to the atmosphere.

$$Na_2O + H_2O \rightarrow 2Na^+ + 2OH^-$$
$$BaO + H_2O \rightarrow Ba^{2+} + 2OH^-$$

```
     H            H
     ..           ..          ..         ..
H : C : H     H : N :     H : O :    H : F :
     ..           ..          ..         ..
     H            H           H
```

FIGURE 7.7
The compounds formed between hydrogen and the second-row elements C, N, O, and F illustrate acid-base trends across the periodic table. In each of these compounds, eight electrons surround the central atom. NH_3 is the most basic of these compounds and HF is the most acidic.

The covalent oxides of many nonmetals react with water to form acids. For example, carbon dioxide and sulfur trioxide react with water to form carbonic and sulfuric acids, respectively.

$$CO_2 + H_2O \rightarrow H_2CO_3$$
$$SO_3 + H_2O \rightarrow H_2SO_4$$

These reactions are examined in detail later in this chapter.

The acid-base behavior of the covalent hydrides of the second row elements, C, N, O, and F, illustrates a trend occurring across the periodic table. Methane, CH_4, has no acid-base chemistry except under the most exotic conditions. Ammonia, NH_3, reacts as a weak base with water. Hydrofluoric acid, HF, reacts as a weak acid with water. (Note a convention. In an acidic compound containing hydrogen, the hydrogen is usually written first in the formula. If the compound is not as acidic as water, the hydrogen is not written first.)

In the Lewis structure for each of these compounds, there are eight electrons about the central atom. Methane cannot act as a base because it has no unshared pair of electrons to accept a proton. The other compounds have unshared pairs of electrons and could act as bases. The presence of lone pairs of electrons does not explain the differences between NH_3 and HF.

Going across the periodic table from nitrogen to fluorine, the charge on the kernel of the central atom increases. Because a more positive kernel pulls electrons closer, lone-pair electrons become less available for sharing. Ammonia is more basic than water because the lone pair on nitrogen extends further and is more available for sharing than a lone pair on oxygen or fluorine. (Note that acid-base chemistry is not statistical in nature. It is

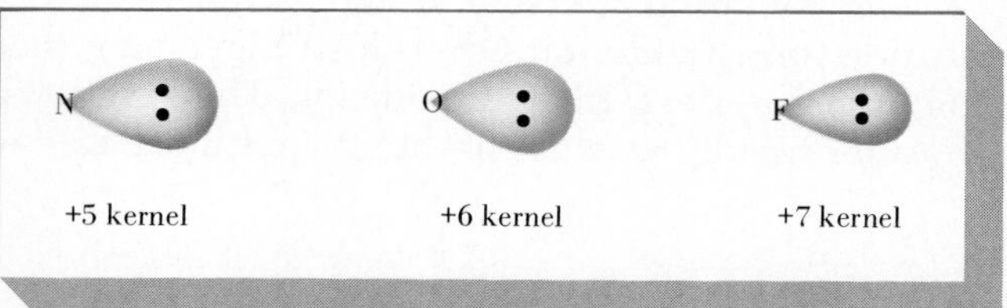

FIGURE 7.8
The higher the kernel charge, the closer the unshared pair of electrons is held. Ammonia is the strongest base and HF is the weakest because the unshared pair on nitrogen is the most available for sharing and that on fluorine is the least available.

wrong to think that water is a better base than ammonia because it has two unshared pairs and ammonia has only one.)

To explain the difference in acidity across the periodic table, a similar argument can be made. The higher the kernel charge, the stronger the acid. The +7 kernel charge on a fluorine atom pulls electrons from the hydrogen in HF more than the +6 kernel charge on an oxygen atom does in water. The hydrogen in HF is more positive, more exposed, and more available to react with a base than is a hydrogen in H_2O.

LEWIS ACIDS

The reaction of copper ions with ammonia is an example of the reaction of transition metal ions with bases. When ammonia is added to a pale-blue solution of copper sulfate, the color changes to deep blue. If alcohol is added to the solution, an intensely colored, deep-blue compound precipitates. This compound has the formula $[Cu(NH_3)_4]SO_4$. Ammonia has reacted with copper ions to form a new species responsible for the deep color.

$$Cu^{2+}(aq) + 4NH_3 \rightarrow Cu(NH_3)_4{}^{2+}$$

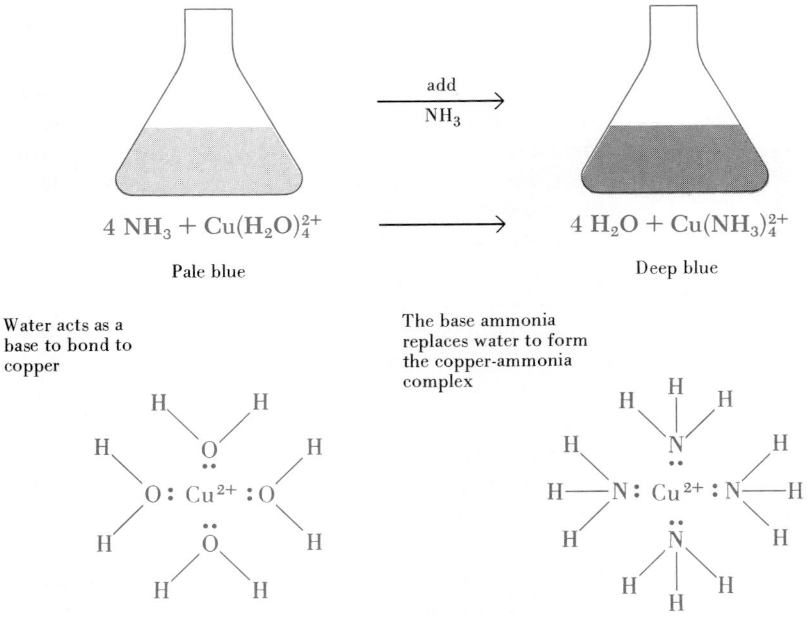

add NH_3 →

$$4\ NH_3 + Cu(H_2O)_4^{2+} \longrightarrow 4\ H_2O + Cu(NH_3)_4^{2+}$$

Pale blue — Deep blue

Water acts as a base to bond to copper

The base ammonia replaces water to form the copper-ammonia complex

FIGURE 7.9
Both ammonia and water molecules can act as bases and bond to 2^+ copper ions. Lone pairs of electrons on the bases are used to form covalent bonds to the Lewis acid.

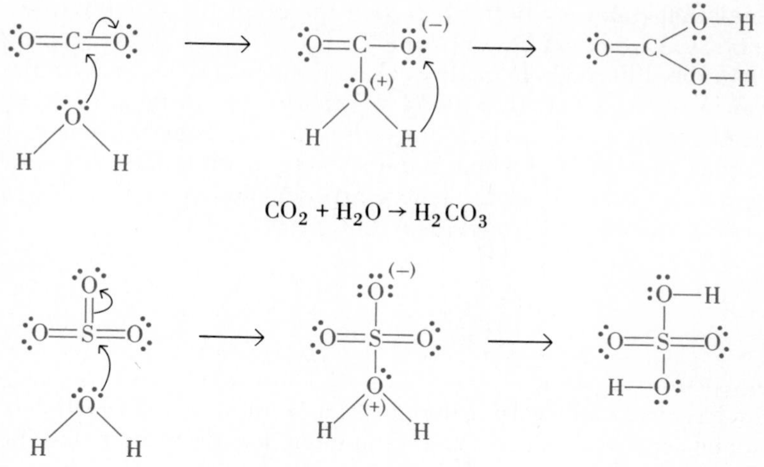

$$CO_2 + H_2O \rightarrow H_2CO_3$$

$$SO_3 + H_2O \rightarrow H_2SO_4$$

FIGURE 7.10
Molecules of CO_2 and SO_3 act as Lewis acids in their reactions with water. The base H_2O furnishes an electron pair to form a new covalent bond. After shifts of protons, the acids H_2CO_3 and H_2SO_4 are formed.

If a strong acid is added to the copper-ammonia solution, the color fades to a pale blue. The hydronium ions of the acid react with ammonia to form ammonium ions so that ammonia molecules are no longer bonded to copper.

$$Cu(NH_3)_4^{2+} + 4H_3O^+ \rightarrow Cu^{2+} + 4NH_4^+ + 4H_2O$$

The reaction of Cu^{2+} with NH_3 is called a *Lewis acid-base reaction.* Ammonia has an unshared pair of electrons. Copper ion has empty orbitals. In the reaction, the base ammonia donates a share in its unshared pair to the copper ion to form a new covalent bond. This reaction is similar to the acid-base reactions discussed earlier in the chapter, for *the base donates an electron pair to form a new covalent bond.* Only the acid is different. Rather than a proton donor, *the acid accepts a share in an electron pair donated by the base.* (Brønsted-Lowry acids become a subset of Lewis acids, namely those that donate protons to react with the unshared electron pairs of bases.)

The molecules CO_2 and SO_3 react as Lewis acids with water. With CO_2, a water molecule furnishes an electron pair to form a new covalent bond with the carbon atom of carbon dioxide. Then one of the protons from H_2O shifts to one of the oxygens on the carbon, giving the structure H_2CO_3. The reaction of water with sulfur trioxide is similar. First H_2O donates an electron pair to the S atom, and then a proton shifts to give the structure H_2SO_4.

Chemists enlarged the class of acid-base reactions to include the reactions of Cu^{2+} with NH_3 and CO_2 with H_2O rather than consider these reactions to be a completely separate class. The choice to expand the concept of acid-base reactions is characteristic of science. By broadening definitions, scientists seek to emphasize similarities in phenomena and to develop broad unifying models.

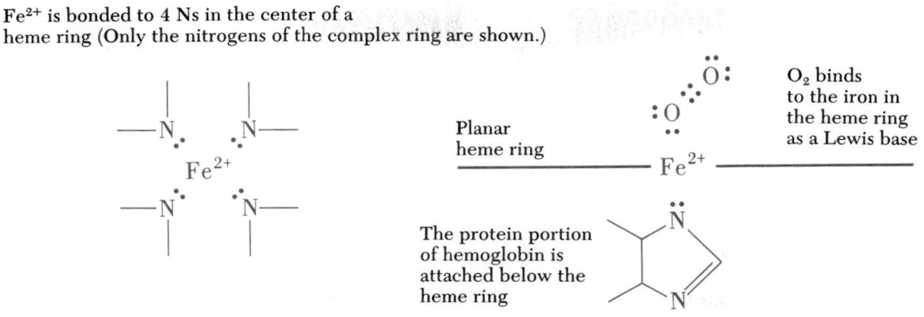

Fe^{2+} is bonded to 4 Ns in the center of a heme ring (Only the nitrogens of the complex ring are shown.)

Planar heme ring

O_2 binds to the iron in the heme ring as a Lewis base

The protein portion of hemoglobin is attached below the heme ring

FIGURE 7.11
Hemoglobin transports oxygen by binding it to a Lewis acid site on iron that is at the center of a rigid heme ring.

A S I D E

ACIDS AND BASES IN BLOOD CHEMISTRY

The transport of oxygen by the protein hemoglobin in blood is an important application of acid-base chemistry. Hemoglobin is the best-known member of a family of proteins that contains an iron ion, Fe^{2+}, in the center of a large, rigid porphyrin ring. The iron is bonded to four basic nitrogens that are part of the ring. The Lewis acid Fe^{2+} has room to bond to six bases. Positions above and below the ring are open. One of the open positions on iron is used to bond the iron-porphyrin complex to the protein portion of hemoglobin. Oxygen is transported at the open sixth position on iron.

The oxygen molecule is weakly bound to the iron in hemoglobin. Oxygen becomes attached when the concentration of oxygen is high in the lungs, and is released as the blood circulates to regions of lower oxygen concentration. (Carbon monoxide acts as a respiratory poison because it competes with oxygen by bonding to the iron in hemoglobin.)

Changes of pH occurring as blood circulates increase the efficiency of oxygen transport. A buffer in the blood consists of carbonic acid and bicarbonate ion. In the blood, CO_2 and H_2CO_3 are rapidly interconverted. When we exhale CO_2, we also

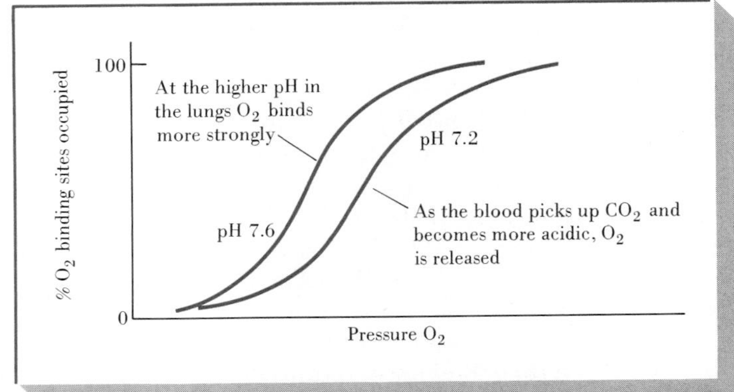

FIGURE 7.12
Acid-base reactions occurring in the blood and on hemoglobin increase the efficiency of oxygen transport. Hemoglobin loads oxygen in the lungs and releases this oxygen as it circulates in the blood.

decrease the amounts of H_2CO_3 and H_3O^+. The acidity of the blood drops slightly so that arterial blood coming from the lungs has a higher pH than venous blood. Then as blood circulates, carbon dioxide accumulates, the acidity of the blood increases, and its pH decreases.

The small pH changes occurring as the blood circulates affect the affinity of hemoglobin for oxygen. At the lower pH in the capillaries, a basic group on the protein accepts a proton. This causes a small change in the shape of the hemoglobin. Then, in turn, more oxygen is released to the tissues. When the pH rises again in the lungs, the hemoglobin protein donates a proton. The protein returns to its former shape, with the iron of hemoglobin better able to attract oxygen again.

QUESTIONS

1. Draw Lewis structures for the reactants and the products of the following acid-base reaction.

$$HCl + H_2O \rightarrow H_3O^+ + Cl^-$$

2. A solution of hydrogen bromide, HBr, is a good conductor of electricity. Hydrogen bromide is a (weak, strong) acid in water. The ions conducting electricity in solution are ________ and ________.
3. HCN is a weak acid in water. A solution of NaCN is (acidic, neutral, basic) because CN^- is (a weak acid, neither an acid nor a base, a weak base).
4. What is the pH of each of the following solutions?
   (a) 1.0 *M* hydrochloric acid
   (b) 1.0×10^{-5} *M* hydrochloric acid
   (c) 1.0 *M* sodium hydroxide
   (d) 1.0×10^{-4} *M* sodium hydroxide
5. In the titration of hydrochloric acid solution (initial pH = 1.0) with sodium hydroxide, what fraction of the original concentration of hydronium ions remains when the pH is 2.0? 3.0? 4.0?
6. A 50.0 mL sample of a base is neutralized by 30.0 mL of 0.100 mol/L hydrochloric acid. What is the concentration of the base?
7. What volume of 0.200 mol/L acid would neutralize 40.0 mL of 0.550 mol/L base?
8. Does a green apple or a ripe apple contain more acid? Why?
9. If a container of sodium hydroxide is left open to the air for a long time, the pH decreases. What is the reaction of the strong base with a minor component of air that causes the pH to decrease?
10. When lemon juice is added to tea, the color of the tea lightens. Does tea contain an indicator? Is the indicator molecule an acid, or is it a base? Explain your answer.
11. Write the reaction, if any, that occurs when hydrochloric acid is added to each of the following solutions. When a solution of sodium hydroxide is added? Which of these solutions is a buffer?

(a) NH_3
(b) NH_4^+ (a solution of ammonium chloride)
(c) NH_3 and NH_4^+

12. Which of the following oxides react with water to form acidic solutions? Basic solutions?
K_2O MgO SO_2 P_4O_{10}
13. Use arguments based on Lewis structures and the periodic table to identify the stronger base in each of the following pairs of compounds.
(a) PH_3 and H_2S
(b) SiH_4 and PH_3
14. Use arguments based on the position of elements in the periodic table to identify the stronger acid in each of the following pairs.
(a) H_2S and HCl
(b) NH_4^+ and H_3O^+
15. When young, the author's children delighted in making "volcanoes" by adding vinegar to sodium bicarbonate. What chemical reaction did they observe?
16. In the following acid-base reactions, underline the acid. Which of the acids are Lewis acids?
(a) $H_2CO_3 + H_2O \rightarrow HCO_3^- + H_3O^+$
(b) $CO_2 + H_2O \rightarrow H_2CO_3$
(c) $SO_2 + H_2O \rightarrow H_2SO_3$
(d) $H_2SO_3 + NH_3 \rightarrow HSO_3^- + NH_4^+$
(e) $Zn^{2+} + 4NH_3 \rightarrow Zn(NH_3)_4^{2+}$
(f) $Zn(NH_3)_4^{2+} + 4H_3O^+ \rightarrow Zn^{2+} + 4NH_4^+ + 4H_2O$
17. Compare hydrogen bonding between water molecules to the acid-base reaction of water molecules to produce H_3O^+ and OH^-. What is the role of an unshared pair of electrons in each case? Of a partially positive hydrogen atom?

CHAPTER

EIGHT

OXIDATION-REDUCTION REACTIONS

A chemical reaction can be a source of energy. People use fires to cook food and battery-powered flashlights to light their way. The reactions occurring in fires and those used to produce electricity in flashlight batteries are called oxidation-reduction reactions. This important class of chemical reactions includes such diverse phenomena as the rusting of an abandoned car and the metabolic conversion of sugar to carbon dioxide and water. In this chapter, we shall explore various ways to describe and classify these chemical reactions, and we will consider a variety of their applications.

Lavoisier was the first to recognize that the reactions occurring in combustion involved chemical combination with oxygen. The importance of these reactions led to the use of the term oxidation to describe reactions in which molecular oxygen combined with other elements.

$$CH_4 + 2O_2 \rightarrow CO_2 + 2H_2O$$
$$4Fe + 3O_2 \rightarrow 2Fe_2O_3$$

The term reduction was introduced to describe reactions with hydrogen or carbon that converted metal oxides to the free metals.

$$H_2 + CuO \rightarrow H_2O + Cu$$
$$3C + 2Fe_2O_3 \rightarrow 3CO_2 + 4Fe$$

SOME CHEMICAL REACTIONS INVOLVE ELECTRON TRANSFER

When finely divided zinc is added to a solution of copper sulfate, the blue color characteristic of copper ions in solution fades, and a red deposit of copper metal forms. When the reaction is complete, zinc sulfate can be isolated from the solution by evaporating the water. The reaction of zinc with copper sulfate is described by the following equations.

$$Zn(s) + CuSO_4(aq) \rightarrow ZnSO_4(aq) + Cu(s)$$
$$Zn(s) + Cu^{2+}(aq) \rightarrow Zn^{2+}(aq) + Cu(s) \text{ net ionic equation}$$

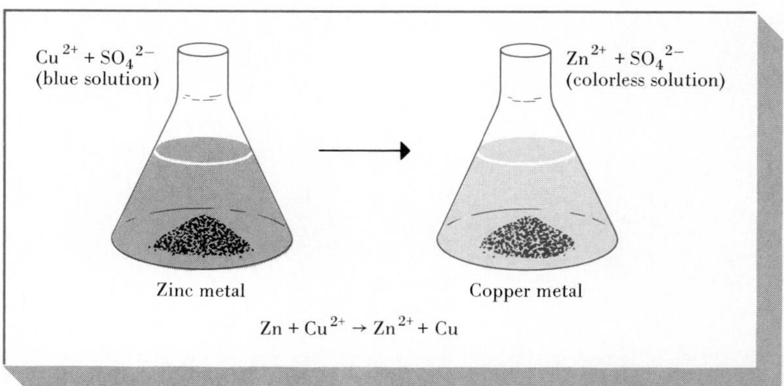

FIGURE 8.1
Zinc metal reacts with copper ions to form zinc ions and copper metal in a redox reaction.

The reaction of zinc metal with a solution of copper sulfate is an electron transfer reaction. Zinc metal donates electrons to the copper ions, and copper ions act as electron acceptors to form copper.

A reaction in which electrons are transferred is called an *oxidation-reduction* reaction or *redox* reaction. No student who can remember the mnemonic "Leo, the lion, says Ger" should confuse the definitions of oxidation and reduction. *L*oss of *e*lectrons is *o*xidation. *G*ain of *e*lectrons is *r*eduction. Zinc, the electron donor, is oxidized to form zinc ions, and the copper ions are reduced.

Because the zinc brought about the reduction of copper ions, zinc is said to be the *reducing agent.* Note that the reducing agent was oxidized in the reaction. Copper ions acted as the *oxidizing agent* for the zinc and were, themselves, reduced.

The reaction of zinc with a strong acid to produce hydrogen is also a redox reaction.

$$Zn + 2H_3O^+ \rightarrow Zn^{2+} + H_2 + 2H_2O$$

Since zinc loses electrons, it is oxidized. Since every electron donor must give its electrons to an electron acceptor, hydronium ions must be reduced.

OXIDATION NUMBERS

The use of oxidation numbers, a formalism, provides a more general system for describing redox reactions. Oxidation numbers are assigned to elements themselves and to elements in compounds. Reactions that involve changes in oxidation numbers are then classified as oxidation-reduction (redox) reactions.

A simplified list of rules for assigning oxidation numbers follows.

1. The oxidation number of an uncombined element is zero.
2. The oxidation number of a monoatomic ion is the charge on the ion.
3. Oxygen has an oxidation number of -2 in its compounds.
4. Hydrogen has an oxidation number of $+1$.
5. The sum of the oxidation numbers of the atoms in a molecule is zero.
6. The sum of the oxidation numbers of the elements in a polyatomic ion is equal to the net charge on the ion.

SAMPLE CALCULATION

Find the oxidation number of N in HNO_3.

First note that according to the simplified list of rules, the oxidation numbers of both H (rule 4) and O (rule 3) are fixed. Since HNO_3 is neutral, the sum of the oxidation numbers of the atoms is 0 according to rule 5. We can write an equation for the sum of the oxidation numbers in HNO_3 and solve for the oxidation number of nitrogen.

$$\text{Sum of the oxidation numbers} = 0$$
$$H + N + 3 \times O = 0$$
$$+1 + N + 3 \times (-2) = 0$$
$$N - 5 = 0$$
$$N = +5$$

Some examples of oxidation numbers are:

Ion or Compound	Oxidation Numbers	Rules Applied
Cu	Cu = 0	1
Cu^{2+}	Cu = +2	2
H_2O	H = +1, O = −2	3, 4
H_3O^+	H = +1, O = −2	3, 4, 6
NO_3^-	N = +5, O = −2	3, 6
HNO_3	H = +1, N = +5, O = −2	3, 4, 5
$KCl(K^+/Cl^-)$	K = +1, Cl = −1	2, 5
$K_2SO_4(2K^+/SO_4^{2-})$	K = +1, S = +6, O = −2	2, 3, 5
Fe_3O_4	Fe = $+\frac{8}{3}$, O = −2	3, 5

REDOX REACTIONS

By definition, *a reaction involving a change in oxidation numbers is a redox reaction. Oxidation is an increase in oxidation number,* and *reduction is a decrease in oxidation number.* The use of oxidation numbers serves to broaden the definition of oxidation and reduction, for reactions in which there is no apparent electron transfer may also be classified as involving oxidation and reduction.

Consider the reactions first used to introduce the terms oxidation and reduction. In the burning of methane, carbon is oxidized from the −4 oxidation state to the +4 state, and oxygen is reduced from the 0 oxidation state to the −2 state.

$$CH_4 + 2O_2 \rightarrow CO_2 + 2H_2O$$

In the rusting of iron, the metal goes from the 0 oxidation state to the +3 state, and oxygen goes from the 0 oxidation state to the −2 state.

$$4Fe + 3O_2 \rightarrow 2Fe_2O_3$$

The burning of methane in natural gas and the rusting of iron are both redox reactions. The conversion of metal oxides (in which metallic elements have positive oxidation numbers) to the free metals (with 0 oxidation numbers) are also redox reactions.

$$H_2 + CuO \rightarrow H_2O + Cu$$
$$3C + 2Fe_2O_3 \rightarrow 3CO_2 + 4Fe$$

Note that the name of a reactant may not indicate whether a given reaction is redox or acid-base. For example, dilute sulfuric acid may react either as an acid or as an oxidizing agent. The following equations illustrate the two types of reactions.

$$H_3O^+ + OH^- \rightarrow 2H_2O \qquad \text{acid-base}$$

In the acid-base reaction of H_3O^+ with OH^-, there is no change in oxidation number.

$$2H_3O^+ + Zn \rightarrow H_2 + Zn^{2+} + 2H_2O \qquad \text{redox}$$

In the redox reaction of H_3O^+ with Zn, hydrogen changes from the +1 oxidation state in hydronium ion to the 0 oxidation state in H_2, and zinc goes from the 0 oxidation state to the +2 state.

REDOX REACTIONS AND THE PERIODIC TABLE

The elements at the far left side of the periodic table readily donate electrons. The outer electrons of these metals are weakly held, for they are relatively far from kernels with charges of +1 or +2. Sodium, potassium, barium, and calcium are all excellent, but expensive and dangerously reactive, reducing agents.

Among elements, the strongest oxidizing agents are fluorine, oxygen, and chlorine, all found in the upper right portion of the periodic table. Atoms of these elements are small and have high kernel charges. These elements readily accept additional electrons from reducing agents.

The chemistry of the halogen family can be used to illustrate redox trends within a family of the periodic table. Chlorine water, a solution of chlorine in water, can be prepared by mixing dilute hydrochloric acid with common household bleach, a solution of sodium hypochlorite.

$$2H_3O^+ + Cl^- + OCl^- \rightarrow 3H_2O + Cl_2$$

When chlorine water is added to a solution of sodium bromide, the chlorine oxidizes bromide ions to bromine and is, itself, reduced to chloride ions.

$$Cl_2 + 2Br^- \rightarrow 2Cl^- + Br_2$$

When either chlorine water or bromine water is added to a solution of sodium iodide, the iodide ions are oxidized to elemental iodine.

$$Cl_2 + 2I^- \rightarrow 2Cl^- + I_2$$
$$Br_2 + 2I^- \rightarrow 2Br^- + I_2$$

Color changes accompanying these reactions can be enhanced by extracting the halogen from chloroform or carbon tetrachloride. In these solvents, chlorine is a faint yellow, bromine is red, and iodine is purple.

The smaller a halogen molecule, the more readily it accepts added electrons. Fluorine, F_2, is the strongest oxidizing agent of the halogens. The larger the halide ion, the less tightly the outer electrons are held. Thus I^- is more easily oxidized than Br^- or Cl^-.

VOLTAIC CELLS

In a voltaic cell, the oxidation and reduction steps of a redox reaction are separated. In the cell shown (Figure 8.2), the solution containing Cu^{2+} is not in contact with the zinc, and the reaction of Cu^{2+} with Zn cannot occur directly. Reduction occurs at the cathode, and oxidation occurs at the anode.

$$Cu^{2+} + 2e^- \rightarrow Cu \qquad \text{cathode reaction}$$
$$Zn \rightarrow Zn^{2+} + 2e^- \qquad \text{anode reaction}$$

Electrons produced at the anode by the oxidation of zinc go through an external circuit to the cathode where they reduce copper ions and plate out copper.

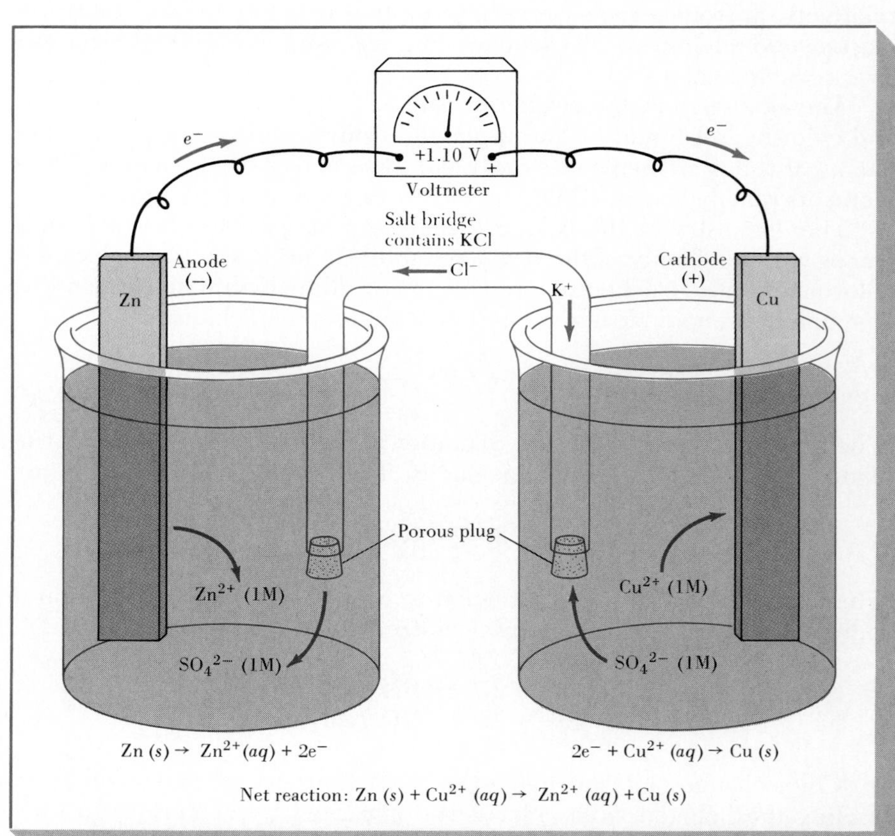

FIGURE 8.2
In a voltaic cell, the oxidation and reduction steps of a redox reaction are separated. Electrons from the oxidation of zinc flow through the external circuit to reduce cupric ions and plate out copper. A connection through a salt bridge is necessary to complete the electric circuit.

A connection through a salt bridge is necessary to complete the electric circuit. A solution of a salt such as KCl permits the motion of ions between the two solutions to balance the flow of electrons through the external circuit. Cations always flow toward the cathode, anions toward the anode.

When the electrodes of a cell are connected, the voltage read is a measure of the combined strengths of the oxidizing and reducing agents. For example, the connection of an electrode in contact with a readily reduced metal ion in solution to an electrode in contact with an easily oxidized metal produces a high voltage. Because the voltage of a cell depends on the concentrations of ions in solution, standard conditions are chosen for comparisons. When the idealized concentrations of copper sulfate and zinc sulfate are both 1.00 *M*, the voltage of the illustrated cell is +1.10 V.

DRY CELLS

Familiar flashlight batteries and the small disk batteries used to power calculators and watches are examples of dry cells. In a dry cell, redox reactions produce an electric current at a nearly constant voltage. A large variety of dry cells are in use; two common cells are discussed here.

The best-known dry cell is the common flashlight battery, a Leclanche cell. A zinc jacket serves as the anode and is oxidized. The cell is filled with a paste containing manganese dioxide and ammonium chloride. A carbon rod serves as an electrode, and manganese dioxide is reduced on discharge. The Leclanche cell has a constant potential of 1.5 V.

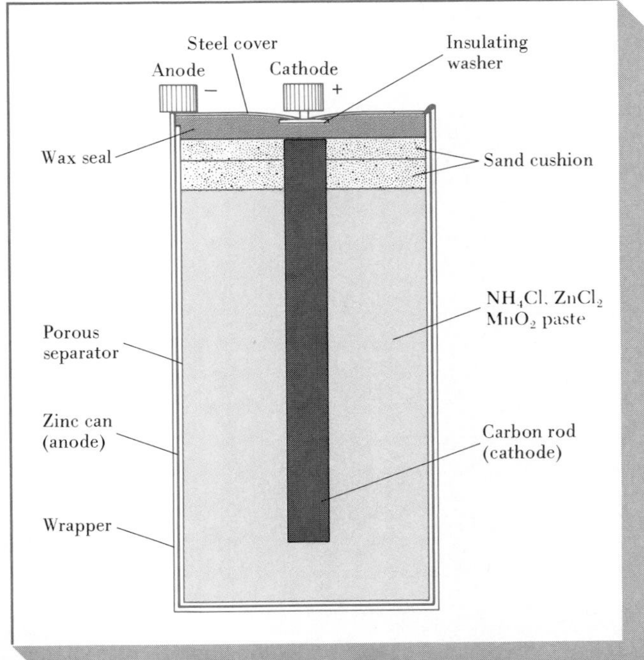

FIGURE 8.3
The Leclanche dry cell is widely used in flashlight batteries. The voltage developed by this cell is +1.5 V.

$$\text{Zn} \rightarrow \text{Zn}^{2+} + 2e^-$$
$$2e^- + 2\text{MnO}_2 + 2\text{NH}_4^+ \rightarrow \text{Mn}_2\text{O}_3 + 2\text{NH}_3 + \text{H}_2\text{O}$$

These equations are a simplification, for one would expect a buildup of ammonia and Zn^{2+} ions. Ammonia and zinc ions combine in a Lewis acid-base reaction, and insoluble zinc compounds are produced in reactions not involving redox.

The small button-shaped batteries have HgO and Zn as electrodes and KOH as the electrolyte. These compact cells deliver current at 1.35 V.

$$\text{Zn} + 2\text{OH}^- \rightarrow \text{ZnO} + \text{H}_2\text{O} + 2e^-$$
$$2e^- + \text{HgO} + \text{H}_2\text{O} \rightarrow \text{Hg} + 2\text{OH}^-$$

RECHARGEABLE ELECTROCHEMICAL CELLS—LEAD STORAGE BATTERIES

Lead storage batteries are widely used as an energy source in automobiles. Current from the battery is used to start the engine, and part of the energy from the running engine is then used to recharge the battery. Lead storage batteries provide a high current at a constant voltage. Because they are rugged and can be recharged, they have a long, useful lifetime.

A lead storage battery consists of a series of cells. Each cell has a 2.0-V potential, so that a 12-V battery consists of six cells. In a cell, there are parallel plates of large surface area in which a honeycomb grid is packed with either lead or lead dioxide. The anode plates are filled with lead, and the cathode plates are filled with PbO_2. A dilute aqueous solution of sulfuric

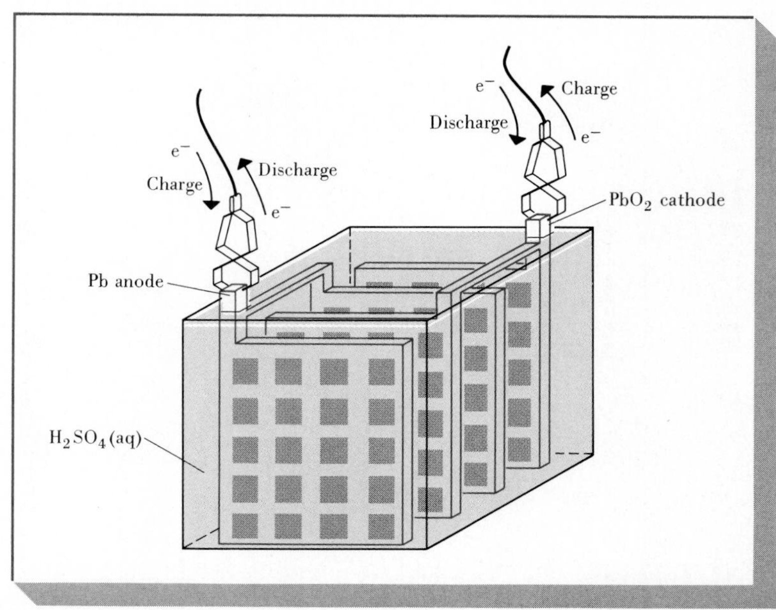

FIGURE 8.4
A lead storage battery consists of a series of cells. Each cell produces a 2.0-V potential. In each cell, there are parallel plates of large surface area in which a honeycomb grid is packed with either lead or lead dioxide. A solution of sulfuric acid provides ions to carry a current through the solution. As the battery discharges, lead sulfate builds up on the surface of the plates.

acid fills the cells. As a cell discharges producing current, the following half-reactions occur.

$$Pb + SO_4^{2-} \longrightarrow PbSO_4 + 2e^- \qquad \text{anode reaction}$$
$$2e^- + PbO_2 + 4H_3O^+ + SO_4^{2-} \longrightarrow PbSO_4 + 6H_2O \qquad \text{cathode reaction}$$
$$Pb + PbO_2 + 2SO_4^{2-} + 4H_3O^+ \longrightarrow 2PbSO_4 + 6H_2O \qquad \text{net reaction}$$

As a battery discharges, lead sulfate forms on both sets of plates and the concentration of sulfuric acid decreases. The density of the sulfuric acid solution in a cell can be used as a measure of the charge on a battery, for as the battery discharges, the acid solution becomes more dilute and less dense.

By applying an external voltage greater than 2.0 volts per cell, the flow of electrons and the chemical reactions occurring at each electrode can be reversed. As long as the lead sulfate clings tightly to each electrode, the battery can be recharged. With age and use, some of this solid flakes and falls to the bottom. When this occurs in a battery, not as much Pb and PbO_2 can form on recharging, and the battery cannot supply as much energy between recharges.

Because heat is produced during the charging and discharging of a battery, some water may evaporate from the cells. Distilled water should be used to restore the fluid level, for some tap waters contain impurities that can contribute to the loss of charge on a battery.

The use of water containing iron impurities can lead to the loss of charge in a battery. Iron, a transition metal, can have either the +2 or +3 oxidation state in solution. In a lead storage battery, ferric ion, Fe^{3+}, gains an electron to form ferrous ion, Fe^{2+}.

$$2Fe^{3+} + Pb + SO_4^{2-} \longrightarrow 2Fe^{2+} + PbSO_4$$

Fe^{2+} can carry that electron through solution to the lead dioxide electrode where it is deposited.

$$2Fe^{2+} + PbO_2 + SO_4^{2-} + 4H_3O^+ \longrightarrow 2Fe^{3+} + PbSO_4 + 6H_2O$$

The regenerated ferric ion can then return to pick up another electron. Because the redox reactions of Fe^{2+} and Fe^{3+} can occur over and over again, small amounts of iron salts will discharge a battery. This is known as the "ferriboat" process for battery discharge.

THE PHOTOGRAPHIC PROCESS

The light-induced reduction of silver halides is the basis of the photographic process. Photographic film contains an emulsion of insoluble, fine-grained silver bromide. On exposure to light, the silver bromide is activated for reduction. (Photons of light cause a redox reaction which produces a small number of silver atoms and bromine that leaves AgBr.) Activated silver bromide crystals, symbolized by $AgBr^*$, are more easily reduced to silver metal than are unactivated crystals. The greater the number of photons absorbed by a grain of AgBr, the more readily it is reduced.

$$AgBr + light \rightarrow AgBr^* \qquad \text{activation}$$

In the development process, the latent image is converted to a negative. By using a very mild reducing agent, selective reduction of only the activated silver bromide produces a dark image consisting of finely divided silver. (Silver atoms present in activated crystals catalyze the further reduction of AgBr.)

$$AgBr^* + \text{developer} \rightarrow Ag + Br^- + \text{oxidized developer}$$

To preserve the image, unreacted AgBr must be removed, for it would darken on continued exposure to light. Photographers use a solution of sodium thiosulfate, $Na_2S_2O_3$, to stop the development reactions and to fix the image on the negative. A Lewis acid-base reaction with $S_2O_3^{2-}$ ions dissolves unreacted AgBr so it can be washed away. (An old name for sodium thiosulfate is sodium hyposulfite, and the fixing solution is sometimes called hypo.)

$$AgBr + 2S_2O_3^{2-} \rightarrow Ag(S_2O_3)_2^{3-} + Br^-$$

The photographic process is repeated to make a print that reverses the image on the negative. When light is passed through the negative onto film, the dark regions of the negative block light and give rise to light regions on the print. The development and fixation steps are repeated to complete the process.

REDOX REACTIONS FOR MAKING SULFURIC ACID AND SODIUM HYDROXIDE

Sulfuric acid is the most widely used strong acid; indeed, it is the largest volume product of the chemical industry. The manufacture of sulfuric acid starts with the direct oxidation of sulfur by oxygen to give sulfur dioxide.

$$S + O_2 \rightarrow SO_2$$

Sulfur dioxide is then oxidized under more forcing conditions to form SO_3.

$$2SO_2 + O_2 \rightarrow 2SO_3$$

The reaction of SO_3 with water gives sulfuric acid.

$$SO_3 + H_2O \rightarrow H_2SO_4$$

In practice, sulfur trioxide is first dissolved in concentrated sulfuric acid. The resulting solution is cooled and then reacted with water.

The most widely used strong base is sodium hydroxide. Its manufacture requires large amounts of electrical energy, for it is prepared by the electrolysis of brine, a concentrated salt solution. Chlorine, an important oxidizing agent, is produced at the anode, and sodium hydroxide and hydrogen gas are produced at the cathode.

$$2Cl^- \rightarrow Cl_2 + 2e^- \qquad \text{anode reaction}$$
$$2e^- + 2Na^+ + 2H_2O \rightarrow H_2 + 2Na^+ + 2OH^- \qquad \text{cathode reaction}$$

It is desirable to keep the products of electrolysis separated in order to obtain pure hydrogen and chlorine and to avoid dangerous mixtures of these gases. A second reason for separation is that sodium hydroxide and chlorine react with one another to form sodium hypochlorite, NaOCl, or bleach.

$$2Na^+ + Cl_2 + 2OH^- \rightarrow Na^+ + OCl^- + Na^+ + Cl^- + H_2O \qquad \text{bleach formation}$$

One way this is accomplished involves the use of liquid mercury as a barrier. In these electrolytic cells, sodium ions are reduced at the mercury surface to produce sodium amalgam, a solution of sodium metal in liquid mercury. Sodium amalgam then reacts with freshwater to form NaOH and hydrogen gas.

$$2Na^+ + 2e^- \rightarrow 2Na(Hg)$$
$$2Na(Hg) + 2H_2O \rightarrow 2Na^+ + 2OH^- + H_2$$

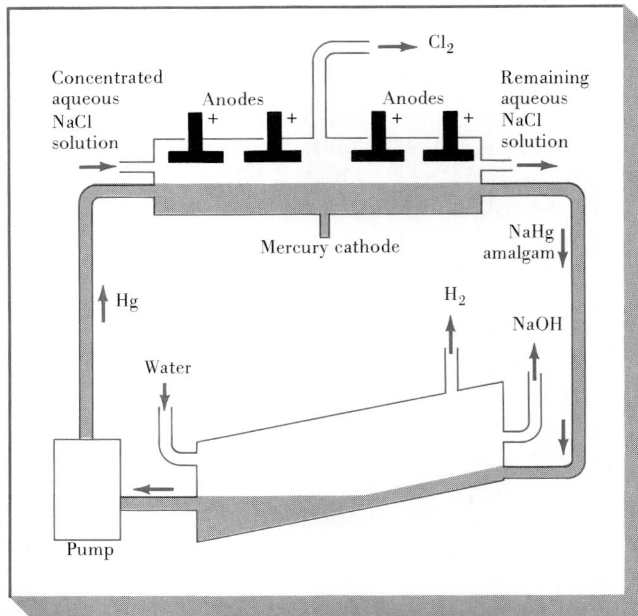

FIGURE 8.5
The mercury cell for the electrolysis of brine separated the production of chlorine from the production of sodium hydroxide by using liquid mercury as a barrier. The loss of mercury from these cells was followed by the production of toxic dimethyl mercury by microorganisms living in sediments where waste water was discharged.

ASIDE

METHYL MERCURY—A TOXIC BY-PRODUCT OF TECHNOLOGY

It was found over many years that mercury was lost from the electrochemical cells used to produce chlorine and sodium hydroxide. Periodically, more mercury was added. While this was of some economic concern, people were not alarmed even though mercury compounds were known to be poisons. It was common knowledge that mercury is an unusually unreactive element and that mercurous chloride, an expected product of the reaction of chlorine and mercury, is exceedingly insoluble.

It was a most unpleasant surprise when high mercury levels were found in fish caught downstream from chlor-alkali plants. Because of mercury pollution, fish caught in these streams and lakes are unfit for human consumption. Microorganisms at the bottom of rivers and lakes had converted the mercury to a soluble compound, methyl mercury, that entered the food chain and accumulated in the fat tissue of fish.

The story of methyl mercury is a sobering one. A seemingly benign technology was suddenly discovered to have a substantial adverse environmental impact. It was not enough to assume a process was safe because there were no known risks. The rejection of technology was also not the answer, for it was only through scientific inquiry that the injection of mercury compounds into the food chain was discovered. Neither ignorance nor the rejection of scientific inquiry is an acceptable approach to controlling technology.

QUESTIONS

1. Calculate the oxidation number of the indicated element in each of the listed compounds.
   (a) P in PF_5, PO_4^{3-}, PCl_3
   (b) S in SO_4^{2-}, H_2SO_3, $Na_2S_2O_3$, H_2S
   (c) Mn in Mn, Mn^{2+}, MnO_2, MnO_4^-
   (d) Cl in HCl, Cl^-, OCl^-, ClO_4^-
2. In the following equations for redox reactions, underline the species being oxidized. Circle the species being reduced. (The equations for the reactions are not balanced.)
   (a) $CH_4 + O_2 \rightarrow CO_2 + H_2O$
   (b) $Cl_2 + Fe^{2+} \rightarrow Cl^- + Fe^{3+}$
   (c) $H_3O^+ + Fe^{2+} + MnO_4^- \rightarrow Fe^{3+} + Mn^{2+} + H_2O$
   (d) $Br_2 + I^- \rightarrow Br^- + I_2$
   (e) $Cl_2 + OH^- \rightarrow Cl^- + OCl^- + H_2O$
3. When Na and Cl_2 react to form NaCl, Na is (oxidized, reduced) and Cl_2 is (oxidized, reduced).
4. The compound sodium hydride is ionic; it is a solid consisting of both Na^+ and H^- ions. Sodium hydride reacts with water to form hydrogen gas.

$$NaH + H_2O \rightarrow NaOH + H_2$$

The reaction of sodium hydride with water can be classified as ______.
(a) An acid-base reaction
(b) A redox reaction
(c) Both an acid-base and a redox reaction

5. Which of the coinage metals, copper, silver, and gold, is most easily oxidized? Least easily oxidized?
6. Which of the following reactions involves oxidation-reduction?
   (a) $Al + H_3O^+ \rightarrow Al^{3+} + H_2 + H_2O$
   (b) $Zn^{2+} + NH_3 \rightarrow Zn(NH_3)_4{}^{2+}$
   (c) $H_2CO_3 + NH_3 \rightarrow HCO_3{}^- + NH_4{}^+$
   (d) $Fe_2O_3 + CO \rightarrow FeO + CO_2$
   (e) $SO_2 + H_2O \rightarrow H_2SO_3$
   (f) $H_2SO_3 + O_2 \rightarrow H_2SO_4$
7. Alchemists found that *aqua regia,* a mixture of hydrochloric acid and nitric acid would oxidize and dissolve gold, even though neither acid acting alone would do so.

$$Au + 4H_3O^+ + 4Cl^- + NO_3{}^- \rightarrow AuCl_4{}^- + NO + 6H_2O$$

   Answer the following questions concerning this reaction.
   (a) What is the oxidation state of gold in $AuCl_4{}^-$?
   (b) Does the nitric acid or the hydrochloric acid act as the oxidizing agent?
   (c) What ion acts as a base toward gold in its positive oxidation state?
8. Why is oxygen a stronger oxidizing agent than sulfur?
9. Why are iodide ions more easily oxidized than chloride ions?
10. Why would current stop flowing in the voltaic cell shown in Figure 7.2 if the salt bridge were removed?
11. In a Leclanche dry cell, current is carried by the movement of $NH_4{}^+$ and Cl^- ions. Immediately after discharge, the concentration of ______ ions is higher near the zinc electrode and the concentration of ______ ions is higher near the MnO_2.
12. As a lead storage battery discharges, the quantity of lead in the anode (decreases, increases), the lead oxide in the cathode (decreases, increases), the lead sulfate at each electrode (decreases, increases), and the concentration of sulfuric acid (decreases, increases).
13. Batteries differ in the quantity of lead and lead dioxide in each cell. Cells containing greater amounts of these compounds should provide (a greater voltage, more current at constant voltage).
14. What would be observed if too weak a reducing agent were used in an attempt to develop a photographic negative? If too strong a reducing agent were used?
15. Sulfur compounds occur in some petroleums and coals. What chemistry might occur after sulfur oxides are introduced into the atmosphere when these fuels are burned?

16. When an electric current is passed through molten NaCl, ______ is oxidized and ______ is reduced.
17. Brine, a concentrated solution of sodium chloride, can be used for the production of hydrochloric acid or for making sodium hydroxide.
    (a) What is the reaction for making hydrochloric acid from brine?
    (b) What is the reaction for making sodium hydroxide from brine?
    (c) Which of these reactions requires the larger energy input?

CHAPTER

NINE

REACTION RATES AND PATHWAYS

heat is oxidized, but the speed with which the reaction occurs varies greatly. Ground into flour or stored in elevators, wheat is insurance against lean years and famine. When eaten, its oxidation occurs by metabolic pathways, fueling our bodies under mild physiological conditions. Yet as a finely divided dust in air, it has sometimes exploded, leveling multistory grain elevators.

What factors govern the speeds at which reactions occur? The design of explosives and the preservation of food present practical problems in the control of rates of reactions. The solutions to these and similar problems often depend on an understanding of factors that affect rates of reactions.

How do reactions occur? The study of reactants and products tells no more about the path of a reaction than the study of building materials and the finished house tells about how a house is built. Scientists study changes in the speeds of reactions as they vary the quantities of reacting substances to gain insight into reaction pathways.

This chapter opens with a consideration of the rates of nuclear reactions. The rates of nuclear reactions illustrate a simple form of reaction kinetics. In addition, the rates of nuclear decay provide us important knowledge concerning the age of our planet and of human history on earth.

The rates of chemical reactions are often more complex than those of nuclear reactions. In this chapter, factors that influence rates of chemical reactions are presented. The connection between reaction rates and reaction pathways is illustrated. The chapter closes with insights pertaining to the relationship between the actions of pharmaceutical drugs and the study of rates of chemical reactions.

NUCLEAR DECAY REACTIONS

Radioactive Isotopes Decay with a Characteristic Half-life

Scientists can measure the production of α particles, β particles, or gamma rays produced by nuclear decay. For example, a beta particle can trigger an electric impulse in a detector known as a Geiger counter. By counting the electric impulses in a time interval and dividing by the time of the interval, a rate can be calculated. For example, if 10,800 counts were recorded in 4 min, the rate of decay would be proportional to 2700 counts per minute (cpm). (A counter does not detect all decay events, but scientists seek to arrange samples and detectors so that decays are detected with the same efficiency for any comparison.)

Though the rate of decay of a radioactive isotope is sometimes fast and sometimes slow, there does exist a unifying feature in all cases. *The rate of decay is proportional to the quantity of isotope present;* it is not affected by changes in temperature, pressure, or chemical state. A graph for the rate of decay of a radioactive isotope against time follows a curve characteristic of *exponential decay*.

The half-life of an isotope is the time required for one-half the nuclei present in the sample to undergo decay. The half-life is independent of the

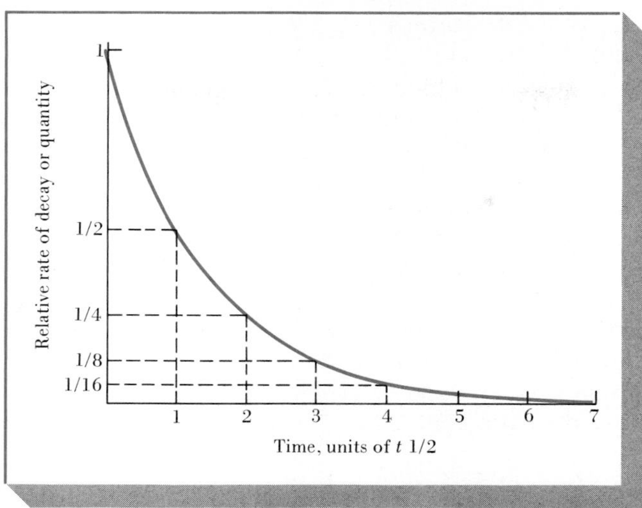

FIGURE 9.1
For nuclear decay reactions, the graph of rate of decay vs. time follows the same characteristic curve. The half-life of an isotope is the time required for one-half the nuclei present in the sample to undergo decay.

original quantity of the isotope, so that after two half-lives, one-fourth of the original quantity remains, and after three half-lives, one-eighth remains. Half-lives of radioisotopes differ greatly. Some are as short as 10^{-9}s and, therefore, require fast methods to measure them. In contrast, the half-life of an isotope of uranium is 4.5 billion years, a number close to the age of the earth. With improvements in the ability to detect and count radioactive decay reactions, both shorter-lived and longer-lived isotopes have been detected.

We can speak of the half-life of a sample of a radioisotope, but we cannot predict the lifetime of an individual nucleus. The half-life describes average behavior, but any particular nucleus may decay after 0.1, 1, or 100 half-lives. With a very small sample or with a radioisotope with a long half-life, fewer decays are observed and the uncertainty in measurements is greater.

Estimating the Age of the Earth

The most abundant isotope of uranium, uranium 238, has a half-life of 4.5×10^9 years. The element formed by the alpha decay of $^{238}_{92}U$ is radioactive as are the products of several subsequent nuclear decay reactions. (The radioactivity which Becquerel discovered in a compound of uranium was due not only to the slow decay of uranium but also to the decay of such radioactive "daughter" isotopes as the polonium and radium, later isolated by the Curies.) The decay series that begins with $^{238}_{92}U$ ends with a stable isotope of lead, $^{206}_{82}Pb$. Because the daughter isotopes have much shorter half-lives than uranium, most, but not all, of the $^{238}_{92}U$ that has decayed over time is now $^{206}_{82}Pb$.

To estimate the age of rocks, scientists measure the ratio of $^{238}_{92}U$ to the sum of $^{206}_{82}Pb$ and $^{238}_{92}U$ present in samples. The sum of the $^{206}_{82}Pb$ and $^{238}_{92}U$ present in a sample is equal to the amount of $^{238}_{92}U$ present when a mineral solidified and cooled. From the fraction of the original $^{238}_{92}U$ remaining, scientists calculate the fraction of a half-life that has elapsed, or the age of the sample. The oldest rocks found on earth solidified about 3×10^9 years ago. Some rocks found in meteorites and some of those brought back from the moon are still older. These rocks, formed about 4.5×10^9 years ago, enable scientists to estimate the age of our solar system.

Prior to cooling, lead and uranium present in molten rock may have separated by differences in migration patterns and by chemical reactions. Since the isotope $^{206}_{82}Pb$ constitutes less than 25 percent of the natural abundance of lead, the absence of other isotopes of lead is evidence that a rock does not contain lead from sources other than radioactive decay.

Carbon-14 Dating in Archaeology

Archaeologists use the decay of $^{14}_{6}C$, an isotope of carbon, to gain new insight into our early history and prehistory. The half-life of this isotope is 5760 years, a time comparable to that for the existence of written history. The ages of a wooden beam from an early castle, a papyrus scroll from the time of the pharaohs, and the charcoal from a prehistoric camp fire can be determined because they contain a small but measurable fraction of the $^{14}_{6}C$ that is found in plants alive today.

Carbon 14 is produced by nuclear reactions occurring in the upper atmosphere. Cosmic rays bombarding the earth's atmosphere constantly produce neutrons; the neutrons react with nitrogen to produce $^{14}_{6}C$.

$$^{14}_{7}N + ^{1}_{0}n \rightarrow ^{14}_{6}C + ^{1}_{1}H$$

In the atmosphere, newly formed $^{14}_{6}C$ reacts with oxygen to form CO_2.

Plants incorporate CO_2 into compounds by reactions occurring in photosynthesis. Leaves and newly formed wood contain the same fraction of $^{14}_{6}C$ as the atmosphere. In living matter as in the atmosphere, the amount of $^{14}_{6}C$ present reflects the balance between the rate of its radioactive decay and the rate at which it is replenished. When a plant dies or a tree is cut down, the $^{14}_{6}C$ present continues to decay while no new $^{14}_{6}C$ is incorporated.

$$^{14}_{6}C \rightarrow ^{14}_{7}N + ^{0}_{-1}e$$

The age of an artifact can be computed by burning the carbon to CO_2, measuring the fraction of the total carbon present that is $^{14}_{6}C$, and comparing that fraction to the fraction of $^{14}_{6}C$ found in the atmosphere today.

The dating of artifacts using carbon 14 raised new questions. The belief that civilization spread by cultural diffusion was challenged by the results of such dating. Dates calculated on the basis of the fraction of $^{14}_{6}C$ remaining supported an alternative hypothesis that certain technologies arose independently in widely separated places. Were the assumptions made in carbon-14 dating wrong, or was it necessary to reconsider ideas about how culture arose and spread?

The discovery of long-lived, well-preserved bristlecone pines provided a way to test the assumption that the atmospheric concentration of carbon 14 is constant over time. The growing seasons recorded in tree rings provided an independent way to document age, for the inner rings of these long-lived trees were formed more than 4000 years before the outer rings. Scientists measured the ages of the inner wood samples of these trees both by counting rings and by using carbon-14 dating.

The ages determined by the two methods were very close but not identical. Scientists concluded that there have been small variations of carbon-14 levels in the atmosphere that reflect variations in the sun's output of energy. However, the corrections to ages obtained by radiochemical dating were in the wrong direction to support those arguing for cultural diffusion as the single mechanism for the appearance of civilization in distant regions.

ASIDE

HAVE COLLISIONS WITH METEORS CAUSED MASS EXTINCTIONS?

Both geology and radioactive dating provide evidence that the earth is very old. A part of the geological evidence is a fossil record preserved in sedimentary rocks. In particular, limestone is formed in shallow seas by the slow deposition of the skeletal remains of microscopic marine organisms. Even a small piece of limestone contains vast numbers of microscopic fossils. Layers of limestone hundreds of feet thick were deposited over millions of years. By sampling deeper layers of limestone, scientists can sample more ancient seas.

When a core is drilled from a thick bed of limestone and the fossils are examined, it is found that the numbers and kinds of fossils present change very slowly throughout most of the core. However, in some cores there are abrupt changes in the relative numbers and kinds of marine fossils occurring in narrow zones. In these regions of discontinuity, many species present in lower parts of the core disappear to be replaced by species found only in the higher segments. Paleontologists refer to these abrupt changes in fossil composition as mass extinctions.

What events could trigger the mass extinctions found in the fossil record? Scientists now believe that a large meteor striking the earth may have been responsible. If a large meteor had struck the earth, dust from the meteor could have darkened the skies for long periods, causing temperatures to drop and photosynthesis to slow. This large and sudden change in climate is believed to have triggered the mass extinction of vulnerable species. (The absence of a large crater on the surface of the earth suggests that the collision must have occurred in the ocean.)

Evidence for this hypothesis first came from measurements of trace amounts of osmium and iridium, elements found with platinum in the periodic table. A technique known as neutron activation is used to detect and measure traces of these elements. First, samples are bombarded with neutrons, and then the gamma-ray emission from the samples is studied. When a neutron is absorbed by a nucleus of an atom, a new, unstable isotope of the same element is formed. These unstable isotopes lose some of their energy by emitting gamma rays. The energies of emitted gamma rays and the half-lives of the gamma-emitting isotopes are measured and compared to the emissions from known samples.

Osmium and iridium are very scarce in terrestrial rocks, but they are more abundant in known samples of meteors. Tiny meteors are constantly bombarding the earth and are burning up as they pass through the atmosphere. Because dust from these meteors is constantly being deposited, traces of osmium and iridium occur throughout limestone. However, in limestone cores taken from widely separated regions of the globe, it has been found that the abundance of osmium and iridium in samples at the boundaries of mass extinctions is more than twenty times as high as it is in other parts of the cores. The greater abundance of these elements at the boundaries is evidence for a large meteor striking the earth.

CHEMICAL REACTIONS

Concentration and Temperature Affect Rates of Chemical Reactions

In contrast to the rates of nuclear reactions, the rates of chemical reactions depend on the physical states of the reactants, and they vary greatly as concentrations and temperatures change. The rates of nuclear decay reactions depend only on the relative instabilities of nuclei comprising the inner cores of atoms, but the rates of chemical reactions depend on the contact between reacting particles.

Chemists classify reactions that take place at surfaces and those that take place in solution. *Heterogeneous reactions* occur at surfaces, and *homogeneous reactions* occur in solution or in the gas phase.

The effects of different surface areas on the rates of heterogeneous chemical reactions can be readily observed. For example, wood shavings used as kindling are easier to light than a yule log. The more surface area of wood in contact with air, the speedier the combustion. The effects of different concentrations on the rates are also apparent. For example, carbon dioxide fire extinguishers slow combustion by providing a smothering blanket of CO_2 gas that lowers the concentration of oxygen at a burning surface.

Increases in concentration also speed homogeneous chemical reactions. For example, the reaction of oxygen and nitrogen to form nitrogen oxides occurs much more rapidly in the cylinder of an automobile engine where the hot gases are compressed than in an open fire where both nitrogen and oxygen are more dilute.

Increases in temperature increase the speed of most reactions. A Fisher burner used in chemistry laboratories provides a dramatic example of temperature effects. The mixture of air and natural gas coming out of the barrel of a burner burns rapidly with a very hot flame. Why doesn't the flame jump back into the barrel? A metal screen carries away heat from the flame above so that the mixture of gases below does not reach ignition temperature. Safety lamps operating on the same principle permitted the mining of coal in the nineteenth century. Wire screens that dissipated heat kept the illuminating flames of miners' lamps from igniting flammable gases in the mine.

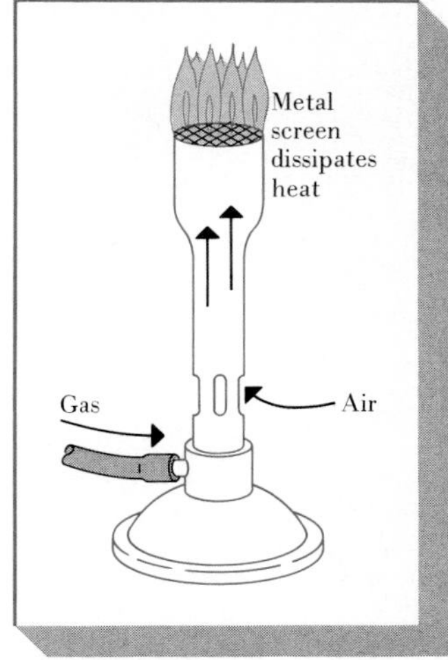

FIGURE 9.2
A Fisher burner used in chemistry laboratories provides a dramatic example of the effects of temperature on reaction rates. Even though the gas and air are mixed at the base of the burner, combustion occurs only at high temperatures over the burner.

A Collision Model for Reaction Rates

A simple model to explain reaction rates is based on collisions. This model provides an explanation for the effects of changing concentrations on rates. The greater the concentration of reacting species, the more frequent the collisions between reacting molecules and the faster the reaction.

$$A + B \rightarrow C$$

According to this model, doubling the concentrations of both A and B, two reacting species, would quadruple the rate of reaction, because collisions of A and B would occur with four times the original frequency.

A collision model also helps explain the effect of temperature on reaction rates. Molecules can react only when they collide, but not all collisions lead to reactions. Reactions involve both the breaking of old bonds and the formation of new bonds. Productive collisions are high-energy collisions;

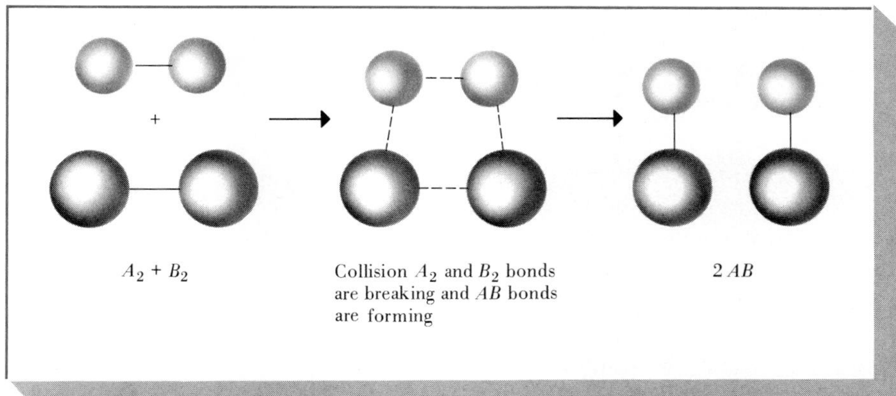

FIGURE 9.3
Collision theory provides a simple model to describe reaction rates. The rate of a chemical reaction depends both on the frequency of collisions between reacting species and the energy of the collisions.

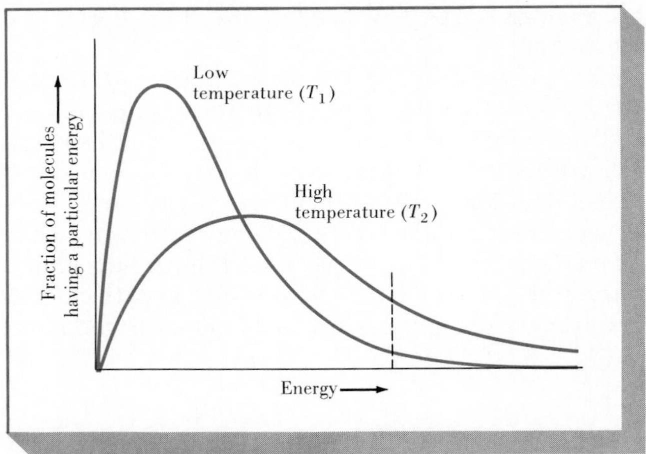

FIGURE 9.4
At a low temperature, only a small fraction of the molecules in a sample have high kinetic energies. The fraction of high-energy molecules increases as the temperature increases.

energy from collisions is needed to weaken or break bonds in reacting molecules. Because the fraction of high-speed molecules in a sample increases rapidly as the temperature increases, the number of high-energy collisions and the reaction rate increase rapidly with increasing temperature.

Most collisions do not involve enough energy to lead to reaction; therefore, chemists speak of a barrier to reaction. For a slow reaction, there is a high barrier to reaction; for a fast reaction, the barrier is lower. The barrier to reaction is called the *activation energy*. The magnitude of the activation energy may determine whether or not a reaction occurs. For example, at 20°C the combustion of gasoline occurs at a negligible rate because the activation energy for this reaction is too high to overcome at this temperature.

If two or more reactions are possible, the reaction with the lower activation energy occurs faster. By analogy, although Denver is higher in altitude than both San Francisco and New Orleans, the barrier for water to flow west across the mountains is high, so the streams, instead, flow east, crossing the lower manmade dams to empty into the Gulf of Mexico.

Catalysts

Catalysts are substances that accelerate the rates of chemical reactions without being consumed by the reaction. A catalyst does not occur as a reactant or as a product in the net equation for a chemical reaction. It may react in one step of a reaction and be regenerated in another, but it is present both at the beginning and at the end of the reaction.

A catalyst provides an alternative lower-energy pathway for a reaction

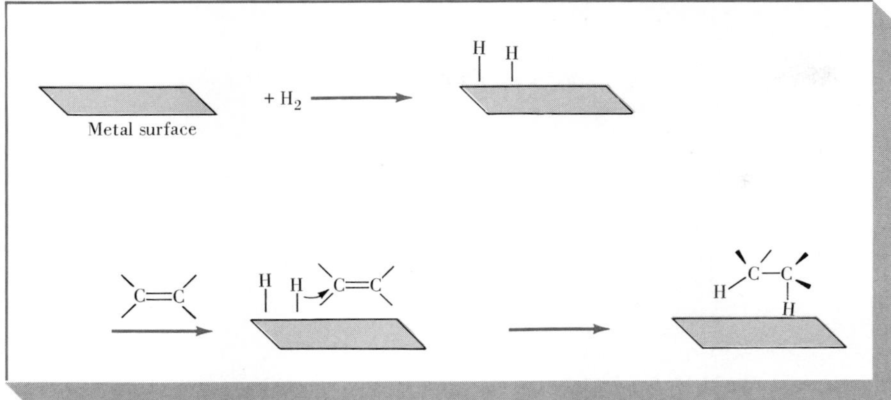

FIGURE 9.5
Finely divided platinum metal is used to catalyze the addition of hydrogen to carbon-carbon double bonds.

to occur. At a fixed temperature, a greater fraction of the collisions between molecules have sufficient energy to surmount the lower barrier, and the reaction is faster.

Some catalysts speed reactions by providing surfaces at which concentrations of reacting species are high and at which bonds in one or more reactant may be weakened or broken. In the production of gasoline from petroleum, finely divided platinum and palladium catalyze the addition of hydrogen to carbon-carbon double bonds. These metals absorb a large amount of hydrogen, and there is a high concentration of hydrogen at their surfaces. At the metal surface, the bond between hydrogen atoms in H_2 is weakened, lowering the activation energy for the addition reaction. Since a catalyst is not used up in a reaction, metals as expensive as platinum and palladium can be used as catalysts even though the gasoline produced sells for a relatively low price.

A Rate Law Describes a Chemical Reaction

The rate of a reaction is the rate of change of concentration. To measure a reaction rate, chemists measure either the rate of disappearance of a reactant or the rate of appearance of a product. One way to measure the rate is to withdraw and analyze samples at regular intervals during a reaction. The rate of a reaction is the slope of the graph of concentration vs. time.

To investigate the relationship between concentration and rate, the concentration of one reactant at a time is varied, and the rate of reaction is measured. It is found that the results of these measurements can be expressed in an equation called the rate law.

$$A + B \rightarrow C + D \qquad \text{hypothetical reaction}$$
$$\text{Rate} = k[A]^a[B]^b \qquad \text{form of rate equation}$$

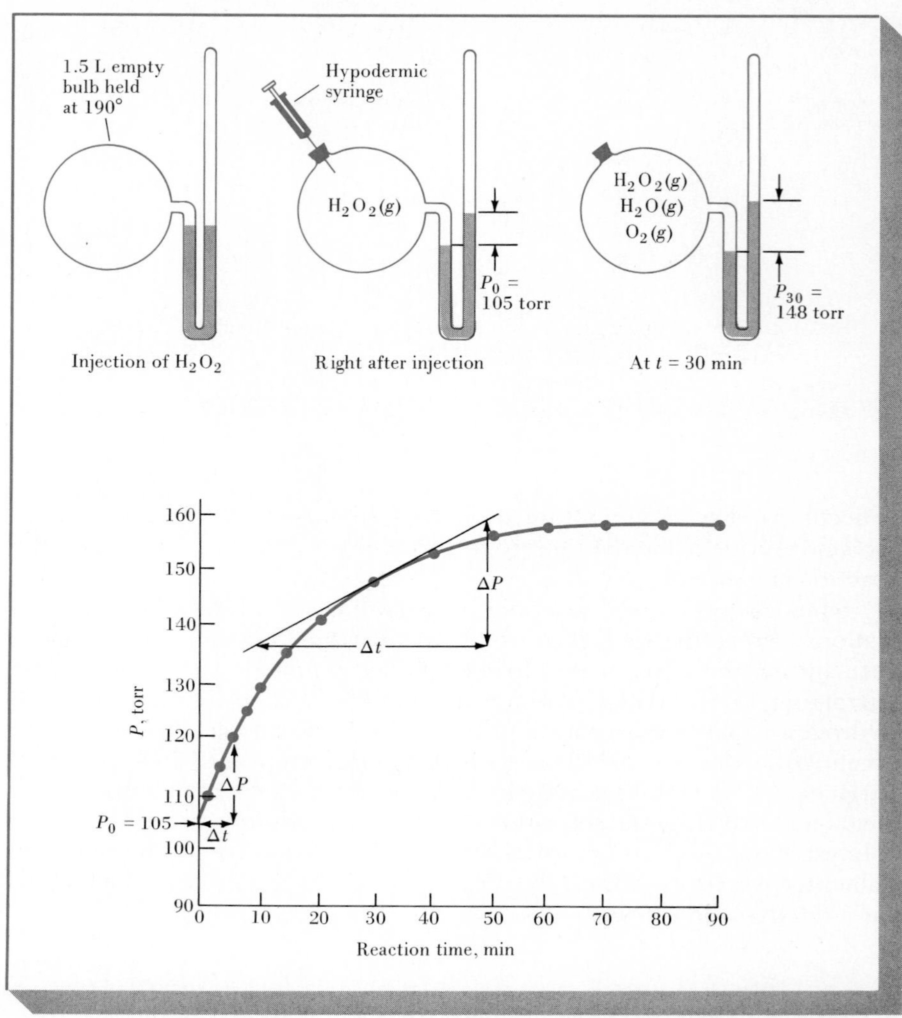

FIGURE 9.6
One way to determine the rate of a chemical reaction is to measure the concentration of a product at different times. The rate of the reaction at a given time is the slope of the graph of concentration vs. time.

In the rate equation, [A] and [B] are concentrations of reactants. (The concentration of a catalyst can appear in a rate equation.) The exponents, a and b, are determined by experiment. The exponent of a concentration in the measured rate law is the order of the reaction. For example, if $a = 1$ and $b = 1$ for a reaction, the reaction rate is first-order in A, first-order in B, and second-order overall. A rate constant k is another quantity that is determined for a reaction.

To illustrate the application of a rate law, consider a hypothetical reaction in which A goes to B. To determine the reaction order in A, a chemist doubles the concentration of A and then measures the effect on the rate. Three possible outcomes of this experiment are given in the accompanying table.

$$\text{A} \rightarrow \text{B}$$

Effect of doubling [A]	Rate Law	Order in A
No increase in rate	Rate $= k[A]^0$	Zero
Rate of reaction doubles	Rate $= k[A]^1$	First
Rate of reaction quadruples	Rate $= k[A]^2$	Second

Many reactions occur in a series of steps. For such a reaction, the slow step of the overall reaction acts as a kinetic bottleneck. Just as a narrow neck on a large bottle limits the speed of emptying it, the rate of the slowest step limits the overall rate of a reaction.

An Application of Kinetics to Mechanism

Chemists study reaction rates to learn how reactions occur. Some reactions occur in a single step; others occur by a multistep pathway. A reaction may or may not go through intermediates. Since we cannot see movements of atoms or electrons during reaction, evidence for how a particular reaction occurs is indirect. Rate studies provide a test that is used to disprove some proposed reaction pathways. When a single pathway that fits all experimental data remains, it is called the mechanism of the reaction.

Substitution reactions in which one carbon-halogen bond is broken and a new carbon-halogen bond is formed have been widely studied. The reaction pathway depends on the groups attached to the carbon. Chemists use rate studies to help determine the correct mechanism for the substitution.

For example, consider two pathways for the replacement of iodine by chloride ion in the compound CH_3I. In pathway I, the net reaction occurs in two steps. First the carbon-iodine bond cleaves in a slow step to generate I^- and CH_3^+, a cation that is an intermediate in the reaction. Then CH_3^+ reacts with Cl^- to form the product CH_3Cl in a fast step. In pathway II, the entering group displaces the leaving group in a single reaction. (A third pathway in which the new bond to carbon forms before the old one breaks would violate the rules of bonding for Lewis structures by having ten electrons about carbon in the intermediate structure.)

Reaction studied

$$CH_3I + Cl^- \rightarrow CH_3Cl + I^-$$

Pathway I

$$CH_3I \rightarrow CH_3^+ + I^- \quad \text{slow}$$
$$CH_3^+ + Cl^- \rightarrow CH_3Cl \quad \text{fast}$$

Expected rate law:

$$\text{Rate} = k[CH_3I]$$

$$\text{Pathway II}$$
$$Cl^- + CH_3I \rightarrow CH_3Cl + I^-$$
$$\text{Expected rate law:}$$
$$\text{Rate} = k[CH_3I]\,[Cl^-]$$

Experimentally, the rate doubles when the concentration of chloride is doubled and the concentration of CH_3I is kept constant. The rate also doubles when the concentration of CH_3I is doubled at constant chloride concentration. Therefore, the reaction is first-order in each reactant and second-order overall. Because the reaction rate depends on the concentration of Cl^-, pathway I cannot be the mechanism. The superior explanation of the observed rates is provided by a mechanism in which chloride ion displaces the iodide ion in a single step (pathway II).

Enzyme Kinetics

Enzymes are proteins that catalyze metabolic reactions. Early in the twentieth century, Michaelis and Menten, working at the University of Chicago, investigated the kinetics of reaction for the digestive enzyme invertase. Invertase cleaves sucrose or cane sugar into simpler sugars. A small amount of the enzyme catalyzes the reaction of a large quantity of its *substrate,* sucrose. (An enzyme catalyzes the reaction of one or more compounds that are called substrates.)

The rate equations for enzymatic reactions differ from those found for simple chemical reactions. At low substrate concentrations, the rate doubles when the substrate concentration is doubled, but at high substrate concentrations, adding more substrate has no effect on the rate. The reaction changes from first-order to zero-order with increasing substrate concentration. A plot of the initial rate of the enzyme-catalyzed reaction vs. the substrate concentration at constant enzyme concentration is shown. With

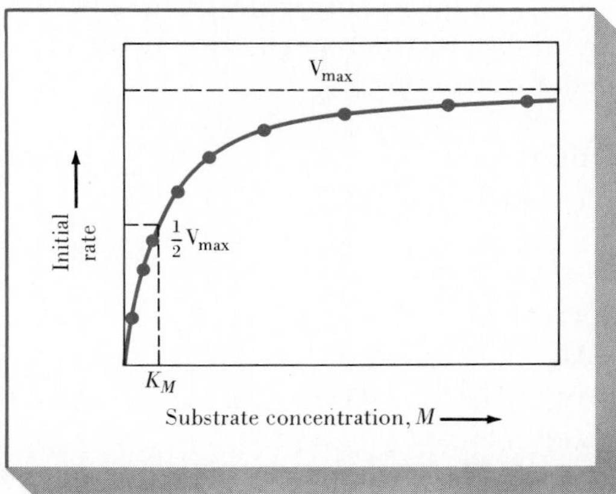

FIGURE 9.7
For an enzyme-catalyzed reaction, the rate of reaction first increases then levels off as the substrate concentration increases.

increasing substrate concentration, the reaction rate first increases and then becomes constant.

To fit the experimental curve, Michaelis and Menten proposed a mechanism in which enzyme-catalyzed reactions occur in steps. In the first step, an enzyme combines with a substrate to form an enzyme-substrate complex. This step is reversible; the enzyme-substrate complex can dissociate to reform the enzyme and substrate. Alternatively, it can react in a second step to give free enzyme and products.

$$\mathrm{E + S \rightarrow ES} \qquad \text{step 1}$$
$$\mathrm{ES \rightarrow E + P} \qquad \text{step 2}$$

Either the first step or the second step of their mechanism can be the slow step that is rate-limiting. The rate of the formation of the enzyme-substrate complex depends on the concentration of both enzyme and substrate while the rate of the second step depends only on the concentration of the complex ES. When the concentration of the substrate is low, reactions of substrate and enzyme are less frequent, and step 1 is slow and rate-limiting. When the substrate concentration is high, the rate of formation of the enzyme-substrate complex is rapid, and step 2 becomes rate-limiting.

ASIDE

MATCHING A MECHANISM TO AN EXPERIMENTAL CURVE IS A TEST FOR CORRECTNESS

Chemists practicing kinetics seek to explain the rate behavior of chemical reactions. Sometimes they propose mechanisms by which complex reactions occur in simpler steps. To test a possible mechanism, they examine its consequences. A correct mechanism should predict a curve identical to that observed.

Michaelis and Menten derived the following equation from their assumed two-step mechanism.

$$\text{Rate} = \frac{V_{\text{max}}\,[\mathrm{S}]}{K_m + [\mathrm{S}]}$$

The rate and the substrate concentration, [S], are measured quantities. V_{max} and K_m are constants that are evaluated graphically. Their equation fits the experimental curve for a wide range of substrate concentrations, confirming their proposed mechanism for catalysis by enzymes. Had their mechanism not fit the experimental curve, the mechanism would have been rejected.

To see how the equation derived by Michaelis and Menten fits the graph of the observed rate, consider the low and the high extremes of substrate concentration. At very low substrate concentration, the denominator of the rate equation is approximately equal to K_m, and the rate is proportional to the substrate concentration. Here the reaction is first-order with respect to [S].

$$\text{Rate} = \frac{V_{\max}\,[\text{S}]}{K_m} \qquad \text{if } [\text{S}] \ll K_m$$

At very high concentrations of substrate, the denominator of the rate equation is approximately equal to [S], and the rate remains constant. Here the rate is zero-order with respect to [S].

$$\text{Rate} = \frac{V_{\max}\,[\text{S}]}{[\text{S}]} = V_{\max} \qquad \text{if } [\text{S}] \gg K_m$$

Enzymes Have Active Sites

The formation of the enzyme-substrate complex plays a very important role in interpreting enzymatic catalysis. The site on the enzyme that binds the substrate is called the *active site*. The substrate fits the active site as a key fits a lock. In the active site, the substrate is aligned with reactive groups on the enzyme or on a second substrate. The binding of the substrate to the enzyme in the enzyme-substrate complex generates a high concentration of reactive groups in an orientation favorable for reaction. The selective and oriented binding of the substrate to an active site is a major way that enzymes speed reactions.

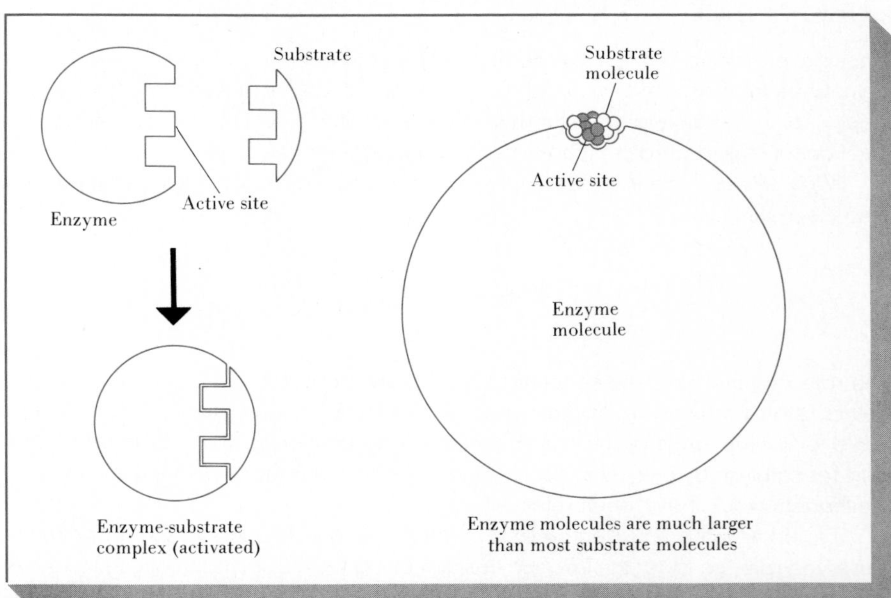

FIGURE 9.8
In an enzyme-catalyzed reaction, the substrate first binds to an active site on the enzyme to form an enzyme-substrate complex.

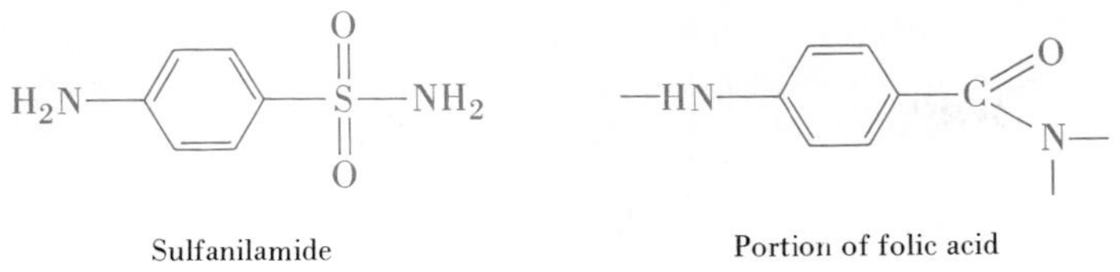

FIGURE 9.9
Sulfanilamide resembles a portion of the folic acid vitamin and it binds to the active site of an enzyme catalyzing the production of folic acid. By occupying the active site, sulfanilamide blocks the production of folic acid required for bacterial growth and reproduction.

The active-site hypothesis has been extended to some proteins that are not enzymes. These proteins selectively bind small molecules as part of a chemical communication system between cells. Some of these proteins provide binding sites for hormones; others, for molecules that transmit nerve impulses between cells.

Sulfa and Penicillin Act at the Active Sites of Enzymes

Most drugs act by affecting one or more enzymes. Some drugs bind at the active sites of enzymes. To do so, their molecular structures must resemble those of normal substrates. Drugs that interfere with the metabolism of invading microorganisms, while not affecting the host, are particularly desirable for use in medicine. When there is a biochemical pathway that is unique to the invading organism, the enzymes on that pathway are potential sites for drug action.

Sulfa drugs, introduced in the 1930s, were the first drugs effective against a wide spectrum of bacteria. Folic acid is required for the manufacture of new DNA and is, therefore, required for cell division. Bacteria must make their own folic acid, a vitamin that is absorbed from a person's diet. Sulfanilamide resembles one of the molecules needed to assemble folic acid; it binds to the active site of the bacterial enzyme that synthesizes folic acid. When sulfanilamide is bound to the active site of the enzyme, it blocks the binding of the substrate which it resembles. A patient treated with sulfa is not affected, while the growth of infecting bacteria is stunted by a vitamin deficiency.

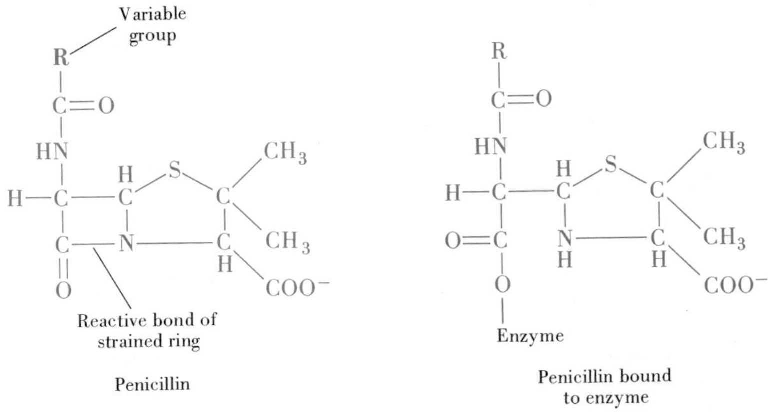

FIGURE 9.10
Penicillin binds to the active site of an enzyme that catalyzes reactions in the formation of cell walls of certain bacteria. While bound at the active site, penicillin reacts to form a covalent bond to the enzyme and render it unable to catalyze additional reactions.

Penicillin is an enzyme inhibitor that acts like a Trojan horse. It binds to the enzyme that completes the construction of cell walls for certain bacteria. Because the penicillin molecule resembles a portion of the cell wall, it is bound to the active site of the enzyme. While bound, it reacts with the enzyme to form a new covalent bond. (Penicillin has a highly reactive four-membered ring that is cleaved by a reactive group on the enzyme.) The modified enzyme no longer can act as a catalyst. The cell walls remain incomplete and leaky, and cell division remains incomplete. Penicillin slows the growth of invading bacteria and allows the body's defenses the time to mount a counterattack.

QUESTIONS

1. During the first 10 min of counting, the rate of decay of a radioisotope dropped from 1600 to 800 cpm. How much time will be required for the decay rate to drop from 800 to 200 cpm?
2. The atmospheric testing of nuclear weapons stopped in 1963. A radioactive isotope of strontium ${}^{90}_{38}Sr$, released in the tests, decays with a half-life of 29 years. In what year will the level of radioactive decay of ${}^{90}_{38}Sr$ released in 1963 drop to one-half its original value? To one-fourth its original value?
3. Why is it necessary to measure the quantities of both ${}^{238}_{92}U$ and ${}^{206}_{82}Pb$ and not just the quantity of ${}^{238}_{92}U$ to determine the age of a rock?
4. The relative rate of decay of carbon 14 in CO_2 prepared from a charcoal artifact is one-eighth that of atmospheric CO_2. What is the approximate age of the artifact?
5. Criticize the following explanation for carbon-14 dating. While a tree is alive, it incorporates carbon 14 from the atmosphere. When the tree dies, the carbon 14 begins to decay. By comparing the fraction of carbon 14 in the wood to the fraction of carbon 14 in the atmosphere, the age of the tree can be calculated.
6. By comparing the fraction of carbon 14 in inner layers of long-lived bristlecone pines to the fraction in outer layers and by determining the age of the tree by counting tree rings, it has been found that the fraction of carbon 14 in the atmosphere has not been constant. Why might the amount of carbon 14 in the atmosphere vary with incoming solar radiation?
7. Table salt and rock salt are samples of sodium chloride that differ in crystal size. Which dissolves faster in water? Why?
8. Platinum catalysts are often prepared by reducing a slurry of PtO_2 powder. Why might the catalyst prepared by this reaction be far superior to one prepared by grinding platinum metal?

$$PtO_2 + 2\ H_2 \rightarrow Pt + 2\ H_2O$$

9. In many gas storage cans, there is a small hole away from the spout. Does the presence of an open vent "catalyze" the emptying of gasoline from the can? Why or why not?

10. Would the rate of reaction of two gases increase with increasing temperature if the volume of the gas were increased to keep the frequency of collision of molecules of the gases constant? Give a reason for your answer.
11. A petroleum refining company advertises that they use platinum in the production of their gasoline. Should you pay a premium price for the gasoline? What is the likely role of the platinum?
12. Ammonia for use in fertilizer is produced by the reaction of hydrogen and nitrogen at high temperature and pressure.

$$N_2 + 3\ H_2 \rightarrow 2\ NH_3 + \text{heat}$$

How do the following changes in conditions affect the rate of formation of ammonia?
(a) Increase in pressure
(b) Increase in temperature
(c) Introduction of a catalyst while holding temperature and pressure constant

13.

$$A + B + C \rightarrow D$$

The above reaction follows the following rate law.

$$\text{Rate} = k[A][B][C]$$

Argue which of the following mechanisms might be more likely, based on analogy to the relative frequencies of two- and three-car collisions.
(a) Simultaneous collision of molecules of A, B, and C to give D
(b) Two-way collision of A and B to give an intermediate I that collides with C in a second step to give D

14. Although methyl alcohol, CH_3OH, does not react with aqueous sodium chloride, it does react with aqueous hydrochloric acid to form CH_3Cl. What is the role of hydronium ions in this reaction?
15. Why are reptiles less active and more slow-moving in cold weather?
16. What forces might be involved in binding of sulfanilamide to the enzyme involved in folic acid synthesis?
17. Do the antibiotics sulfa and penicillin kill germs or do they only slow the rate of reproduction of invading bacteria? Explain.
18. How do the rates of enzyme-catalyzed reactions respond to temperature? (Consider the effects of hard-boiling an egg or of storing it in a refrigerator for the time required for an egg to spoil.)

CHAPTER

TEN

ENERGY, ENTROPY, AND CHEMICAL CHANGE

ater runs downhill, and fuels burn. Both the dripping of a measured volume of water and the burning of a measured length of a candle have been used to measure the passage of hours during the night. We can observe the passing of seasons by noticing the blossoms of early spring, the greens of summer, and the bright hues of autumn. We observe and experience a direction of change with time.

What accounts for the direction of chemical change? Scientists believe that processes of biology, physics, geology, and chemistry are governed by the same laws. To examine chemical changes, we need to consider a part of science called *thermodynamics.* Thermodynamics deals with changes in energy and disorder.

HEAT, WORK, AND ENERGY

Early in the nineteenth century, Benjamin Thompson, Count Rumford (1753–1814) studied the quantitative relationship between *heat* and *work.* He measured the heat produced during the drilling of the bores in brass cannons. (Water was used to cool the hot metal.) He found that the heat produced by the turning of a drill was proportional to the work done in turning the drill. Heat and work are two forms of energy.

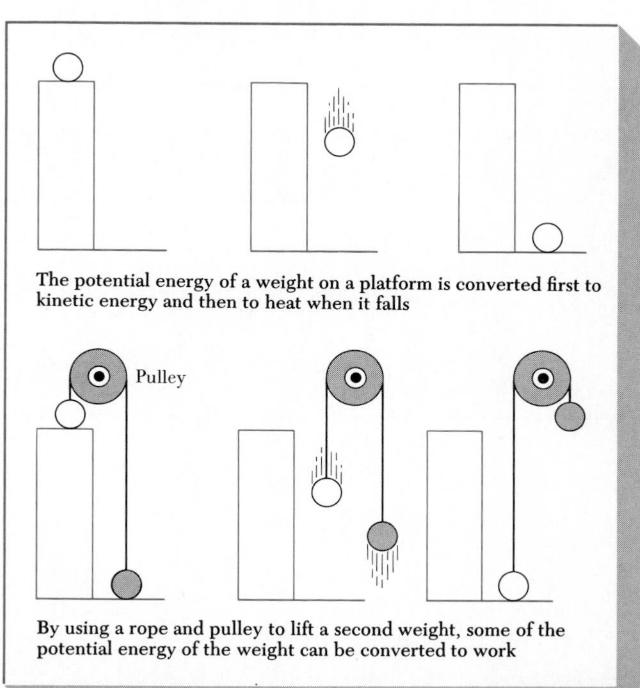

The potential energy of a weight on a platform is converted first to kinetic energy and then to heat when it falls
Pulley
By using a rope and pulley to lift a second weight, some of the potential energy of the weight can be converted to work

FIGURE 10.1
Potential energy can be converted into heat or it can be converted into work. Heat and work are two forms of energy.

Thompson formulated the law of conservation of energy. This relationship, also known as the *first law of thermodynamics,* states that the energy of a system can only be increased by heating it or by doing work on it. Conversely, the energy of a system can only be lowered when the system loses heat or does work.

$$E_{\text{final}} - E_{\text{initial}} = \Delta E = q + w$$

Change in energy = heat added to system + work done on system

Although the total energy change depends only on the initial and final states, the amount of either heat or work can vary. For example, an object on a platform has potential energy relative to the ground. (The potential energy is the product of the gravitational force acting on the object and its height.) The change in potential energy is the same whether heat is produced or work is done. If the object falls, no work is done. The potential energy is converted first to kinetic energy as the object falls and then to heat when it strikes the ground. If the object is lowered using a pulley, the potential energy can be used to raise another object.

HEAT AND WORK FROM CHEMICAL REACTIONS

When we use heat from a fire for cooking or an electric current from a lead storage battery to start an automobile, we look to chemical reactions as a source of energy. Chemists apply the first law to the study of chemical reactions. In the case of a reaction, the initial state describes the reactants and the final state, the products.

Reactions that produce heat are called *exothermic reactions,* and reactions that absorb heat from their surroundings are called *endothermic reactions.* The burning of natural gas is an exothermic reaction, and the formation of nitric oxide from nitrogen and oxygen is an endothermic reaction.

$$CH_4 + O_2 \rightarrow CO_2 + 2H_2O + \text{heat}$$
$$\text{Heat} + N_2 + O_2 \rightarrow 2NO$$

Sometimes the direction of a reaction can be reversed. Note that the reversal of an endothermic reaction would turn it into one that is exothermic, and vice versa.

The energy of a reaction may appear partly as heat and partly as work, or it may appear entirely as heat. Consider the voltaic cell shown in Figure 8.2. If the reaction of zinc with a copper salt is harnessed to drive an electric motor, then some of the energy from the chemical reaction is converted to work. If, however, the zinc and copper salt solution are simply mixed, the redox reaction occurs directly on the zinc surface where copper plates out and zinc is oxidized. No work is produced, and the energy of the reaction is converted entirely to heat.

MEASURING HEATS OF REACTIONS

Under certain restricted conditions, the heat liberated or absorbed by a chemical reaction depends only on the initial and final states of a system. When a reaction occurs in a container with rigid walls to keep the volume constant, and no device is used to do work, then no work is done. Under these restricted conditions, the change in energy can be most easily determined, for it is equal to the heat of the reaction,

$$\Delta E = q \qquad \text{when } w \text{ is zero}$$

Chemists measure heats of reactions using a calorimeter. For example, the heat of combustion of an organic compound is measured by burning it in an atmosphere of oxygen in a bomb calorimeter. Heat from the reaction is absorbed by the calorimeter and the surrounding water bath. The heat of combustion of the compound can be calculated from the temperature increase of the calorimeter and bath and the heat capacity of the calorimeter and bath.

When dieticians speak of the caloric content of a food, they refer to the heat of combustion of that food. For example, the combustion of 1 mol of the sugar glucose yields 673.0 kcal.

$$C_6H_{12}O_6 + 6O_2 \rightarrow 6CO_2 + 6H_2O + 673.0 \text{ kcal/mol}$$

The energy content of glucose is the same whether it is burned in the body or in a calorimeter. Dieticians, however, use different units. A dietician says

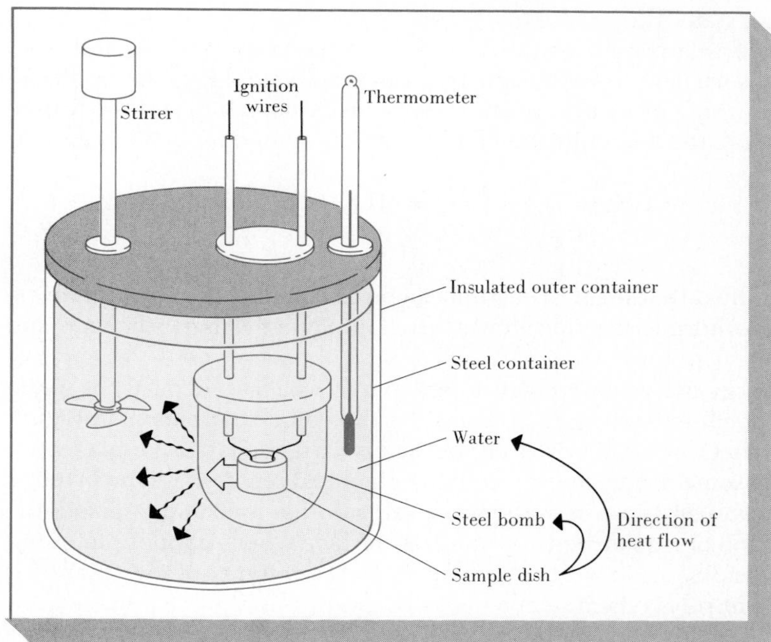

FIGURE 10.2
A bomb calorimeter is used to measure the heat of combustion of many compounds. The heat of the reaction flows into the bomb and then into the surrounding water. By measuring the temperature increase, the heat of reaction can be determined.

that glucose contains 3.74 cal/g. [By multiplying the calories in 1 g of glucose by 180 g (the molecular weight of glucose), one obtains 673 cal/mol.] A calorie unit for food in the diet is 1 kcal of energy. In the body, that energy may be converted to heat or to a combination of heat and work, or it may be stored chemically as fat.

A CLOSER LOOK AT WORK AND HEAT

Work and heat are two forms of energy in transit. Both the acceleration of a rocket and the thawing of snow involve the transfer of energy. Work is directional; we can lift weights or throw a ball. Electric energy is carried by a current of electrons, and light energy is carried by a beam of photons. Heat is a less organized form of energy. The transfer of heat is a transfer of random motion.

The conversion of work into heat involves a loss of organization, for energy with direction is converted to random molecular motion. Because of friction in mechanical systems and resistance in electric circuits, some energy is lost as heat in every energy transfer.

Heat can be converted to either mechanical or electrical work. For example, heat is used to produce steam, and steam is used to power an engine or to turn a generator. The processes, however, are not efficient. The energy available from the expansion of steam depends upon the difference in temperature between the boiler and the exhaust. Usually only slightly more than 30 percent of the caloric value of a fuel can be captured and used to do electrical work.

ENTROPY—DISORDER OR RANDOMNESS

Some processes that take place spontaneously involve no net energy change. Reversals of these spontaneous changes are never observed. For example, gases mix but they do not spontaneously separate. No person is known to have suffocated because oxygen and nitrogen in the air separated and he or she stepped into a pocket of nitrogen. Hot and cold objects in contact come to have the same temperature. No one has seen a warm cup of coffee begin to freeze at the rim and boil in the center.

Neither the separation of gases nor the simultaneous generation of high and low temperatures would violate the law of conservation of energy. Nonetheless, these events do not occur. Energy considerations alone do not explain why gases mix and why objects in contact come to have the same temperature. To account for the direction of spontaneous change, an additional concept is needed.

Entropy is a quantitative measure of the disorder of a state. A highly ordered state such as a crystalline solid has a low entropy. A highly disordered state such as a gas has a high entropy. An increase of disorder is an increase of entropy, for a more random state has a higher entropy.

The direction of spontaneous change is toward greater randomness. Consider examples using coins and dice to illustrate this concept. If a large number of coins are flipped, nearly equal numbers of heads and tails can be expected. We are more confident of that prediction than we are in predicting the result of a single coin flip. If a single die is rolled, the likelihood of it being a 1 or a 6 is equal. However, if 100 dice are shaken and cast, we expect some dice to roll 1s, others 2s, and so forth. We would expect the rolled total of the 100 dice to be much closer to 350 than to either 100 or 600. Change favors disorder; extreme differences disappear, and a more random or averaged state is observed.

ASIDE

DYNAMITE AND THE NOBEL PRIZE

An explosive can be considered a textbook case of applied thermodynamics. A shock from a detonator initiates a rapid exothermic reaction. As the reaction proceeds, the liberated heat causes the remaining explosive to rapidly rise in temperature which, in turn, accelerates the rate of reaction. The hot gaseous products formed in the reaction exert great pressure and expand doing work against their surroundings. With the blast, there is a large increase in entropy for both the gaseous products and for the surroundings.

$$\underset{\text{Nitroglycerin}}{2C_3H_6N_3O_6} \rightarrow 6CO + 3N_2 + 6H_2O$$

Alfred Nobel, the man who instituted and endowed the Nobel prizes in science, was the inventor of dynamite. He found ways to stabilize the explosive nitroglycerin and to make it safer for use. Later, he developed and patented other formulations for explosives. Money from these patents funds the Nobel prizes.

Explosives are used to quarry rock and to construct dams and mountain roads. They are also used in bombs and shells. Nobel's inventions have been used in both constructive and destructive ways. They are among the many technologies that have increased our power for doing both good and bad.

ENTROPY, PHASES, AND THE SECOND LAW OF THERMODYNAMICS

When ice melts or liquid water turns to steam, there is an increase in entropy for the system. In the solid state, water molecules are highly ordered; networks of hydrogen bonds orient each water molecule to its neighbors. In the liquid phase, water molecules remain close to one another and hydrogen-bonded, but there is greater freedom for molecules to move past one another or to tumble. In the gas phase, molecules are much further apart and are not at all oriented to one another. Of the three phases of matter, the solid state has the lowest entropy and the vapor state the highest.

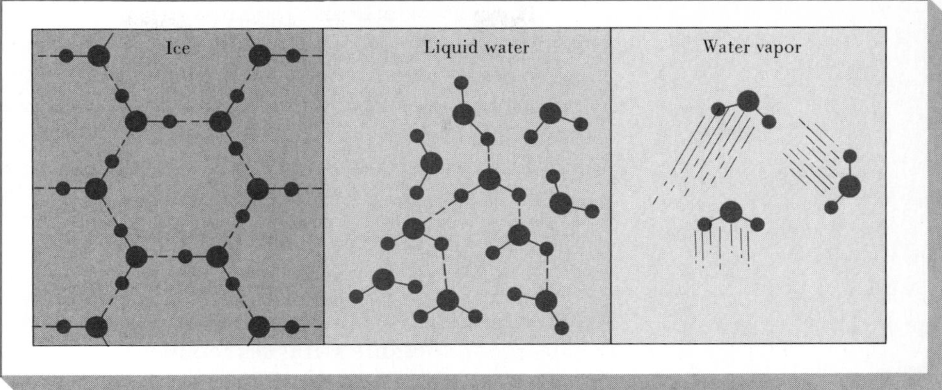

FIGURE 10.3
Disorder increases as ice changes to liquid water and as liquid water changes to water vapor; more hydrogen bonds are broken and molecules begin to move past one another and then become separate.

According to the second law of thermodynamics, only those processes for which the total entropy increases or remains constant are observed to occur. The total entropy is the entropy of both the system and its surroundings.

$$\Delta S_{total} = \Delta S_{system} + \Delta S_{surroundings} > 0$$

That ice melts in surroundings at 20°C and not at -20°C is a consequence of the second law. Although there is an increase in entropy for the system at either temperature, the total change of entropy for the process is positive at the higher temperature and negative at the lower temperature. However, the calculation of entropies for a system or for its surroundings is beyond the scope of this course.

A S I D E

LIFE AND THE SECOND LAW

How does life exist in a universe moving toward greater randomness? Life processes continuously create order in the synthesis of proteins and nucleic acids, in the organized movements of muscle, and in the processing of information in the brain. How is the low entropy of living organisms possible when the universe is moving toward greater disorder?

Plants make complex sugars from carbon dioxide and water. The removal of gaseous CO_2 and liquid H_2O from the environment to be stored as sugars in a fruit is a tremendous increase in order. However the second law asserts that the entropy of the plant and its surroundings must be increasing. For green plants, the surroundings include the sun since radiant energy from the sun drives photosynthesis. The sun is aging, burning up its nuclear fuels, and radiating energy into space. The second law is not violated.

For an animal that digests food and eliminates waste, the system and its surroundings do not extend so very far. Food molecules are intermediate in order. Animals can maintain and increase order by exporting disorder. The carbon dioxide and water vapor that are exhaled have the high entropy associated with gases. By exhaling gases of high entropy into the surroundings, an animal maintains its organization. Upon death, the processes that maintain low entropy cease. ''Ashes to ashes; dust to dust.''

EQUILIBRIA

We are familiar with many *equilibrium* phenomena. For example, when a small amount of table salt is stirred in water, it dissolves. Still more salt can be dissolved until the solution becomes saturated, and no further salt will dissolve. The solid salt is then at equilibrium with the salt in solution.

$$NaCl(s) \rightarrow Na^+(aq) + Cl^-(aq)$$

At the molecular level, processes at equilibrium are dynamic. In a saturated salt solution, ions in a salt lattice continue to go into solution, but an equal number of ions leave the solution for the crystal.

That salt may have been obtained by the evaporation of seawater. As seawater evaporates in the sun, the concentration of salt increases until the solution becomes saturated. Upon further evaporation of water, salt crystals form, but the concentration of dissolved salt remains constant, and the same equilibrium exists between the solid and ions in solution.

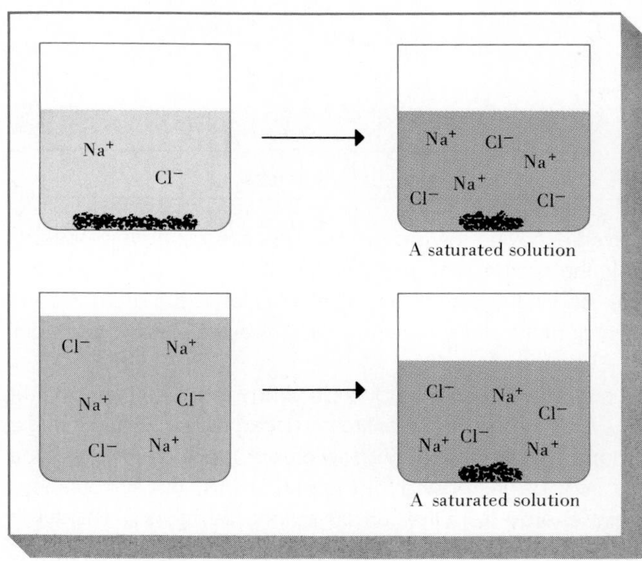

FIGURE 10.4
The composition of a saturated solution of sodium chloride is the same whether the solution is prepared by dissolving salt or it is prepared by concentrating a dilute solution of salt.

In another familiar example, ice and water can remain in equilibrium at 0°C. If excess ice is added to warmer water, some of the ice melts and the water cools to 0°C. If excess water at 0°C is in contact with ice below 0°C, some of the water freezes and the ice warms to 0°C.

Chemical reactions can also proceed to equilibrium. At equilibrium, the rates of the forward and reverse reactions are equal, and no net change of concentrations occurs. Equilibrium represents the balance of competing dynamic processes. (Rates of reaction may be very slow. Discussions of equilibria always assume that there is a process for getting to equilibrium.)

The acid-base reaction of two water molecules to form hydronium ions and hydroxide ions is an example of chemical equilibrium.

$$H_2O + H_2O \rightarrow H_3O^+ + OH^-$$

The product of the concentration of hydronium ions times the concentration of hydroxide ions is a constant K_w.

$$K_w = [H_3O^+] \times [OH^-] = 1 \times 10^{-14}$$

When an acid is added, the concentration of H_3O^+ goes up and that of OH^- goes down. Conversely, when a base is added, the $[H_3O^+]$ goes down and the $[OH^-]$ goes up.

REACTIONS DEPEND ON ENERGY AND ENTROPY CHANGES

The position of equilibrium, and indeed the direction of a reaction, can depend on temperature and pressure. For example, when limestone is heated to a high temperature, it is decomposed to form lime and carbon dioxide.

$$\text{Heat} + CaCO_3 \rightarrow CaO + CO_2 \qquad \text{at } 1000°C$$

When lime is used to mark lines on an athletic field, it reacts with carbon dioxide from the air to reform calcium carbonate.

$$CaO + CO_2 \rightarrow CaCO_3 + \text{heat} \qquad \text{at } 25°C$$

The position of equilibrium for a system consisting of $CaCO_3$, CaO, and CO_2 changes with temperature. At low temperatures, the exothermic reaction to form $CaCO_3$ is favored. As the temperature is increased from 25 to 1000°C, the direction of the reaction is reversed. At high temperatures, entropy considerations are more important and the endothermic reaction forming CO_2 gas as one of the products is favored.

MANIPULATING CHEMICAL EQUILIBRIA—LE CHATELIER'S PRINCIPLE

Both energy and entropy influence the position of equilibrium in a system. For example, the reaction of nitrogen with oxygen to form nitric oxide takes place only at high temperatures. It occurs in lightning strikes, in fires, and in automobile engines. The reaction is endothermic, for nitrogen molecules and oxygen molecules have stronger bonds than those found in molecules of nitrogen oxides. However, a mixture containing only nitrogen and oxygen molecules is more ordered and is at a lower entropy than is a mixture of all three gases. An increase in entropy accompanies the reaction to form some NO.

$$\text{Heat} + N_2 + O_2 \rightarrow 2NO$$

When conditions such as temperature or pressure are changed, the composition of a system at equilibrium changes. Often the direction of change can be predicted by applying Le Chatelier's principle. This states

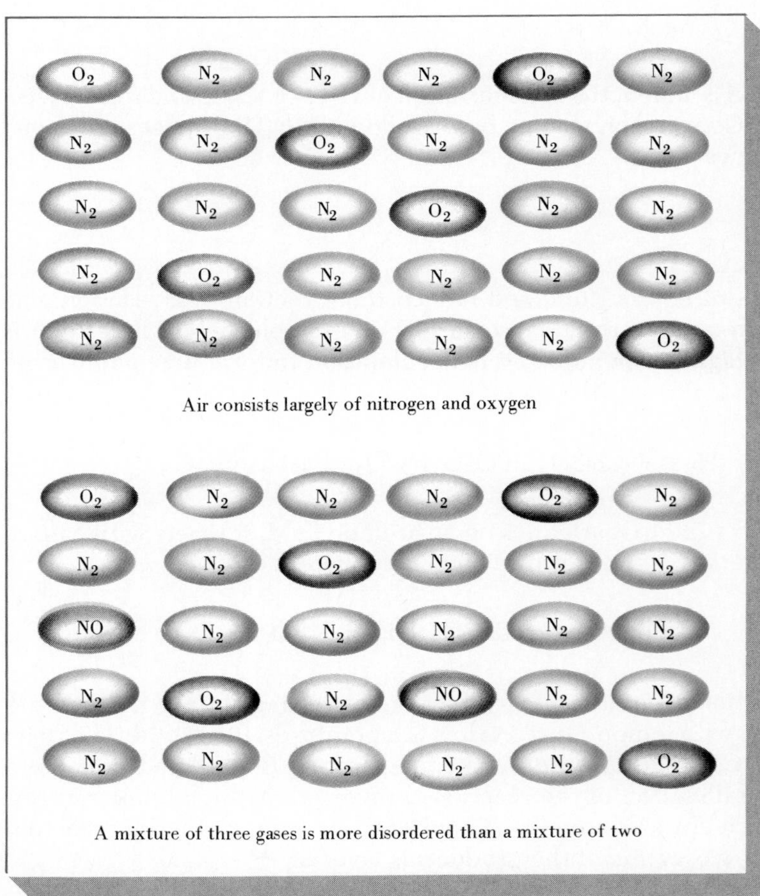

FIGURE 10.5
An increase in entropy accompanies the formation of a small amount of nitric oxide from nitrogen and oxygen.

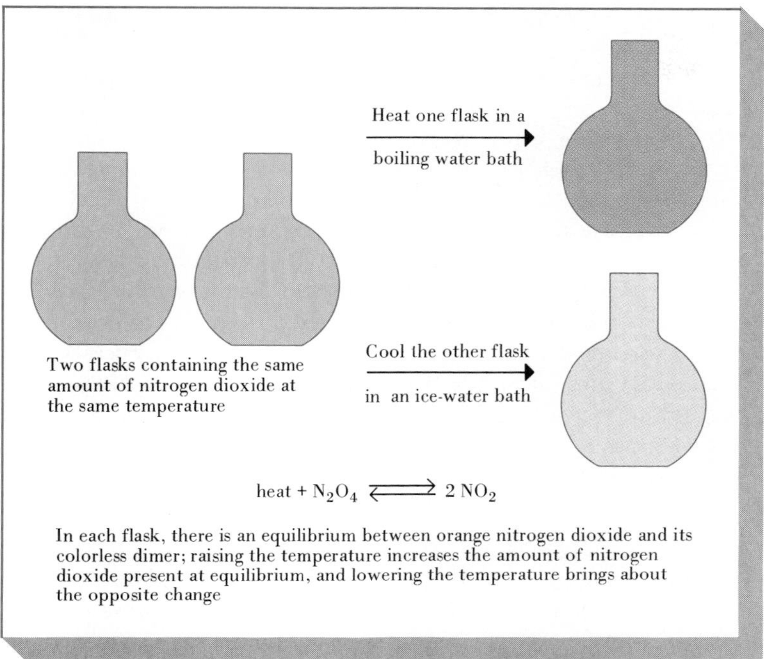

Heat one flask in a
boiling water bath
Two flasks containing the same amount of nitrogen dioxide at the same temperature
Cool the other flask
in an ice-water bath
heat + N2O4 ⇌ 2 NO2
In each flask, there is an equilibrium between orange nitrogen dioxide and its colorless dimer; raising the temperature increases the amount of nitrogen dioxide present at equilibrium, and lowering the temperature brings about the opposite change

FIGURE 10.6
Le Chatelier's principle can readily be demonstrated by showing the effect of temperature change on the amount of nitrogen dioxide in equilibrium with its dimer.

that *when conditions are changed, the position of equilibrium shifts to resist external changes.* For example, an increase in temperature favors the reaction that absorbs heat. Hence, the formation of NO becomes more favorable as the temperature is increased. (By converting some heat to chemical potential energy, this equilibrium shift resists a temperature rise.)

The formation of nitrogen dioxide from N_2 and O_2 is favored by increases in both pressure and temperature.

$$\text{Heat} + N_2 + 2O_2 \rightarrow 2NO_2$$

To resist a pressure increase, the volume decreases as 3 mol of reactant gases forms 2 mol of product gases. A shift toward more nitrogen dioxide at equilibrium would also resist a temperature increase, for its formation is endothermic.

A S I D E

REDUCING HARMFUL AUTO EMISSIONS

Both nitrogen oxides and unburned hydrocarbons from fuel contribute heavily to photochemical air pollution or smog. The reduction of nitrogen oxide and hydrocarbon emissions from automobiles has been mandated by federal law. One approach to reducing these emissions involved changes in engine design based in part on Le Chatelier's principle.

Although gases in an automobile cylinder do not reach equilibrium, the change in concentrations accompanying reaction is in the direction of equilibrium. As gases leave the cylinders, they rapidly cool and expand, stopping reactions and trapping nitrogen oxides that were formed at the elevated temperatures and pressures. To decrease nitrogen oxide emissions, the compression ratio was lowered at the expense of engine power. Lowering the compression ratio lowered the combustion temperatures and decreased the formation of nitrogen oxides.

Unfortunately, these changes in engine design resulted in less complete combustion of fuel. To further reduce emissions, exhaust gases are passed over a catalyst. The catalyst promotes the further oxidation of unburned hydrocarbons to form carbon dioxide and water. These catalysts also shift the amounts of nitrogen oxides in exhaust gases toward quantities in equilibrium at lower temperatures so they help limit NO and NO_2 emissions.

ENERGY, ENTROPY, AND RESOURCES

Thermodynamics provides an overview of mineral and energy resources. Whether it is petroleum or iron ore, a resource can be defined as a useful material found at high concentration (low entropy). Resources vary in the amount of energy needed for recovery. For example, the greater the amount of dirt and rock covering a coal seam, the greater the cost to remove the overburden and mine the coal.

As we exhaust supplies of our most highly concentrated ores, we will need to use less-concentrated, lower-grade ores. The amount of energy and the cost of the energy needed to recover lower-grade resources will necessarily increase. (Occasionally, the development of new, more efficient recovery processes will offset these cost increases.) In the future, the cost of energy will be an increasingly large amount of the total cost of producing many materials. Since energy resources are limited, the conservation and efficient use of energy will become increasingly important to our economic well-being.

We need to assess both the economic and the environmental impacts of industrial processes; the processing and use of resources can greatly affect our environment. Chemical wastes are a necessary part of the recovery of all minerals, but the amounts and types of wastes depend on the resource and on the particular processes used. The production and use of energy generates waste as well, since heat is a by-product of all energy production and use. Both the short-term cost of pollution control and the long-term cost of the failure to control pollution need to be taken into account.

QUESTIONS

1. Why does a person perspire more during exercise than when at rest? Is the chemical energy in food converted completely into muscular work?

2. The heat produced when a piece of zinc is dropped into a solution of cupric sulfate is (less than, the same as, more than) the heat produced when Zn reacts with Cu^{2+} in a galvanic cell. Justify your answer.
3. Calculate the calories in 1.00 g of the sugar glucose. The heat of combustion for glucose is 673.0 kcal/mol.
4. Is the energy required to vaporize 1 mol of water at 100°C more than or less than the energy required to vaporize 1 mol of a nonpolar liquid that has the same boiling point? Justify your answer.
5. Is the increase in entropy for the vaporization of 1 mol of water (less than, the same as, more than) the increase in entropy for the vaporization of 1 mol of a nonpolar liquid having the same boiling point? (The final volume of each gas would be the same; consider differences in the structure of the liquids.)
6. If the flow of water in a waterfall were spontaneously reversed, would it violate the first law of thermodynamics? The second law?
7. Why do the clothes on a line dry, even though the temperature is below the boiling point of water?
8. Why does a gram of water vapor have a higher entropy than a gram of liquid water?
9. Some eighteenth century scientists thought that air was a compound because the relative amounts of oxygen and nitrogen did not vary. They thought the lighter nitrogen should be on top of the heavier oxygen if these elements were not chemically combined. Why do nitrogen and oxygen mix, and why is the composition of air nearly constant on the surface of the earth?
10. It is much more difficult to determine the half-life of a radioactive sample when the number of decays occurring is small. Why is it easier to determine this average quantity with larger samples or shorter half-lives?
11. Some natural gas contains a relatively high amount of helium. What is the probable source of the helium? People concerned with future supplies of helium advocate separating helium from natural gas and storing it in abandoned mines. Why would it be much easier to recover helium from natural gas now than from the atmosphere in the future?
12. Why does it cost more to obtain iron from a low-grade ore than from ore with a higher iron content?
13. Gunpowder is a mixture of saltpeter (KNO_3), sulfur, and charcoal. What products of the reaction of gunpowder make it act as a propellant? The oxidation of carbon and sulfur is exothermic. Why is that important for the action of gunpowder?
14. Ammonia for use as a fertilizer is made by the reaction of hydrogen and nitrogen at high temperature and pressure.

$$N_2 + 3H_2 \rightarrow 2NH_3 + \text{heat}$$

(a) How does an increase in the temperature affect the amount of ammonia produced at equilibrium?

(b) How does an increase in the pressure affect the amount of ammonia produced at equilibrium?

15. In a bottle containing a soft drink, the carbon dioxide in the vapor and the carbon dioxide in the liquid are in equilibrium. Apply Le Chatelier's principle to account for the loss of carbon dioxide from the soft drink upon standing in an open glass.
16. More NO and NO_2 are formed when an excess of air is used to burn fuel. Apply Le Chatelier's principle to show that nitrogen oxide emissions can be reduced by decreasing the amount of air so that there is just enough oxygen to burn the fuel.

CHAPTER

ELEVEN

ENERGY FOR THE TWENTY-FIRST CENTURY

In preindustrial societies, the use of energy was limited and the density of population was low. Feed for domestic animals was grown on the farm. Trees were cut for fuel and lumber. Small towns might be located at a site where a waterfall powered a mill, and people who worked in a mill would live in nearby houses.

With the industrial revolution there began a rapid growth in the use of energy. Tractors and automobiles replaced mules and horses; steam engines replaced waterwheels and windmills. In many places, energy in the form of electricity became available at the flick of a switch.

New sources of energy were developed to meet the needs of industrialization. The industrial revolution was first fueled by the burning of wood, threatening to denude the forested countryside of England. The development of the technology for mining coal helped avert this early environmental crisis; coal replaced wood as the fuel for industrial growth. In the twentieth century, petroleum products have grown to supplant coal as the principal source of fuel for producing energy.

Changes in the organization of society accompanied the use of increased amounts of energy. The generation and distribution of electricity to homes and businesses and the construction of pipelines to distribute natural gas enabled the production of energy to be separated from its end use. The use of energy increased and its production became concentrated as industry and cities grew. Transportation systems were developed to let workers commute and to facilitate the wide distribution of mass-produced goods. Modern industrial society came into being.

ENERGY USE AFFECTS ALL ASPECTS OF MODERN LIFE

In the twentieth century, the cost of energy greatly affects our standard of living. Energy costs may be readily apparent as in the cases of home heating and transportation, or the costs may be hidden. The price of food rises with increased costs of energy to make fertilizers, to run farm machinery, and to process and transport food. The price of clothing varies with the price of the petroleum-based chemicals used to make synthetic fibers. Energy costs influence all areas of modern life.

Small changes in the price of oil have large impacts on the economies of both developed and developing nations. Oil-producing nations use revenues to develop at home and to invest abroad. For developed oil-importing nations, trade balances swing widely with changes in the price of petroleum and increased costs for energy slow economic expansions and hasten economic recessions. Undeveloped nations with few resources have gone into debt to pay for needed oil and gasoline.

An Arab oil embargo following a war between Israel and Arab countries led to short-term shortages in gasoline and heating oil. A series of rapid price increases by OPEC countries led to the quadrupling of oil prices in 1973. Increases in construction costs and high interest rates have left the

TABLE 11.1

WORLD POPULATION, ECONOMIC OUTPUT, AND FOSSIL FUEL CONSUMPTION, 1900–1986*

	Population (billions)	Gross World Product (trillion 1980 dollars)	Fossil Fuel Consumption (billion tons of coal equivalent)
1900	1.6	0.6	1
1950	2.5	2.9	3
1986	5.0	13.1	12

*Population statistics from United Nations; gross world product in 1900, authors' estimate, and in 1950 from Herbert R. Block, *The Planetary Product in 1980: A Creative Pause?* (Washington, D.C.: U.S. Department of State, 1981), with updates from International Monetary Fund; fossil fuels consumption in 1900 from M. King Hubbert, "Energy Resources," in *Resources and Man* (Washington, D.C.: National Academy of Sciences, 1969); for remaining years, Worldwatch estimates based on data from American Petroleum Institute and U.S. Department of Energy.

Source: Reproduced from *STATE OF THE WORLD, 1987*, A Worldwatch Institute Report on Progress Toward a Sustainable Society, Project Editor: Lester R. Brown. By permission of W. W. Norton & Company, Inc. Copyright © 1987 by the Worldwatch Institute.

nuclear power industry in shambles. The completion dates for some nuclear plants have been delayed, and the construction of others has been abandoned.

In 1986, a rapid drop in the price of oil stimulated the economies of industrialized nations; at the same time it wreaked havoc with the economies of poor oil-producing nations including Mexico. One result in the drop of oil prices is that the prospect of default by a major debtor nation threatens to disrupt the world banking system. In addition, the research leading to the development of new energy sources has faced sporadic and diminished funding.

There is widespread and growing belief that the availability, cost, and distribution of energy will dominate American and world politics in coming decades. Should energy availability and cost again become unpredictable, conflicts between groups within nations, between supplier and user nations, and between rich and poor nations will increase. The world we live in will become a more dangerous place.

At the same time, groups concerned about the waste of finite resources, the safe disposal of nuclear wastes, the effects of acid rain on the environment and, indeed, on the stability of climates throughout the world have become active in the political processes of western nations. People are concerned with the profound environmental impacts that can accompany our use and misuse of energy.

In this chapter, sources of energy and associated environmental concerns will be explored. An underlying theme of the chapter is summarized in the phrase "There is no free lunch." In deciding how to produce and use energy in the twenty-first century, choices that greatly affect both our political and natural environments will be made. Informed and thoughtful decisions are needed.

PETROLEUM, COAL, AND NATURAL GAS ARE FOSSIL FUELS

Radiant energy from the sun captured by plants and by photosynthetic bacteria is stored in carbon compounds and in an atmosphere of oxygen. Photosynthesis, using light from the sun, continually produces sugars from carbon dioxide and water. When plants and animals die and decay, these reduced carbon compounds may be preserved and modified by geological processes involving high temperatures and pressures and long periods of time. These resulting fossil fuels are petroleum, coal, and natural gas.

Coal originated in forests which were submerged and then buried. Peat, which is mined from bogs and burned for fuel, is an intermediate product in the formation of coal from wood. Coals contain fossils of ancient trees; this confirms their biological origin. Coal, like petroleum, consists of carbon, hydrogen, and frequent nitrogen and sulfur compounds.

Compounds in coal have a lower ratio of hydrogen to carbon than those in oil. Many of the compounds in coal have double bonds and rings. Their platelike structures enable efficient stacking, giving coal a solid structure. Because it is a solid, coal is more difficult to extract from the earth and more expensive to transport than petroleum. In the 1980s the use of coal for producing energy in the United States occurs chiefly in very large electricity-generating steam plants.

Petroleum is believed to have originated in microorganisms living in the sea. Because it is liquid, petroleum does not preserve imprints that furnish evidence of fossil origins. Scientists searched for and found molecular fossils, hydrocarbons that would provide evidence for a biological origin. These molecules contain repeating isoprene (C_5H_8) units like those found in vitamin A, rubber, turpentine, chlorophyll, cholesterol, and a variety of other naturally occurring compounds. The presence of these isoprene compounds is evidence that petroleum originated in living organisms.

$$\begin{array}{c} \quad CH_3 \\ \quad | \\ (-CH_2C{=}CHCH_2-) \end{array}$$

Petroleum has a higher hydrogen content than coal; and ranges of petroleum molecular weights are smaller than those found in coal. Petroleums are classed as sweet or sour, light or heavy. Sweet petroleums are low in sulfur; sour petroleums contain several percent sulfur. Light petroleums are free-flowing liquids; heavy petroleums are viscous liquids that are difficult to pump and refine.

Natural gas is a by-product of the geological processes that produce petroleum and coal. Methane, the principal component of natural gas, is responsible for fires and explosions that sometimes occur in underground coal mines. When geological formations are favorable, natural gas may be trapped in domes under impermeable rock formations. Because natural gas is low in sulfur and nitrogen, it is an environmentally cleaner fuel than coal or oil.

SUPPLIES OF FOSSIL FUELS ARE LIMITED

Concern for future supplies of fossil fuels is not new. Both gloomy and optimistic projections of supply and demand abound. These projections make very different assumptions about population growth rates, per capita energy use, and quantities of oil, gas, or coal that remain to be discovered. However, most agree that the supplies of fossil fuel are finite and will run out, and that choices made and implemented in the near future will affect the amount of time available to develop alternative energy sources.

Supplies of easily available oil and natural gas may be exhausted early in the twenty-first century. The known reserves of petroleum have remained nearly constant for many years because oil was being discovered almost as fast as it was being used. However, the average size of new oil fields has been growing smaller and smaller. The future availability of natural gas is less certain, for new discoveries have led to increased estimates of reserves.

When the price of oil rises, it may become attractive to reopen spent oil fields and to extract much of the original oil that remains. New petroleum recovery techniques that include the injection of high-temperature steam or detergents serve to free residual oil from the rocks and sands to which it clings. These tertiary petroleum extraction techniques are expensive, so they are not used while the price of petroleum remains relatively low.

Supplies of coal are much greater than those of oil and gas. It is estimated that there is enough coal in the United States to supply this country's energy needs for 500 years. Mining that coal can have serious environmental impacts. In eastern mountain areas, strip mining has led to the erosion of slopes and the pollution of streams. In the west, abundant coal is found in areas where water is in short supply. Water to mine and wash coal is also needed for irrigation and for use in cities. The cost of coal varies with shipping costs and with the costs of treatments to lessen adverse environmental impacts.

However, it is not only the exhaustion of supplies of finite resources that is of concern. The use of large amounts of fossil fuels contributes to the production of acid rain and it may, in time, seriously disrupt the climate of the world.

ACID RAIN REPRESENTS A GROWING THREAT TO OUR ENVIRONMENT

An unwanted side effect of the burning of large amounts of fossil fuels has been the ejection of large amounts of nitrogen oxides and sulfur oxides high into the atmosphere. (The production of nitrogen oxides from the high-temperature reactions of nitrogen and oxygen was discussed in Chapter 10. Chapter 8 discussed sulfur, present in many untreated fuels, and its oxidation first to form SO_2 and then, in a slower second step, to form SO_3.) Further reactions of these gases with oxygen in the air and with water lead to the formation of the strong nitric and sulfuric acids. Rain carries these acids back to earth and often hundreds of miles downwind of their source.

Higher smokestacks, once thought to minimize the effects of pollution by diluting these gaseous by-products of power production, have only

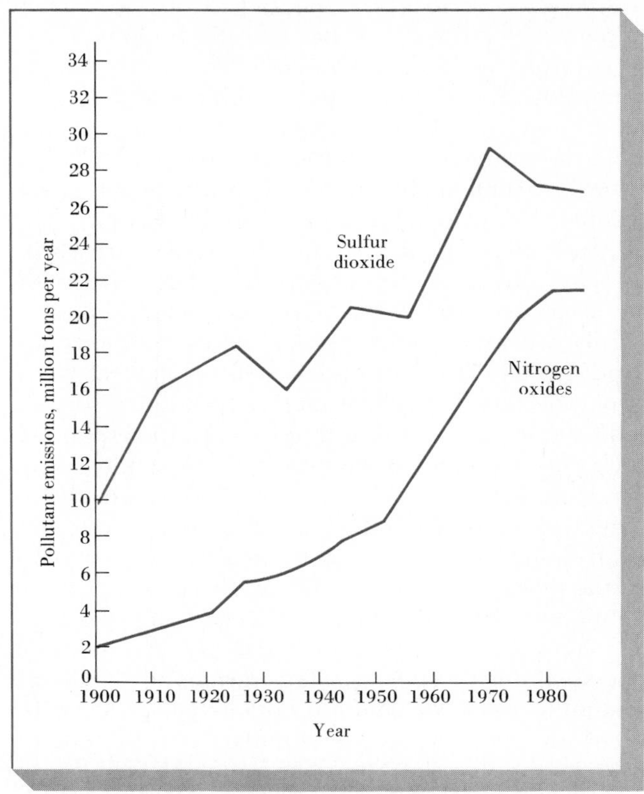

FIGURE 11.1
Sulfur dioxide and nitrogen oxide emission trends — national totals. *(Adopted from Acid Rain and Transported Pollutants, Office of Technology Assessment Congress of the United States, UNIPUB, New York, 1985.)*

served to increase their spread. In fact, the use of high stacks may have aggravated problems, for the gases remain aloft longer giving more time for reactions that produce strong acids. Because SO_2 and nitrogen oxides dissolve in water, they are concentrated in the tiny water droplets of clouds. Reactions occurring in clouds may hasten the formation of strong acids.

Damaging effects due to acid rain were first noted in Scandinavian lakes and forests and in the northeastern portions of the United States and Canada. Since then, the effects have spread in ever-widening rings. The effects are irregular. Fish may be killed in one lake, while a nearby lake continues to support a full spectrum of aquatic life. In some lakes, basic rocks such as limestone neutralize the acids.

$$H_3O^+ + CaCO_3 \rightarrow Ca^{2+} + HCO_3^-$$

Since the granites of the Adirondacks and the Canadian Shield do not neutralize the acids, these regions are particularly susceptible to damages.

While the initial concern about the effects of acid rain was focused on lakes and streams, attention is turning toward the effects on forests. In the

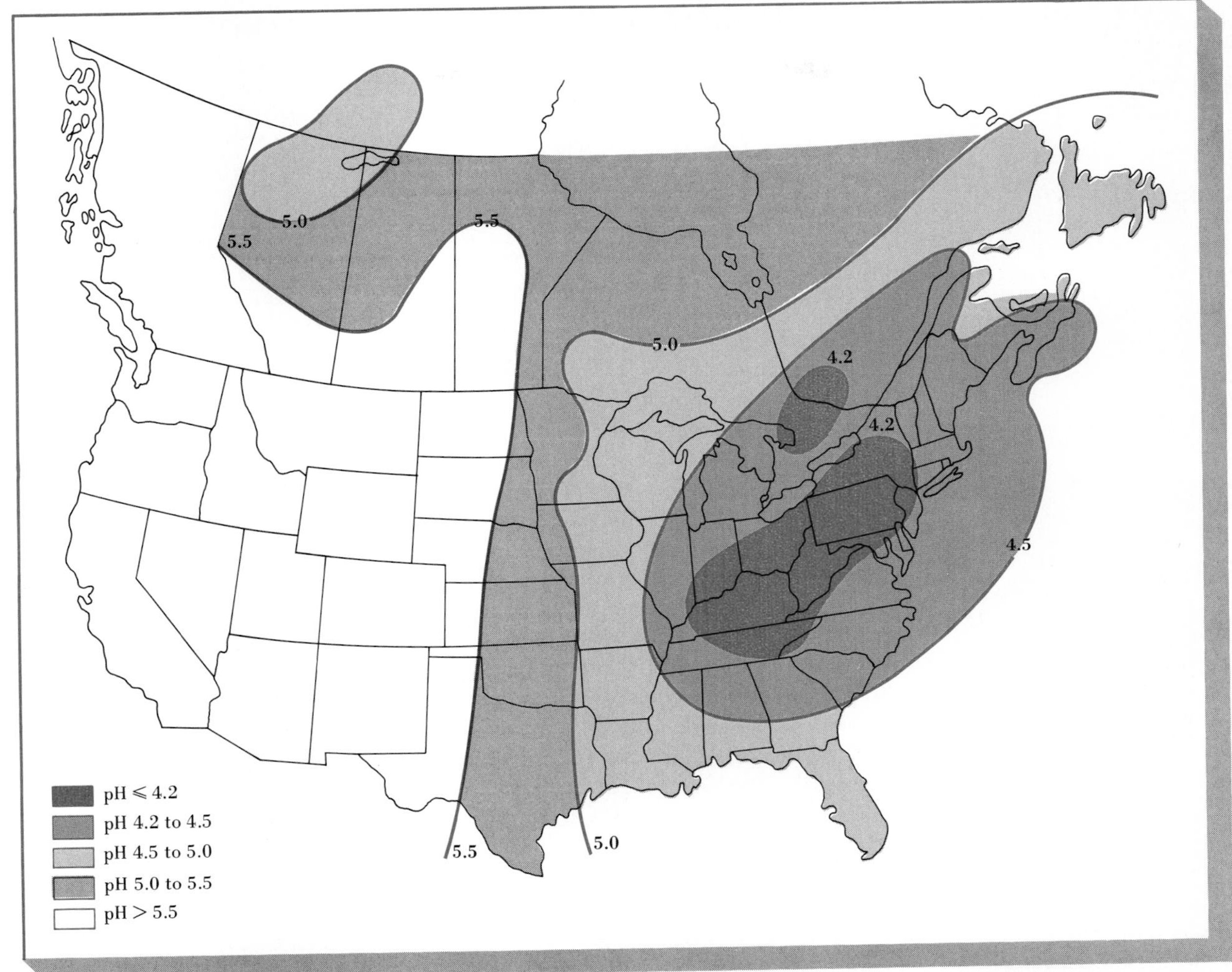

FIGURE 11.2
Precipitation acidity — annual average pH for 1980. *(From Acid Rain and Transported Pollutants, Office of Technology Assessment Congress of the United States, UNIPUB, New York, 1985.)*

higher elevations of New England, growth rates for trees have dropped dramatically during the last 20 years. Concern for loss of vegetation and increased erosion is growing. Far greater problems are found in the Black Forest region of Germany.

While the adverse effects on northern lakes and forests are apparent, the measurement of acid rain presents difficulties. The pH of "pure" rainwater is below 7. Rain is naturally acidic, for a solution of carbon dioxide in water is acidic. It is the presence of small amounts of the strong acids, HNO_3 and H_2SO_4, that is thought to be responsible for damages. Measured pH values may vary since the pH of unbuffered solutions is sensitive to small amounts of impurities and to changes in the amount of dissolved CO_2. It may be more reliable to concentrate samples and to measure the anions,

NO_3^- and SO_4^{2-}, than it is to measure the pH. Indeed, proposed goals for the control of acid rain are stated in terms of the reduction of sulfur emissions.

A S I D E

WHAT ABOUT ACID RAIN CAUSES BIOLOGICAL DAMAGE

What causes the mortality of fish in northern lakes and the deaths of trees at higher elevations? Scientists trying to answer these questions offer no simple answers. It is not yet certain which pollutants are the primary culprits or how they act. Our lack of knowledge about cause-and-effect relationships leading to biological damage effects raises questions for control strategies.

It is not simply the acidity of rain that is damaging. Coal has been used in great amounts for many decades, yet the damages associated with acid rain have appeared more recently. Some evidence points to nitrogen oxides and ozone as major contributing factors. Other evidence suggests that occasional events involving high acidities may be more important than average levels of acidity.

The complexity is illustrated by the damage to forests at higher elevations. It is found that high levels of acids are present in clouds; trees at high elevations are frequently bathed in clouds. Yet why the damage? Do the acids and ozone damage leaves, causing nutrients to be leached? Or does the nitric acid overfertilize trees so that they do not adequately prepare for winter? We do not know. The damage to the forests in Germany, where automobile emissions were not controlled before 1984, is worse than that found in the United States, an observation that supports some role for nitrogen oxides and ozone.

Similarly, evidence for fish kills is mixed. It has been found that the pH of high-elevation streams is lower in the spring than at other times of the year. Perhaps acid deposition accumulates in snow and ice, so that a surge of acidity accompanies the snow melt. When in their life cycle are fish most vulnerable? The timing of peaks of acidity may be important.

Complete answers to questions of cause may be slow in coming. In the absence of this knowledge, the effects of various control strategies on emissions cannot be known with certainty.

STRATEGIES TO REDUCE ACID RAIN

The control of acid rain is less a technical problem than it is a political and an economic one. Because much of the emissions come from large utilities and industries, control is possible. A low-cost approach to achieving modest reductions in SO_2 emissions involves the burning of coal containing less sulfur. Simply washing coal reduces the amounts of impurities high in sulfur such as iron pyrite, FeS_2. To switch from burning higher-sulfur coals mined

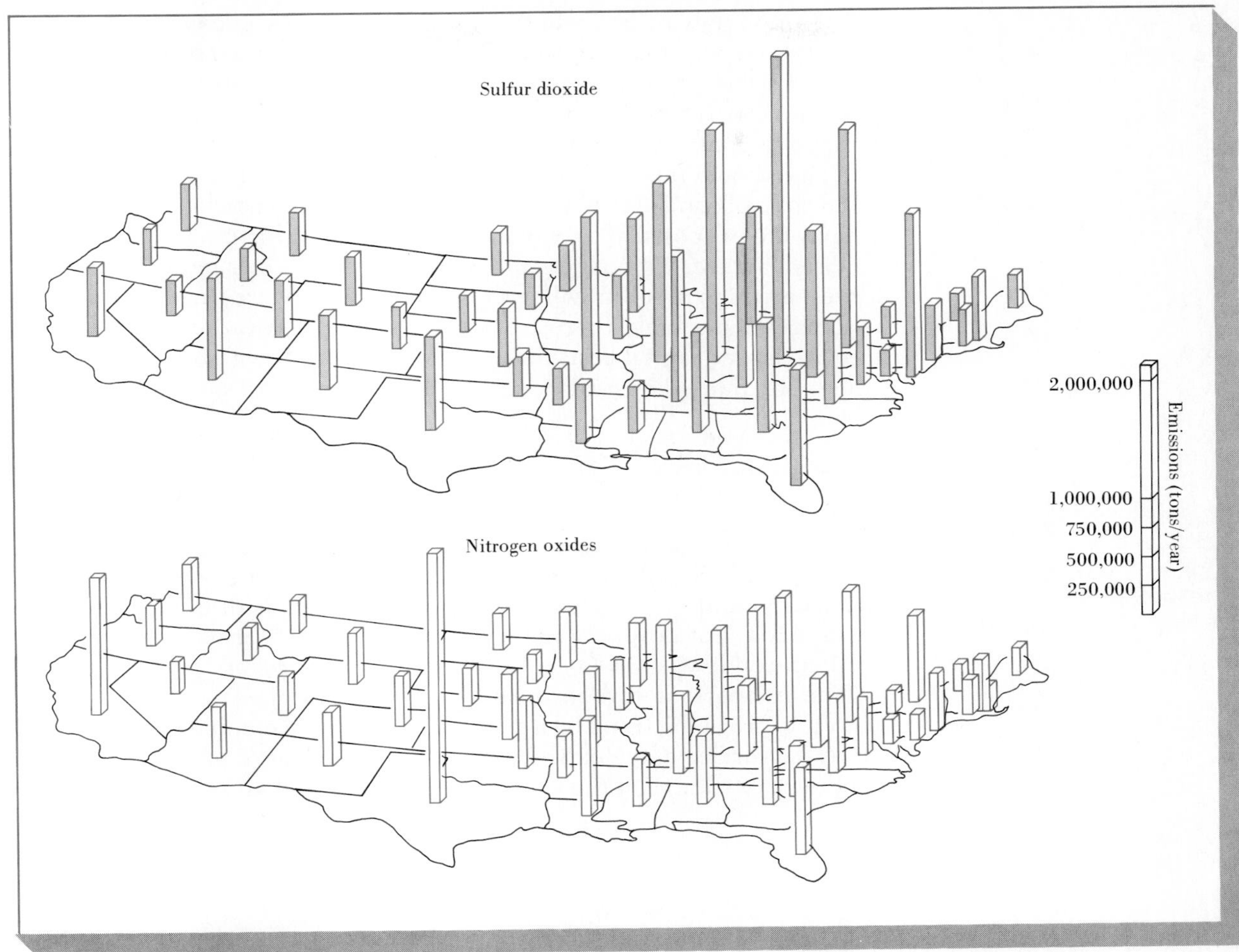

FIGURE 11.3
Sulfur dioxide and nitrogen oxides emissions—state totals for 1980. *(From Acid Rain and Transported Pollutants, Office of Technology Assessment Congress of the United States, UNIPUB, New York, 1985.)*

in the midwest and northern Appalachians to lower-sulfur coals mined in the west or southern Appalachians would further reduce sulfur emissions.

Additional technology to further reduce the emission of oxides of nitrogen and sulfur exists, but it is expensive. The installation of scrubbers to dissolve acidic gases so they can be neutralized is one approach. In another approach, pulverized coal is mixed with crushed limestone and burned. The basic limestone neutralizes much of the acidic gas before it can be released to the atmosphere.

$$SO_2 + CaCO_3 \rightarrow CaSO_3 + CO_2$$

If acid rain is a national and international problem, can its control be approached at a state and local level? Environmental costs of pollution are paid by those downwind from the power producers and consumers. Many Canadians believe that the United States' failure to take steps to control sulfur emissions poses the greatest threat to good relations between these two countries. And in Europe where boundaries are closer, no single nation can hope to control acid rain.

At a national level, who should pay for the control of acid rain? The capital costs of control are high. Should utilities in the northeast and middle west pay the full costs? If so, will industries in those regions be at a competitive disadvantage? To further complicate the issue, coals found in the middle west are high in sulfur, while those found in the west are low. If we shift the sources of our fuels, there are large economic dislocations; if we do not, there are larger capital costs for pollution control. If we postpone difficult choices, still more thousands of acres will become deforested and subject to erosion.

CARBON DIOXIDE EMISSIONS ARE WARMING THE EARTH

The use of fossil fuels has increased dramatically since the start of the industrial revolution. Carbon dioxide fixed by prehistoric plants is being returned to the atmosphere faster than plants can utilize it in photosynthesis. Measured levels of CO_2 in the atmosphere are rising.

The increase of carbon dioxide in the atmosphere contributes to a long-range warming trend by a greenhouse effect. Incoming solar radiation warms the surface of the earth by day. Most of the energy striking the earth is in the visible range of the spectrum. The earth radiates energy back into

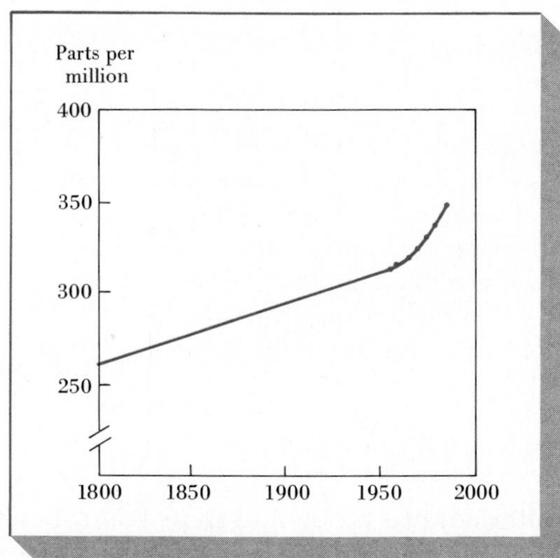

FIGURE 11.4
Atmospheric levels of carbon dioxide, 1800-1986. *(Reproduced from STATE OF THE WORLD, 1987, A Worldwatch Institute Report on Progress Toward a Sustainable Society, Project Editor: Lester R. Brown. By permission of W. W. Norton & Company, Inc. Copyright © 1987 by the Worldwatch Institute.)*

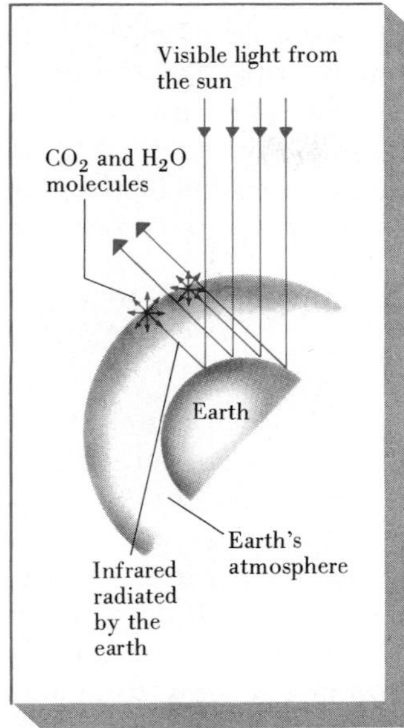

FIGURE 11.5
Increases of atmospheric carbon dioxide can lead to the warming of the earth's atmosphere by a greenhouse effect. The earth cools by sending energy into space in the form of infrared radiation. Molecules such as carbon dioxide and water that absorb infrared radiation cause a portion of that radiation to be retained in the atmosphere and converted into heat.

space. Since the earth is so much cooler than the sun, it radiates longer-wavelength infrared radiation.

Unlike oxygen and nitrogen in the atmosphere, both carbon dioxide and water vapor absorb infrared radiation. (Molecules with polar bonds absorb infrared radiation far more efficiently than nonpolar molecules. The energy absorbed causes the molecules to vibrate at greater energies.) Because of rising levels of carbon dioxide, the atmosphere absorbs more infrared radiation and retains more heat. Energy radiated from the surface of the earth warms the atmosphere rather than going into space.

The shift in the balance of radiation gains and losses is expected to have profound effects on the climate of the earth. Polar ice caps reflect much incoming radiation from the sun. As the atmosphere warms, some ice will melt. In the northern hemisphere, a large polar ice sheet is floating on the Arctic Ocean. When this ice melts, the exposed water will absorb more solar radiation than did the ice sheet. This increase in the absorption of solar energy will amplify the warming of the earth initiated by increases in atmospheric CO_2.

Many scientists believe that if the consumption of fossil fuels at current high levels is continued into the next century, the earth will warm, the sea level will rise several feet, and climates will change in unpredictable ways. If the consumption of fossil fuel is lessened, these changes will occur more slowly. These scientists believe that it is imperative to increase the use of forms of energy other than fossil fuels.

NUCLEAR FISSION IS A SOURCE OF ENERGY

Forces holding protons and neutrons together in atomic nuclei are far greater than those binding electrons to nuclei in atoms. In the fission of isotopes of uranium, the product nuclei have less mass and are more tightly bound than the reacting nuclei. Large quantities of energy are released as kinetic energy of the particles produced by fission and as high-energy photons called gamma rays.

$$^{235}_{92}\mathrm{U} + {}^{1}_{0}n \rightarrow {}^{92}_{36}\mathrm{Kr} + {}^{141}_{56}\mathrm{Ba} + 3\,{}^{1}_{0}n$$
$$^{238}_{92}\mathrm{U} + {}^{1}_{0}n \rightarrow {}^{141}_{55}\mathrm{Cs} + {}^{93}_{37}\mathrm{Rb} + 2\,{}^{1}_{0}n$$

The rate of energy production depends on the rate of fission reactions. If, on average, one or more neutrons from each fission is captured by another uranium-235 isotope, a chain reaction can occur; if fewer neutrons are captured, a chain reaction is not sustained. If the amount of uranium 235 is above a critical mass, a chain reaction does occur and large amounts of energy are released. (In an atomic bomb two subcritical masses of uranium 235 are blown together by a conventional explosive. Then, in a small fraction of a second, the energy of the fission reactions is released before the force of the explosion separates the remaining uranium 235 and stops the chain reaction.)

In a nuclear reactor, the nuclear fuel is less concentrated than that used in the manufacture of weapons. The rate of fission is controlled. Control rods contain cadmium and boron, two elements that absorb neutrons efficiently.

$${}^{1}_{0}n + {}^{113}_{48}\mathrm{Cd} \rightarrow {}^{114}_{48}\mathrm{Cd}$$
$${}^{1}_{0}n + {}^{10}_{5}\mathrm{B} \rightarrow {}^{11}_{5}\mathrm{B}$$

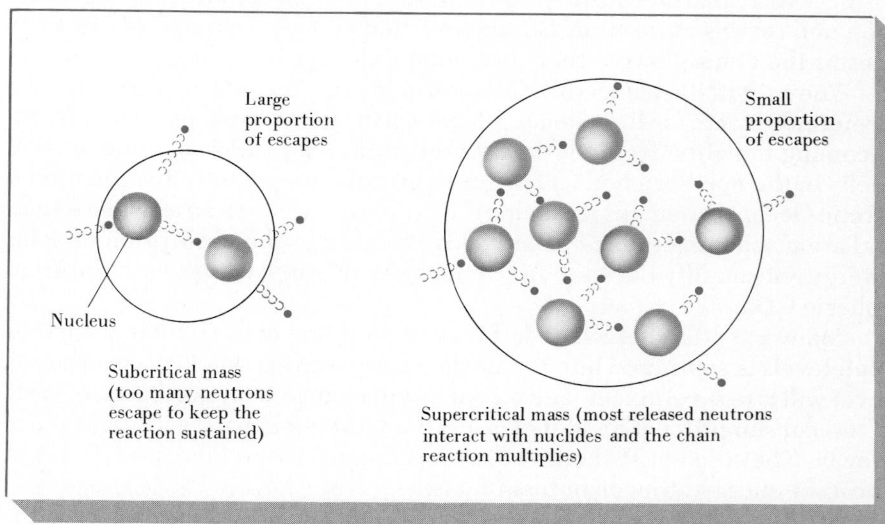

FIGURE 11.6
The rate of nuclear fission depends on the efficiency of capture of neutrons. If the mass of fissionable material is below a critical mass, too many neutrons are lost and a chain reaction is not sustained. If the mass of fissionable material is far above the critical mass, a highly branched chain reaction can occur leading to an explosion. By controlling the mass of fissionable material and the flux of neutrons, a controlled, sustained rate of fission can be maintained.

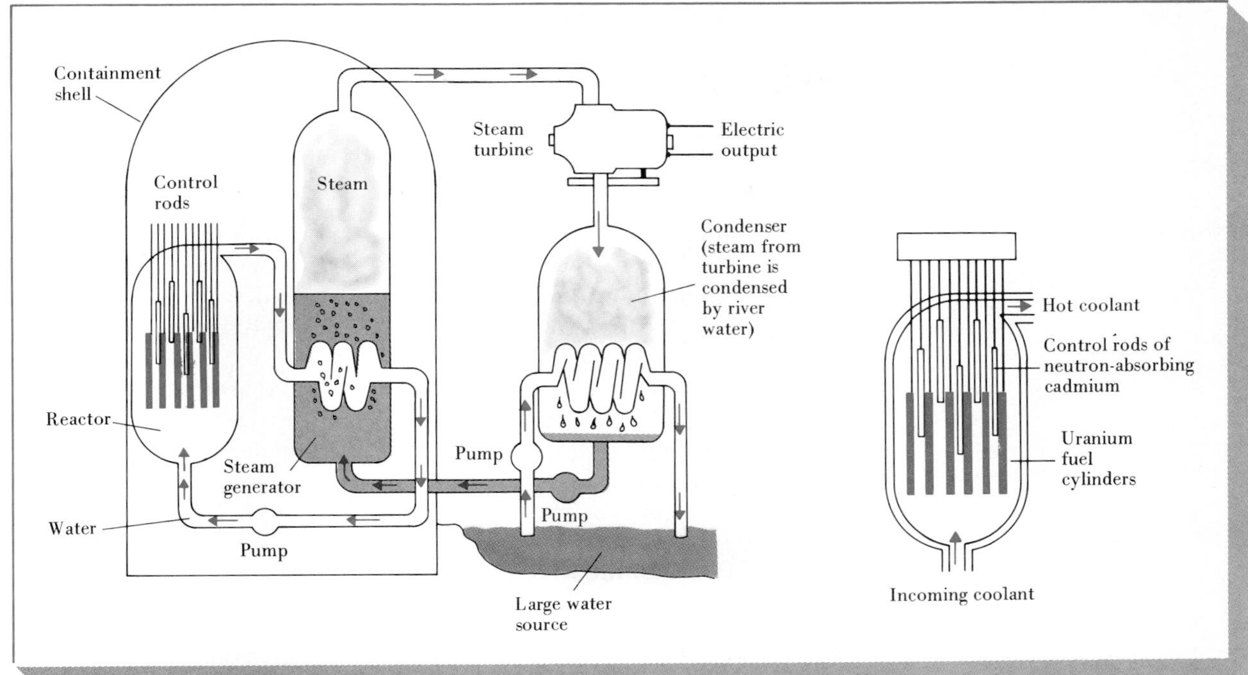

FIGURE 11.7
In a nuclear reactor, the energy released by fission heats water to form steam. The steam is then used to generate electricity in the same way as in fossil-fuel-operated power plants.

The fraction of uranium 235 decreases as the nuclear fuel is consumed. By withdrawing the control rods gradually, the rate of heat production by the reactor can be maintained. Should nuclear fission begin to occur too fast, the control rods can be thrust into the core to quench the reaction by reducing the flux of neutrons.

In addition, a reactor contains a moderator, usually water or graphite, a form of carbon. Slow neutrons are captured more effectively in collisions with uranium 235 than are fast neutrons; collisions with water molecules or carbon atoms slow the high-speed neutrons ejected upon fission.

A heat exchanger is used to transfer thermal energy from the reactor core to a secondary loop for use in the generation of electricity. As in the case of the generation of electricity from the burning of fossil fuels, only a part of the heat energy produced by the nuclear reactions can be converted into electricity.

The natural abundance of uranium 235 is low; it comprises only 0.7 percent of uranium. For weapons use or for the production of power, the fraction of uranium 235 is enriched. It is difficult and expensive to separate isotopes of uranium. (During World War II, uranium enrichment facilities were built at Oak Ridge, Tennessee and Paducah, Kentucky. Small differences in the rates of diffusion of uranium fluoride gas, UF_6, that differed in mass were used to produce weapons-grade uranium. Since the enrichment in one step is small, the enrichment process had to be repeated many times.) If the use of uranium 235 for the production of power grows, the supply of this isotope will be depleted early in the next century.

RADIOACTIVE WASTES POSE TECHNICAL AND POLITICAL PROBLEMS

The development of nuclear power is at a standstill in the United States; construction has not started on any new nuclear plants. (It has, however, proceeded in some other countries, particularly France and Japan.) In this country there are too many uncertainties about the future of nuclear power for utilities to commit large amounts of capital.

The uncertainties center around public fears and around environmental issues associated with the disposal of nuclear wastes. Public fears can have a large impact on policy. The American public perceives that nuclear power is far more dangerous than do most experts. In contrast, the state-run nuclear power industry in France receives broad support. There, the selection of a site for a nuclear plant and the construction of the plant can be completed with a minimum of delay. That is not the case in this country where environmental impact hearings, changes in regulations, and con-

TABLE 11.2

NUCLEAR SHARE OF ELECTRICITY IN SELECTED INDUSTRIALIZED COUNTRIES—1985*

Country	Nuclear Share of Electricity for 1985, %
United States	16
United Kingdom	19
Canada	13
Sweden	42
West Germany	31
Japan	23
France	65
Soviet Union	10

*Nuclear Energy Agency, "Projected Costs of Generating Electricity from Nuclear and Coal-Fired Power Stations for Commissioning in 1995," OECD, Paris, 1986; electricity share from International Atomic Energy Agency, *Nuclear Power: Status and Trends, 1986 Edition* (Vienna: 1986).

Source: Reproduced from *STATE OF THE WORLD, 1987,* A Worldwatch Institute Report on Progress Toward a Sustainable Society, Project Editor: Lester R. Brown. By permission of W. W. Norton & Company, Inc. Copyright © 1987 by the Worldwatch Institute.

struction delays have added years to the completion of some plants and completely halted construction of others.

The products of the nuclear fission reactions occurring in a reactor are themselves radioactive. In addition, radioactivity is induced in many of the components of the reactor that are exposed to a high flux of neutrons. Radioactive wastes differ greatly in their half-lives. High levels of radioactivity of short-lived isotopes present only small, long-range disposal problems, for these materials can be stored until their radiation subsides. After 10 half-lives, radiation from any isotope is about one-thousandth of its original level.

People are concerned about the disposal of longer-lived wastes. To dispose of these wastes, means to contain highly radioactive materials need to be found. For example, plutonium 239 produced by the reactions of neutrons with uranium 238 has a half-life of 24,000 years. This half-life is short enough that the level of radioactive decay is high, but is long enough to present a long-term storage or disposal problem. It is an alpha emitter; plutonium compounds would be highly toxic if ingested, but the metal itself poses little danger.

Proposals for the permanent disposal of radioactive wastes call for the wastes to be concentrated, enclosed in glass or ceramic containers, and buried in geological formations that are impermeable to water. Some environmentalists are concerned that we cannot be certain that these wastes would not reenter the environment at a future time. In the meantime, spent fuels are accumulating at reactor sites. Some operating reactors may be closed when the limit of allowed on-site storage of radioactive materials is reached.

The Nuclear Waste Disposal Act of 1982 provided strategies for the disposal of radioactive wastes. Low-level wastes such as those generated by hospitals were to be buried in specially designated landfills. Permanent disposal sites for high-level wastes are to be chosen and put into operation in the 1990s. However, state governments and the executive and legislative branches of the federal government have been reluctant to make the final difficult choices that will lead to permanent disposal of high-level wastes. The absence of a clear disposal policy makes it impossible to project costs for future nuclear power production. It is necessary to solve the problem of waste disposal if nuclear power is to contribute to meeting future energy needs in the United States.

BREEDER REACTORS WOULD EXTEND THE FUTURE OF NUCLEAR POWER

Nuclear reactors designed to produce both power and fissionable isotopes are called breeder reactors. Breeder reactors produce more fissionable atoms than they consume, for uranium 238 is converted into fissionable plutonium 239. Excess fuel produced by breeder reactors could be used in conventional reactors.

Breeder reactors have the potential to extend the supply of nuclear fuel more than a hundredfold. To increase the yield of plutonium 239, a fast neutron flux is used. Compared to a conventional reactor, the core of a breeder reactor is small and the operating temperatures are high. Liquid sodium is used as the coolant.

$$
\begin{aligned}
{}^{1}_{0}n + {}^{238}_{92}\mathrm{U} &\rightarrow {}^{239}_{92}\mathrm{U} \\
{}^{239}_{92}\mathrm{U} &\rightarrow {}^{0}_{-1}e + {}^{239}_{93}\mathrm{Np} \\
{}^{239}_{93}\mathrm{Np} &\rightarrow {}^{0}_{-1}e + {}^{239}_{94}\mathrm{Pu}
\end{aligned}
$$

Breeder reactors can be designed to be inherently safer in operation than conventional reactors. Should fission speed up and begin to go out of control, the liquid sodium can cool by convection (should pumps fail) without nearing its boiling point. It has been shown in experimental breeder reactors operated to test limits that, as temperatures rise, the core expands lowering the neutron flux and slowing nuclear fission.

The future of breeder technology is uncertain. Because sodium reacts violently with water, the navy was averse to developing breeder reactors to power nuclear submarines. As it developed, the peacetime nuclear industry followed the lead of the navy in designing conventional reactors. However, reactor features desirable on a submarine and those desirable for the peacetime generation of power need not be identical. At present, breeder options are being pursued in Europe but not in the United States. In 1984, Congress discontinued funding for the Clinch River Breeder Reactor. Uncertainties about expense and about the price at which power may be sold have slowed the development of this technology.

RECYCLING NUCLEAR FUELS—OPPORTUNITIES AND RISKS

Spent fuel rods from conventional reactors as well as from breeder reactors contain fissionable plutonium. Since plutonium differs from uranium in its chemistry, it is far easier to separate plutonium from uranium than it is to separate the isotope uranium 235 from uranium 238. Plutonium may serve as a fuel in a nuclear reactor, or it may be used to construct nuclear weapons. If it is not to be used in these ways, it presents an important disposal problem (discussed earlier in this chapter).

The recovery and recycling of plutonium presents risks. One concern is that of possible nuclear proliferation. In addition, shipping spent fuel rods risks accidents and spills. Returning plutonium for fuel presents another sort of risk, because there is fear that terrorists could hijack a shipment and use the stolen plutonium to construct a bomb. These concerns are not entirely new. Nuclear powers have recovered plutonium from the fuel rods of specially designed reactors, have transported that plutonium, and have used that plutonium to construct nuclear weapons.

The United States has chosen not to recycle fuels from nuclear power plants. Again, France has chosen a different policy; it is recycling its own fuels and those from some other nations.

ASIDE

NUCLEAR ACCIDENTS AT THREE MILE ISLAND AND AT CHERNOBYL

Two major nuclear accidents have eroded the public's confidence in nuclear power and shaken the complacency of the nuclear power industry. In 1979, an accident at the Three Mile Island nuclear reactor in Pennsylvania disabled the reactor and melted a part of the core. Some radioactive gases were vented, but the fuel and the majority of the radioactive species did not escape the containment vessel. In 1986, a more serious nuclear accident occurred near Chernobyl in the Soviet Union. Again, there was a meltdown, but in this accident there was a large release of radioactivity.

In each case, a reactor with inherent design weaknesses was operating near its limits as a part of a test procedure. In each case, operators had turned off safety devices contrary to operating procedures. In each case, the accident was due to human error by inadequately trained operators working with too little supervision. In each case, the accident has led to a review of design, of operating procedures, and of training; and in each case, the accident has led to heightened public concerns about nuclear power.

In the Three Mile Island accident, not enough cooling water reached the reactor and the water that was there vaporized to steam. The heat from the nuclear reactions melted the zirconium alloy fuel rods in part of the core. In addition, the hot metal reacted with steam to form metal oxides and hydrogen gas. There was a buildup of pressure that was relieved in part by explosions due to the reaction of hydrogen and oxygen. However, when the water coolant was lost, neutrons were no longer slowed by collisions with water molecules and the efficiency of neutron capture dropped slowing nuclear fission.

At Chernobyl in an experiment designed to take place at low reactor power, the power dropped more than was planned. To correct this condition operators removed additional control rods causing the power to come up too fast to be controlled. In this case, there was not a strong containment vessel and part of the fuel and its fission products were vented into the atmosphere. In the Chernobyl reactor, graphite was used to slow neutrons so the loss of water cooling did not slow fission to nearly the extent it did at Three Mile Island. In addition, the graphite burned for several days in a conventional blaze that carried radioactive materials high into the atmosphere. (As a part of the firefighting procedure, boron carbide was dropped on the blaze to slow the continuing fission.)

NUCLEAR FUSION MAY SOMEDAY PROVIDE ENERGY

The fusion of hydrogen nuclei (the deuterium isotope) to form helium releases far more energy than does the fission of nuclei of heavy elements.

$$^{2}_{1}H + ^{2}_{1}H \rightarrow ^{4}_{2}He + \text{energy}$$

However, the reaction is hard to bring about. As positive hydrogen nuclei approach one another, the repulsive forces are great. In a hydrogen bomb, the explosion of a fission bomb is used to implode the fuel and hold it together long enough for fusion to occur. It is fusion that fuels the sun and stars. In stars, the combination of strong gravitational forces and high temperatures cause nuclear collisions to be sufficiently frequent and violent to sustain fusion.

For the production of power, a means to contain the nuclear fuel needs to be found. The required temperatures exceed the melting point of conventional materials. Two approaches are being taken. One is to contain the ionized hydrogen gas (plasma) magnetically, and the other is to use lasers to rapidly heat a solid nuclear fuel, lithium hydride, to cause the lithium nucleus to fuse with the hydrogen nucleus.

$$^{6}_{3}\mathrm{Li} + {}^{2}_{1}\mathrm{H} \rightarrow 2\,{}^{4}_{2}\mathrm{He} + \text{energy}$$

While research on fusion continues, it is not yet known whether this technology will ever contribute to future energy needs.

SOLAR ENERGY AND ALTERNATIVE ENERGY SOURCES

Solar energy is the largest underdeveloped source of energy available. The energy of the sun can be used to heat water and homes or it can be used to generate electric power. However, the amount of incoming solar energy varies with time of day, location, climate, and season. Although solar power can help to meet our energy needs in some ways in some places, it is not a panacea.

The use of solar energy for direct-heating purposes is called passive solar energy. The use of electric energy, particularly resistance heating, to heat homes or water is wasteful. There is energy loss in the production of electricity, in the transmission of electricity, and in the conversion of electricity back into heat. There is a great potential for saving by using the sun for heating purposes in many locations, and the technology is available.

The conversion of sunlight into electric energy, sometimes called photovoltaic energy, is a technology still under development. In one application, solar voltaics have been used to provide power to satellites in orbit. This method is still expensive, but the cost is dropping.

In another approach to harnessing solar energy, sunlight is used to drive the electrolysis of water to generate hydrogen at one electrode and oxygen at the other. Hydrogen serves to store energy. The metals palladium and platinum can take up large amounts of hydrogen. The hydrogen can then be reacted with oxygen in fuel cells to generate electricity. This approach to solar energy would make solar energy available when there is a need for electrical energy and not just when the sun is shining. However the large-scale use of hydrogen would present difficulties and dangers.

There remain undeveloped a variety of energy sources that can be used to help meet future needs. To the extent that these resources are developed, supplies of finite resources are extended. Energy in motion can do work. The movement of wind and of the tides can be harnessed for energy production. At certain geological hot spots, the heat from geothermal wells can be used to produce energy.

ENERGY EFFICIENCY AND CONSERVATION

Low-cost energy is a thing of the past. We live in a society in which energy is used to grow our foods, to clothe our bodies, to warm and cool our homes and workplaces, and to transport people and goods. A simpler life may be desired and desirable for a few, but it cannot sustain civilization as we know it. At the same time, we are faced with dwindling resources and the destruction of environments caused by by-products of energy production.

"Conservation" is a bad word to some. They associate it with a call to reject an industrial society. Yet third world countries see industrialization and increased energy use as a hope to escape from poverty. Whether or not western societies increase energy use, there may be greater energy demands in the world.

Energy efficiency may be synonymous with conservation for many. Energy is too valuable to waste. The careful use of energy in the present can extend energy supplies and reduce the demands on the environment. At the very least, it buys time.

What will be the energy legacy that we leave to the twenty-first century? This chapter pointed to environmental concerns associated with each of the major sources of energy. Risks need to be evaluated and options chosen carefully. We have a short time in which to plan for the future. A part of that planning should be to keep options open. We need to be developing nuclear power, solar power, and alternative energy sources, or, by default, we will find ourselves relying exclusively on coal. Choices may not be easy, and the effort to keep open currently unpopular options for research and development may not be politically popular, but to close options now may be dangerously short-sighted.

QUESTIONS

1. During the 1970s and early 1980s, the price of natural gas was regulated at a low level. What are the short-term and long-term effects of such a regulatory policy on the availability of natural gas?
2. Because of interruptions to the supply of petroleum from the Middle East, western nations have sought to reduce their dependence on foreign oil. Why did European nations believe that it was not in their national interest to support an American boycott of equipment for the completion of a Soviet pipeline to deliver natural gas from Siberia to the west?
3. Why is natural gas considered a cleaner fuel than petroleum?
4. Which fossil fuel is most nearly free from noncombustible contaminants when it is extracted from the earth? Which is least free?

5. In the early development of the automobile, low-sulfur oil from Pennsylvania was used as a lubricant. What would have been the effect of using a higher-sulfur oil in engines?
6. Statues of marble (a metamorphic rock that is chiefly $CaCO_3$) and bronze (an alloy containing copper and tin) have shown dramatically increased rates of destruction since automobiles have come into wider use. Why?
7. Why does low-sulfur oil demand a premium price on the market?
8. Is there the same need to control auto emissions in Montana and Wyoming as in California? Should the regulation of emission controls be undertaken by the states or by the federal government? Why?
9. Ohio is a state that produces little energy but uses much. How would congressional delegations from Ohio and Vermont differ on the need for control of acid rain and on the formula for cost allotment? Should the regulation of emissions be undertaken by the states or by the federal government? Why?
10. What have been the effects of increases in the price of oil on research directed toward the development of new sources of energy? As the price of oil dropped during the mid-1980s, what happened to the funding for developing alternative energy sources?
11. In mountain regions of the eastern United States, trees at higher elevations are stunted in growth more than the same species at lower elevations. In what ways may acid fog pose a wider threat to the environment than acid rain?
12. Proposals to use electric cars for driving in cities cite reduced pollution as a major benefit. In what ways would this be true? In what ways would this be false?
13. Both nitrogen oxides and sulfur oxides contribute to air pollution. Acid rain west of the Mississippi River has more nitrogen oxides than that east of the Mississippi. Conversely, acid rain east of the Mississippi contains more sulfur oxides than that west of the Mississippi. Why?
14. What countries would benefit most from a lengthened growing season accompanying the warming of the earth's atmosphere?
15. Why is a nuclear power plant better able to respond to changes in electrical demand than a coal-fired, steam plant?
16. Should decommissioned nuclear submarines be scuttled in deep parts of the ocean? Why or why not?
17. What scientific reason was there for the several-year delay in cleaning up the damaged nuclear reactor at the Three Mile Island?
18. Why do the tailings from uranium processing plants present a disposal problem?
19. Third world countries purchasing nuclear reactors from developed countries are concerned with fuel reprocessing capabilities. Why are developed countries hesitant about sharing this technology?
20. Should environmental groups have the right to seek injunctions for halting construction of new power plants for environmental reasons? What is the effect of a 1-year delay on the construction costs? (Add

interest and inflation.) What would be the effect of the loss of this privilege on the right of the public to seek policy reviews?

21. A nuclear fission chain reaction is believed to have occurred hundreds of thousands of years ago in a uranium-containing deposit in Africa. Which of the following lines of evidence would not be needed to support this hypothesis? Give a reason for your choice.
    (a) Lower than normal ratio of uranium 235 to uranium 238
    (b) Occurrence of plutonium
    (c) Observation of products of fission
22. What are the advantages of a breeder reactor over a conventional nuclear reactor?
23. Should the development of commercial breeder reactors be subsidized by the federal government? Why or why not?
24. If nuclear fission were to produce power commercially in the future, plants would need to be located in regions of high population density, for the plants would produce very large amounts of electricity. (Electric power is lost in transport over long distances.) Is fusion a political possibility in the United States?

CHAPTER

TWELVE

METALS

From the dawn of civilization, the use of metals has played a prominent role in human culture. In many places, civilization began with the use of bronze. In the bronze age, specialized artisans used this hard alloy to fashion sharper spears and keener blades than could be made from stone or copper alone. Bronze is made from tin and copper. Since ores of these metals are not found together, trade developed as the use of bronze grew.

The production of iron and hard, tough steel made possible the growth of industry in the nineteenth century. Steam engines powered looms and drove the ships and trains that brought nations closer together. In America, farmers used steel plows to break the prairie sod and laborers spanned the continent with steel rails as the west was settled. Cities grew upward as architects designed and built skyscrapers using steel beams for support.

Flight and the use of aluminum are twentieth-century developments. The fragile biplanes flown in World War I have been replaced by jets flying faster than the speed of sound and by airliners carrying hundreds of passengers across oceans and continents. Light, strong aluminum helped speed the pace of life and lower the costs of transportation.

How did it happen that copper was one of the first metals to be widely used and aluminum was one of the last? Copper is a relatively rare metal in the surface of the earth, while aluminum is the most abundant metal present. An understanding of the chemistry involved in the production of metals can increase our appreciation of history, economics, and geography.

COPPER, TIN, AND LEAD

Copper

Copper, a major component of both bronze and brass, has played important roles in technology. Because it is a relatively scarce metal and is a superior conductor of electricity, its value remains high even though other metals have replaced copper for many applications.

Copper, like the other coinage metals, gold and silver, is sometimes found free in nature. More often, copper is found combined with sulfur in the CuS and $CuFeS_2$ ores, which may have been formed by the precipitation of highly insoluble sulfides of copper from molten rock. Copper ores are found in mountainous regions including the southern Appalachians in Tennessee, the Rocky Mountains in Montana, Utah, and Arizona, and the Chilean Andes.

Copper compounds in the sulfide ores are concentrated by flotation. Oil and water are added to finely crushed rock. Water wets the silicates that compose most of the rock, but it does not wet copper sulfides in the ore. When air is blown through the mixture, a froth containing oil and copper sulfides floats to the surface where it is skimmed.

Copper sulfide is roasted to obtain copper. Oxygen from air has the seemingly paradoxical effect of reducing copper to its metallic state. Oxygen combines with sulfur to form sulfur dioxide, leaving the copper free.

$$CuS + O_2 \rightarrow Cu + SO_2$$

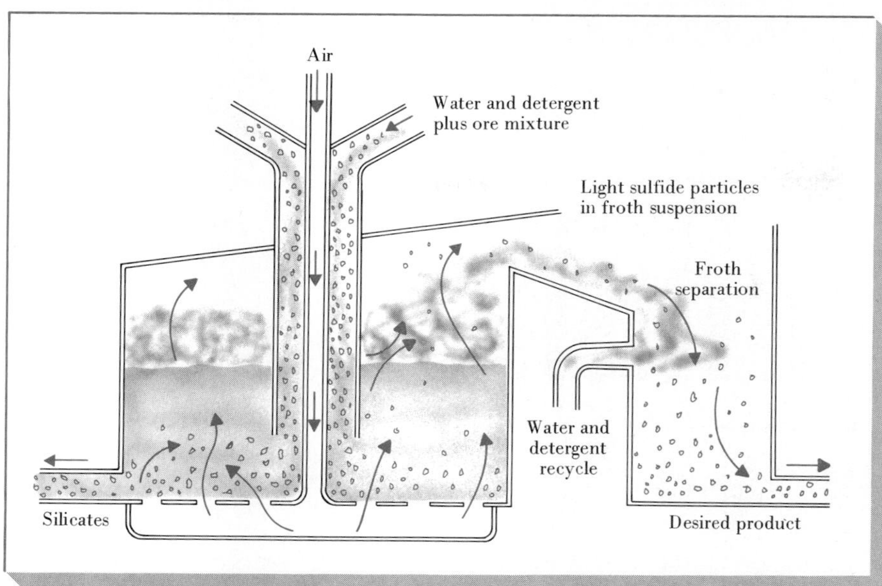

FIGURE 12.1
In the processing of copper ore, CuS is first concentrated by flotation. Water wets the silicate minerals that comprise the major portion of the ore; these minerals sink to the bottom and are separated. Oil coats the surface of the CuS, and a suspension of this mineral is first carried to the surface in a froth and then skimmed.

In the case of $CuFeS_2$, both iron and sulfur are oxidized by O_2 while the copper is reduced.

$$CuFeS_2 + \tfrac{5}{2}O_2 \rightarrow Cu + FeO + 2SO_2$$

The release of sulfur dioxide from early copper smelters led to acid rain that devastated the nearby countryside. At Copper Hill, Tennessee, acid rain killed the forest cover and the heavy rains of the region gouged the steep mountain slopes. Years later, the smelting process was modified to recover the sulfur dioxide and to make sulfuric acid as a by-product of the production of copper. Because copper is refined in the sparsely populated mountain areas where it is mined, the public was not alerted to the destructive effects of acid rain.

Very pure copper is required for all electrical applications. Copper from the smelter still contains small amounts of other metals found with copper in the ore. Because impurities greatly decrease the electrical conductivity of copper, the metal is purified by electrolysis. The impure copper is used as an anode. There, under an applied voltage, copper is oxidized. Copper ions migrate to the cathode where they are reduced to give copper that is more than 99.9% pure.

The electrolysis removes small amounts of iron and silver as well as other metals. Iron is oxidized and goes into solution, but it does not plate out with the purified copper at the cathode since Fe^{2+} is harder to reduce than Cu^{2+}. Because it is harder to oxidize than copper, silver in the impure anode falls to the bottom of the cell as the copper in the anode is oxidized. The

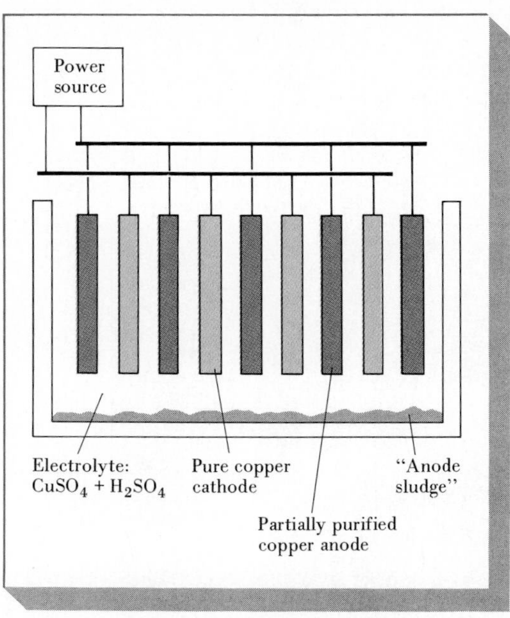

FIGURE 12.2
Electrolytic cell for refining copper.

recovery of silver from the anode sludge helps pay the cost of the electrolytic purification of copper.

Tin and Lead

Tin, like lead, is an unreactive metal in the same family of the periodic table as carbon. It is found as the oxide SnO_2, which is readily reduced by carbon.

$$SnO_2 + C \rightarrow Sn + CO_2$$

Although it is not an abundant metal, tin has properties that have led to its important applications. Tin played an important role in early metallurgy because it was used with copper to make bronze, an alloy of the two metals that is harder than either metal alone. Tin cans have been widely used to preserve and market foods. These cans are actually made from steel that has a thin protective inner lining of unreactive tin.

Lead, like copper, occurs as sulfide ores in mountain regions and is readily recovered from these ores. It is the most abundant heavy metal in the earth's crust. Lead is found most often in the +2 oxidation state, whereas tin occurs in the +4 state in its ores. Like copper, lead is readily recovered from its ores.

$$2PbS + 3O_2 \rightarrow 2PbO + 2SO_2$$
$$PbO + C \rightarrow Pb + CO$$

Lead has been used in a wide variety of applications. Some lead compounds are both relatively insoluble and brightly colored: they have been used in paints and colored inks. The softness and high density of lead made it desirable for use in shotgun shells; lead shot is "easy" on the gun barrel and "hard" on the duck or pheasant. Lead's low-melting point made it desirable for casting hot type for printing.

From the time automobiles came into wide use until the late 1980s, a major use of lead has been in the production of gasoline additives. Tetraethyl lead, a compound with four ethyl (CH_3CH_2—) groups covalently bonded to lead, was added to gasoline to improve its antiknock characteristics. Upon heating in the engine cylinder, the long, weak covalent bonds between carbon and lead cleave to produce carbon species with odd numbers of electrons. These odd electron species react readily with molecular oxygen and help to promote smooth combustion.

The decomposition of tetraethyl lead produces metallic lead. To get the lead out of engines, compounds containing chlorine and bromine were also added to gasoline. An aerosol of PbBrCl and related compounds of lead is emitted in the exhaust. Because lead compounds poison the catalysts used to reduce hydrocarbon and nitrogen oxide emissions, lead additives are not permitted in gasolines for newer cars with emission control devices. They will gradually be removed from all gasolines during the 1980s.

Unfortunately, the widespread use of lead has had adverse health effects. Lead, like many other heavy metals, is poisonous. It binds to sulfur atoms of proteins. Brain damage often results from long-term exposure to lead. In old houses lead-bearing, peeling paint poses a particular problem for small children who often ingest this poisonous compound. Earthenware pots with lead glazes can be a source of lead in the diet. Lead poisoning may have been a hazard as early as the time of the Roman Empire, when lead acetate, "sugar of lead," was sometimes used to sweeten wine and lead was used in plumbing and in pots.

Because of its toxicity, lead is being replaced in paints and in shotgun shells. Like the use of lead in gasoline, the use of hot type for printing is being replaced by newer technologies.

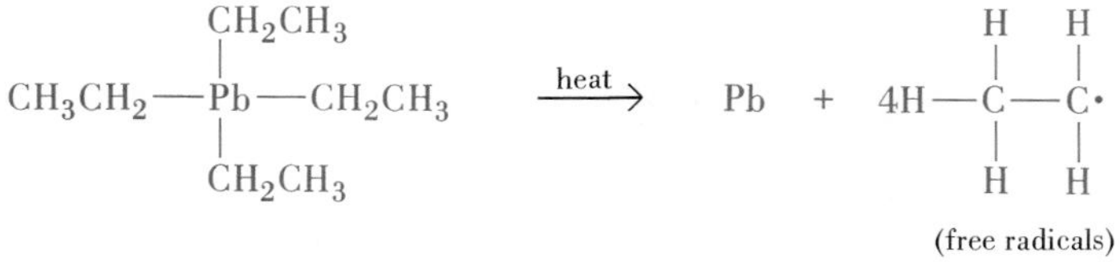

$$(CH_3CH_2)_4Pb \xrightarrow{\text{heat}} Pb + 4\,H{-}\overset{H}{\underset{H}{C}}{-}\overset{H}{\underset{H}{C}}\cdot \quad \text{(free radicals)}$$

Decomposition of tetraethyl lead

$$2CH_3CH_2\cdot \longrightarrow CH_3CH_2CH_2CH_3$$

and (free radical reactions)

$$CH_3CH_2\cdot + O_2 \longrightarrow CH_3CH_2{-}O{-}O\cdot$$

FIGURE 12.3
Tetraethyl lead has been used as a gasoline additive to improve the antiknock characteristics of the fuel. When heated, tetraethyl lead decomposes to form reactive free radicals that help promote smooth combustion.

IRON AND STEEL

Iron

Iron has seen limited use for many centuries. The Bible mentions the early use of iron by the Hittites. Archaeologists have identified sites where iron was produced during Roman times. However, the properties of iron and steel are sensitive to impurities and early steels made from some iron ores were very brittle. Since ores found in different locations contained different impurities, the production of iron and steel remained an art.

Improved methods of refining iron and making steel introduced in the nineteenth century led to a rapid increase in the production of iron. The ability to produce consistently high-grade iron and steel followed developments in chemical analysis and inventions that improved production processes.

Iron is recovered from iron oxides, Fe_2O_3 and Fe_3O_4, that are found in sedimentary deposits. The availability of limestone and coal in western Pennsylvania and the ability to ship ore through the Great Lakes from mines in Minnesota contributed to the early growth of Pittsburgh as a center for steel production in the United States.

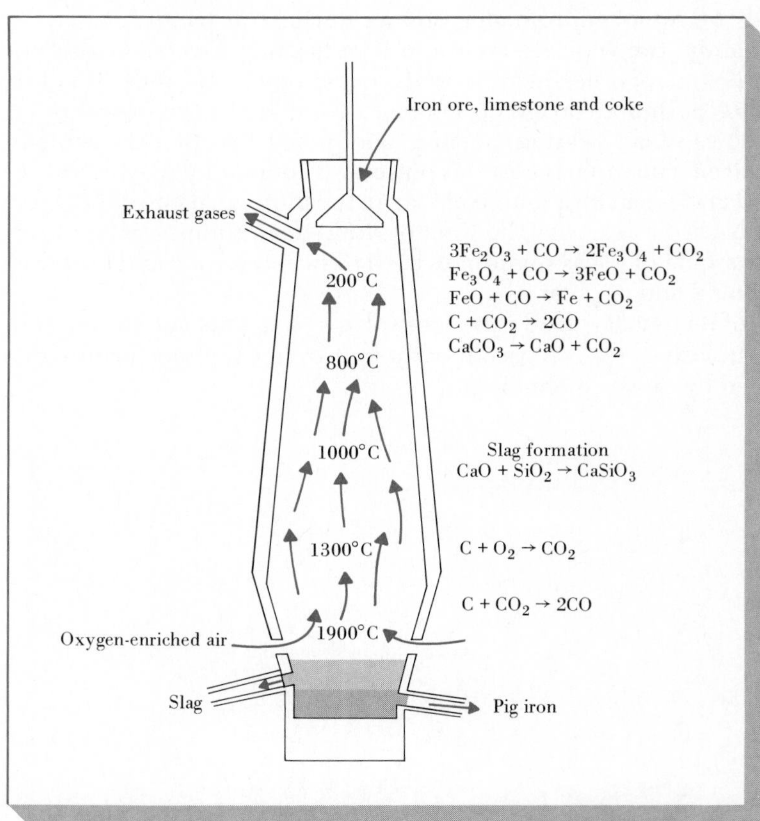

FIGURE 12.4
The blast furnace is used in the production of iron from its ore.

Iron ore is reduced in blast furnaces. A charge of iron ore, coke (C), and limestone ($CaCO_3$) is heated in the furnace, and air is introduced at the bottom. The partial combustion of the coke to carbon monoxide provides heat for the process. Carbon monoxide, not carbon, reduces the ore. (Reactions between coke and iron ore are very slow because there is only a small amount of contact at the surfaces of the two solids.)

The role of the limestone is to lower the melting temperature of silicate impurities in the ore and to remove acidic phosphates that adversely affect the quality of iron and steel. The slag formed in these reactions floats on the molten iron. Periodically the furnace is tapped; slag and cast or pig iron are drawn off. Slag is widely used in the making of concrete; therefore, highway construction provides a useful outlet for some of this by-product.

Cast iron from the blast furnace contains about 94% iron with carbon and traces of other elements comprising the remainder. Blast furnaces are operated continuously for months and years. Ore, coke, and limestone are added at the top and iron and slag are withdrawn from the bottom.

Galvanized Iron and Steel

Iron is subject to corrosion; it reacts with oxygen, particularly when wet, to form rust, Fe_2O_3. Iron can be protected by paint or by coating it with a layer of zinc to produce galvanized iron. (Zinc is more readily oxidized than iron.)

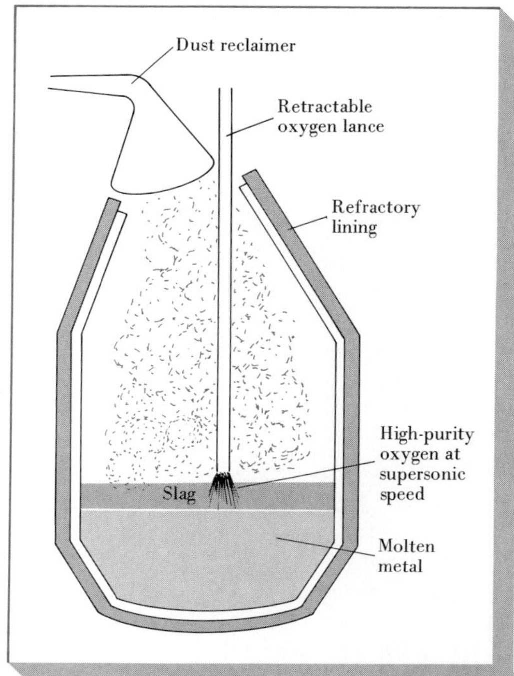

FIGURE 12.5
To make steel, the carbon content of molten pig iron is reduced by a high-temperature reaction with oxygen. A flux of CaO is used to help remove phosphorus and silicon remaining in the iron.

In the first step of the oxidation of iron to form rust, Fe^{2+} is formed. When Fe^{2+} is formed in galvanized iron, it is reduced back to iron by zinc before it can be further oxidized.

$$Fe^{2+} + Zn \rightarrow Fe + Zn^{2+}$$

Zinc reacts as a sacrificial electrode protecting the exposed iron from oxidation.

Steel is made by burning the excess carbon from the iron. Steels differ in carbon content and properties. High-carbon steel can be tempered to hold a sharp edge while low-carbon steels are tougher. Alloy steels are made by introducing other metals. Steel-containing manganese is very hard and tough; it is used in the blades of earth-moving machinery and in safes. Elastic, tough chrome-vanadium steel is used to make wrenches and engine valves. Stainless steel, an alloy containing chromium and nickel, does not rust because its surface is protected by a tightly adhering oxide layer.

ALUMINUM

Aluminum is a light, strong metal widely used in transportation and packaging. Its high strength-to-weight ratio makes its use attractive for planes and increasingly for automobiles as well. As the cost of gasoline rises, the fuel economies of lightweight cars become more attractive. Although aluminum cans cost more to make than steel cans or glass bottles, they are lighter and less bulky. Transportation savings help offset initial costs for aluminum users.

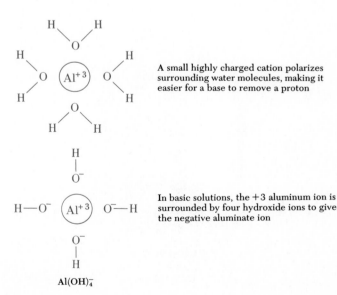

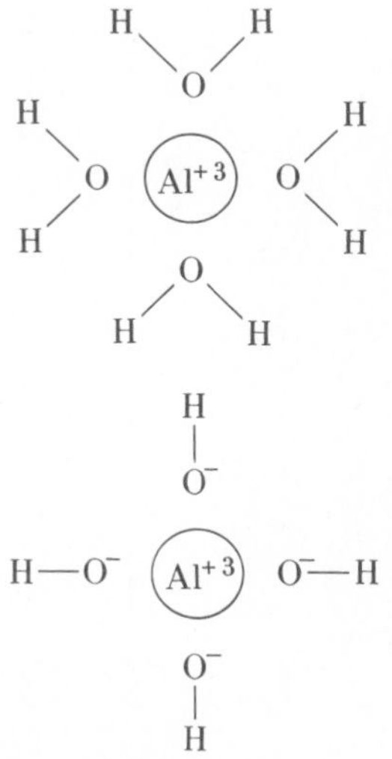

A small highly charged cation polarizes surrounding water molecules, making it easier for a base to remove a proton

In basic solutions, the +3 aluminum ion is surrounded by four hydroxide ions to give the negative aluminate ion

$Al(OH)_4^-$

FIGURE 12.6
The small +3 aluminum ion greatly increases the acidity of solvating water molecules.

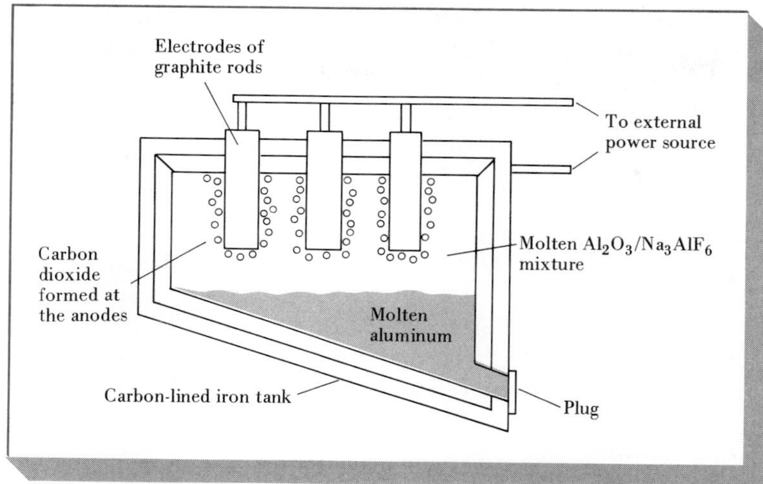

FIGURE 12.7
An electrolytic cell is used for the production of aluminum.

Aluminum occurs widely in granites and clays, but aluminum cannot be recovered economically from these minerals. Bauxite, $Al_2O_3 \cdot 2H_2O$, is a sedimentary ore used for the production of aluminum. Bauxite deposits are found in Arkansas and in Jamaica.

The aluminum in bauxite is separated by extraction from the iron and silica that occur with it. Al_2O_3 dissolves in strongly basic solutions, while Fe_2O_3 is insoluble. When base is added to bauxite, the aluminum oxide dissolves while the iron oxide remains behind.

$$Al_2O_3 + 2OH^- + 3H_2O \rightarrow 2Al(OH)_4^-$$

After the aluminum is leached from the bauxite, CO_2 is used to neutralize the excess base, causing $Al(OH)_3$ to precipitate from solution.

$$2Al(OH)_4^- + 2CO_2 \rightarrow 2Al(OH)_3 + 2HCO_3^-$$

The aluminum hydroxide is collected and heated to form Al_2O_3.

Aluminum oxide, unlike many other metal hydroxides, dissolves in strong base because the high positive charge on the Al^{3+} ion increases the acidity of attached water molecules. The small +3 ion pulls the electrons of the attached water molecules toward the aluminum and away from the hydrogen atoms of water. This electron shift polarizes the oxygen-hydrogen bond so that a hydroxide ion in solution can remove a proton to give the soluble $Al(OH)_4^-$ ion.

Aluminum-oxygen bonds are so strong that aluminum ore cannot be reduced using common chemical-reducing agents. Aluminum is produced by electrolysis. To electrolyze Al_2O_3, a suitable solvent is required. Charles Hall in the United States and Paul Heroult in France found that Al_2O_3 dissolves in the fused mineral cryolite, Na_3AlF_6, to give conducting solutions. Their independent discoveries made in 1886 opened the way to low-cost aluminum production. A current of electricity is passed through a solution of Al_2O_3 in cryolite. Liquid aluminum, collected at the bottom of the cell, serves as one electrode and carbon as the other. The electrode reactions produce aluminum at one electrode and carbon dioxide at the other.

$$Al^{3+} + 3e^- \rightarrow Al \qquad \text{reduction}$$
$$C + 2O^{2-} \rightarrow CO_2 + 4e^- \qquad \text{oxidation}$$

Much energy is required to refine aluminum. To produce aluminum by electrolysis, both a high voltage and a high current are required. In the United States, aluminum refineries were built in the Tennessee valley and in the Pacific northwest, areas that have a tradition of abundant, low-cost electric power.

Aluminum is Protected by an Oxide Coat

Aluminum is protected by a thin, strongly adhering film of aluminum oxide. If the film is scratched, the metal quickly reacts with oxygen in the air or with water to re-form the coating. This coat is impervious to oxygen so it protects the bulk of the metal. An aluminum beverage can lying by the roadside may take hundreds of years to oxidize.

The reaction of aluminum with oxygen can be easily demonstrated. If the aluminum is treated with a small amount of mercury (or a solution of a readily reduced compound of mercury) and a scratch is made, the presence of the mercury keeps the oxide coat from adhering. Aluminum oxide continues to form, and the aluminum becomes deeply etched.

Acids and bases react with the aluminum oxide coating to expose fresh metal. Acids attack oxygen atoms in Al_2O_3, and bases attack aluminum atoms. Freshly exposed metal is then oxidized.

$$2Al + 6H_3O^+ \rightarrow 2Al^{3+} + 3H_2 + 6H_2O$$
$$2Al + 6H_2O + 2OH^- \rightarrow 2Al(OH)_4^- + 3H_2$$

For this reason, highly acidic foods requiring long cooking times should not be prepared in aluminum cookware.

Recycling Aluminum

Much of the energy used to produce aluminum remains stored in a discarded aluminum can or in an obsolete plane. The recovery of aluminum from scrap requires about 5 percent of the energy needed to produce aluminum from ore. Recycled aluminum is melted and the aluminum oxide is skimmed. Energy is required to melt the metal, but the cost of reducing +3 aluminum is avoided.

For the recovery of aluminum to be effective, any iron must be removed. Iron impurities in aluminum would be weak spots where the oxide coating would not cling, and the oxidation of the metal could continue. Magnets are used to remove iron before the aluminum scrap is melted. The wasteful practice of making cans that have aluminum sides and steel tops has been discontinued.

A S I D E

METALLOIDS AND SEMICONDUCTORS

Silicon and germanium lie below the nonmetal carbon and above the metals tin and lead in the periodic table. Silicon and germanium belong to a small group of elements classed as *metalloids,* elements with properties between those of metals and nonmetals. Metalloids conduct electricity, but do so poorly. By adding tiny amounts of other elements, scientists can greatly enhance the conductivity of silicon and germanium. Devices made of "doped" silicon and germanium are widely used to construct circuits in computers.

In metals, conducting electrons are in orbitals that extend throughout the crystal. In the diamond structure of carbon, electrons are in localized bonds and are not free

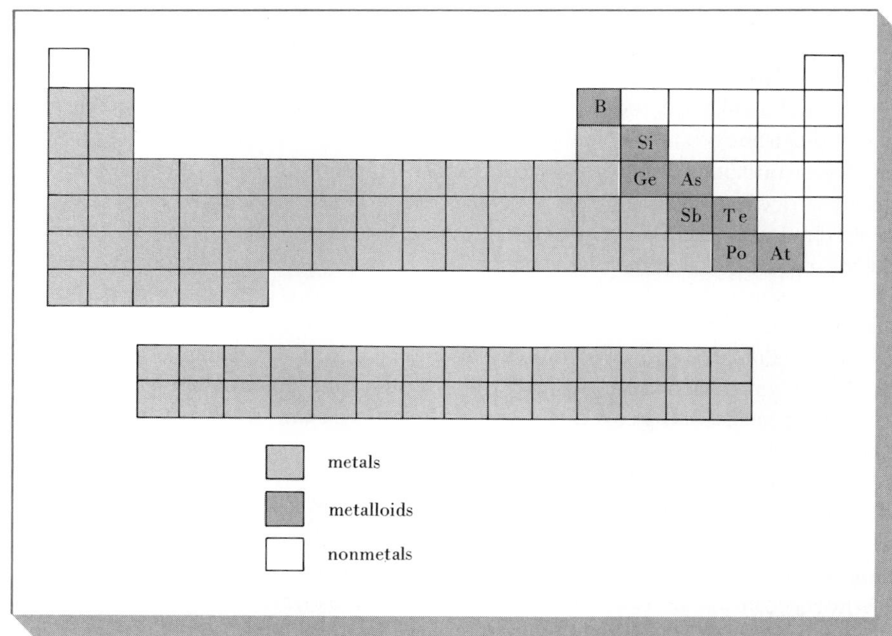

FIGURE 12.8
In the periodic table, metalloids form a narrow band dividing metals from nonmetals.

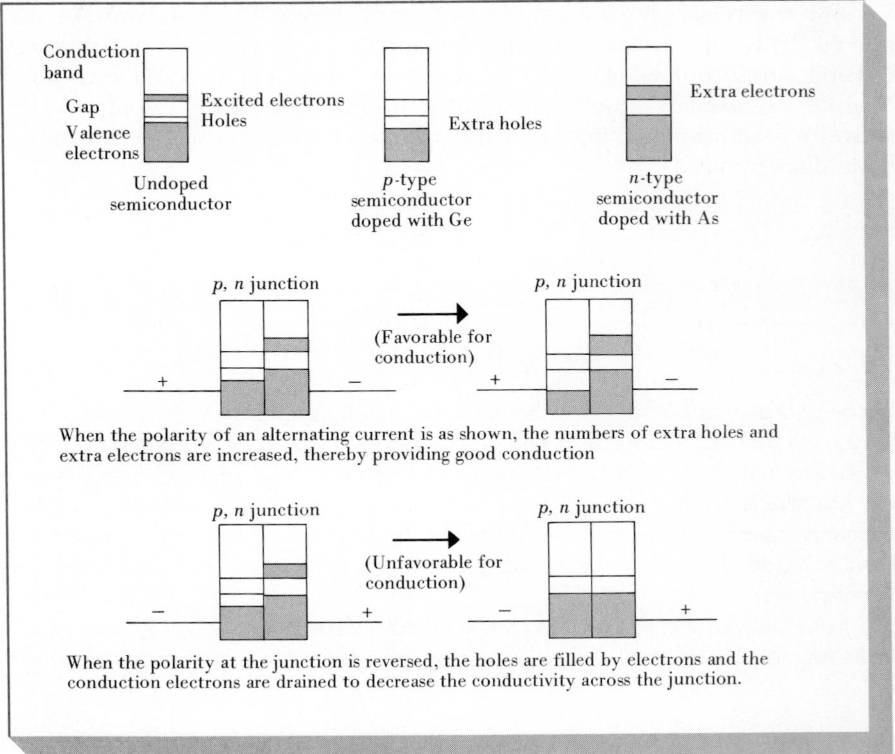

FIGURE 12.9
Semiconductors carry current either by the movement of positive holes or by the movement of electrons in a conduction band. By modifying semiconductors by adding controlled amounts of other elements and by combining different semiconductors in a circuit, electrical properties important for the miniaturizing of circuits can be prepared.

to migrate. The crystal structures of silicon and germanium resemble that of diamond, but not all the electrons are localized. In crystals with the larger silicon or germanium atoms, electrons are not as tightly held, and a few are excited to conducting bands.

Silicon and germanium are *semiconductors*. Current is carried both by conducting electrons and by positive holes. A nearby electron can fill a hole left by electron excitation and thereby leave behind a new hole; this movement of a positive hole is an electric current. Unlike metals, semiconductors become better conductors as the temperature increases, for the number of excited electrons increases with increasing temperature.

Small but highly controlled amounts of impurities are introduced to enhance the conductivity of semiconductors. If gallium with only three outer electrons is added, it produces a lattice with an excess of positive holes (a *p* semiconductor). If arsenic with five outer electrons is added, it produces a lattice with an excess of conducting electrons (an *n* semiconductor).

Transistors are made from sandwiches of *n* and *p* conductors. These have electrical properties important for the miniaturization of complex circuits. For example, electrons can flow in only one direction through an *np* junction. This device turns alternating current into direct current, allowing us to use a common electric outlet in place of batteries.

QUESTIONS

1. The following elements have symbols derived from Latin names. Either the element or its compounds were known in ancient times. Which one of these elements was known only in compounds?
   (a) Ag ("argentum") silver
   (b) Au ("aurum") gold
   (c) Cu ("cuprum") copper
   (d) Na ("natron") sodium
   (e) Pb ("plumbum") lead
2. What is a chemical explanation for finding gold artifacts in Egyptian burial tombs even though gold is a rare element in the crust of the earth?
3. In the electrolytic purification of copper, why do iron impurities remain in solution? Why are silver impurities found in the anode sludge?
4. Copper, silver, and gold are in column 11 in the periodic table. How does the ease of oxidation vary going down the table in this family? If oxidation were identical to removing an electron from an isolated atom, which of these elements should be most easily oxidized? (Consider the relative atomic size.)
5. Both copper and tin were liberated from ores near ancient campfires. Write the reaction for the formation of each metal. Which ore is reduced by carbon?
6. Why would a pulverized mixture of charcoal, ore, and limestone be unsuitable for use in a blast furnace?
7. Why are iron oxides reduced more rapidly by carbon monoxide than by solid carbon?
8. What is the role of $CaCO_3$ in the production of iron?
9. Scrap iron and steel are recycled to make steel. Why would it be important to be able to analyze the metal content of scrap to control the properties of the steel? What complications are posed by the use of a wider variety of metals in auto construction?
10. Why is aluminum used in automobiles since it is weaker than iron and costs more to produce?
11. Magnesium is a lightweight metal with uses similar to aluminum. It is in the same family of the periodic table as the element calcium which is readily oxidized by water or air. Why might metallic magnesium be resistant to oxidation?
12. Why does the solubility of aluminum compounds leached from bauxite with base decrease as CO_2 is absorbed from the air?
13. Why is it more economical to recycle aluminum than iron?
14. Why is the pH of a solution of aluminum sulfate less than 7? Write a reaction of a hydrated aluminum ion, $Al(H_2O)_6^{3+}$, with water to account for the decrease in pH.
15. Account for the observation that solutions of $FeCl_3$, $Fe(NO_3)_3$, and $Fe_2(SO_4)_3$ are acidic. How would Fe^{3+} affect attached water molecules?

16. Carbon is used in the production of both aluminum and iron. Write the reactions of carbon in the production of each metal.
17. For many uses, aluminum and copper need to be free of impurities. Why is copper purified as a metal and aluminum as an ore? (Try to think of chemical differences that would account for this difference.)
18. A student writes that aluminum is more susceptible to oxidation than is iron. In what way is this statement correct? In what way is it wrong?
19. Both aluminum and magnesium are protected from oxidation by an oxide coat. Why should these elements form strong covalent bonds to oxygen when elements below them in columns 2 and 13 bond more ionically to oxygen? (How does the ability to share electrons depend on atomic size?)
20. Which element in the same family of the periodic table as silicon should be most useful as a semiconductor? Briefly defend your choice.

CHAPTER

THIRTEEN

DOWN-TO-EARTH CHEMISTRY

A glass chandelier, a ceramic vase, a concrete dam, and a freshly plowed field seem to have little in common. Yet each is fashioned by nature or by humans from rocks and/or weathered rocks. Potters fashion ceramics from clay, and farmers grow wheat in clay soils. People modify chemical structures by heating limestone with clay to make cement and heating it with silica sand to produce glass.

There are similarities in the structures of ceramics, of concrete and glass, and of soils; they are all built up from a small number of chemical building blocks. There is a straightforward relationship between the structures and the properties of these familiar materials. In this chapter, we explore the everyday chemistry of glass and cement, of ceramics, and of agriculture.

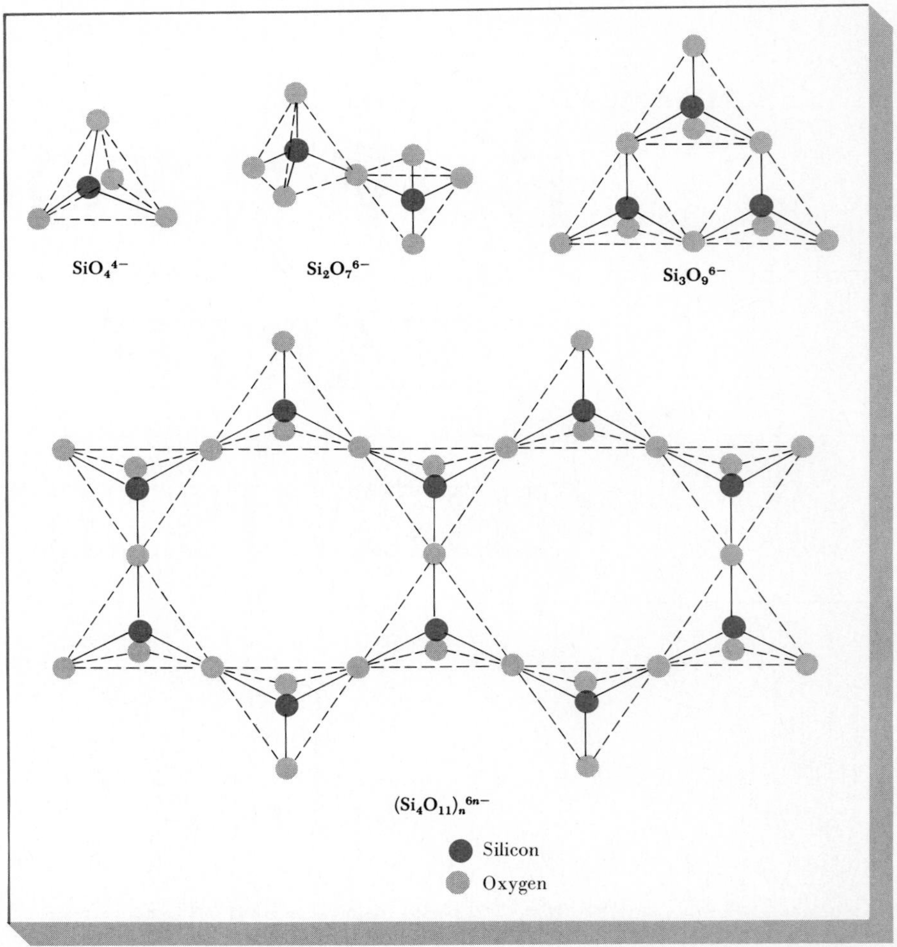

FIGURE 13.1
Some of the wide variety of structures found in silicate minerals are illustrated. These structures can be considered to be built by combining tetrahedral silicate anions.

SILICATES HAVE A VARIETY OF STRUCTURES

Silicon and oxygen are the two most abundant elements in the crust of the earth. Minerals called *silicates* contain these elements in varying ratios combined with a wide range of other elements.

The variety of silicates arises in part from the various ways that the tetrahedral silicate units, SiO_4^{4-}, can combine. Silicate tetrahedra can join to form dimers, trimers, linear chains, two-dimensional sheets, and three-dimensional networks. When tetrahedra are joined in a chain, two oxygen atoms of each tetrahedral unit are shared with neighboring silicon atoms. In a sheet, three of the oxygen atoms bonded to each silicon atom serve as bridges. In sand and quartz, each silicon atom is bonded to four other silicon atoms through oxygen bridges.

The electric charges on silicate species vary. A bridging oxygen atom connected to two silicon atoms carries no charge, but each oxygen atom bonded to a single silicon atom has a -1 charge. In a chain, there are two negative oxygen atoms for each silicon atom; in a sheet, there is only one negative oxygen per silicon; in the three-dimensional structures of quartz and sand, the oxygen atoms carry no charge.

In ionic silicate compounds, the negative charge on silicate anions is balanced by the positive charge on cations, but the identity of the cations can vary. For example, both the minerals $CaMg(SiO_3)_2$ and $LiAl(SiO_3)_2$ have long silicate chains, but in one of these the negative charge is balanced by Ca^{2+} and Mg^{2+} ions, and, in the other, it is balanced by Li^+ and Al^{3+} ions.

The physical appearances of some silicate minerals reflect their underlying molecular structure. For example, the mineral asbestos, $CaMg(SiO_3)_2$, is fibrous. The structure contains parallel silicate chains.

FIGURE 13.2
The structures of some silicate minerals reflect their underlying molecular structure. *(Left: Dr. Jeremy Burgess, Science Library, Photo Researchers. Right: J & L Weber/ Peter Arnold, Inc.)*

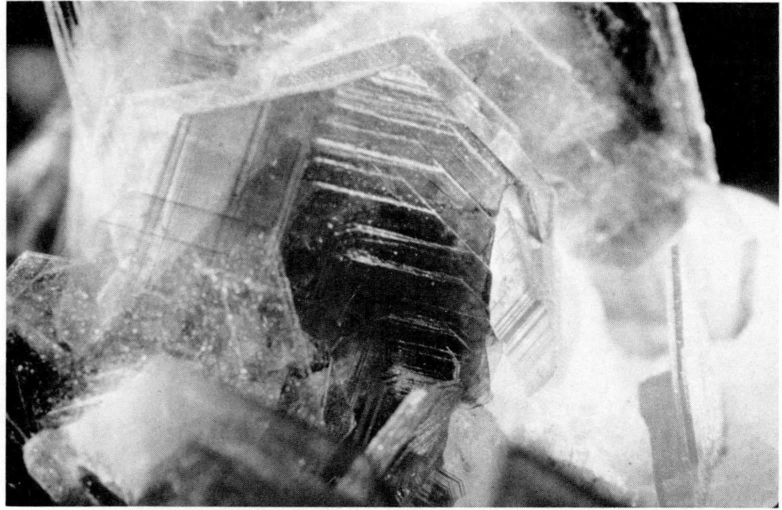

(Asbestos has been widely used for insulation and brake linings. However, small fibers in the dust from the mineral contribute to the formation of lung cancers. The removal of asbestos from schools and other public buildings is proving to be a slow and expensive process.) A different structure is found in the mineral talc, $Mg_3(Si_2O_5)_2(OH)_2$. Talc contains stacked two-dimensional silicate sheets. The layers can slide by one another, giving the mineral a slippery feeling.

GLASS

To make common *soda lime glass*, sand is heated with $CaCO_3$ (limestone) and Na_2CO_3 (soda ash). Heat decomposes the carbonates to give carbon dioxide and the oxides CaO and Na_2O.

$$\text{Heat} + CaCO_3 \rightarrow CaO + CO_2$$
$$\text{Heat} + Na_2CO_3 \rightarrow Na_2O + CO_2$$

Basic oxide ions furnish a pair of electrons to attack a silicon atom (a Lewis acid site) and to cut some, but not all, of the bridges of the SiO_2 network. The product of the reaction is a viscous liquid containing long silicate chains that vary in length. Silicate chains with their high negative charge are stiff and extended, and positive sodium and calcium ions balance the charge.

When glass is heated to 600 – 700°C, it softens and becomes plastic. It can be blown into molds to make bottles, drawn into tubing, rolled into sheets, or fashioned into ornate figures. The silicate chains and cations in glass are arranged somewhat randomly; they do not stack to form a crystal lattice. Upon cooling, the glass hardens and retains its shape. (Glass windowpanes from very old houses are thicker at their bases than at their tops, indicating that the glass flowed slightly over many years.)

Recipes for glass differ, for the properties of glass are altered by the addition of different materials. Glass made with lead oxide is used in crystalware. It refracts light more than ordinary glass and appears more brilliant. Boron oxide, B_2O_3, together with SiO_2 and Na_2CO_3, is used to make *borosilicate* glass. Borosilicate glass expands less than ordinary glass on heating. Because it can better withstand rapid heating and cooling, it can be used for the manufacture of cooking utensils.

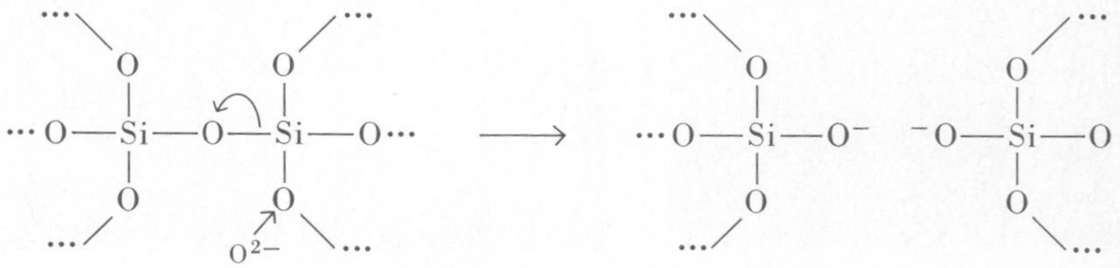

FIGURE 13.3
In the formation of glass, basic O^{2-} ions attack Si atoms to convert the SiO_2 network to silicate chains.

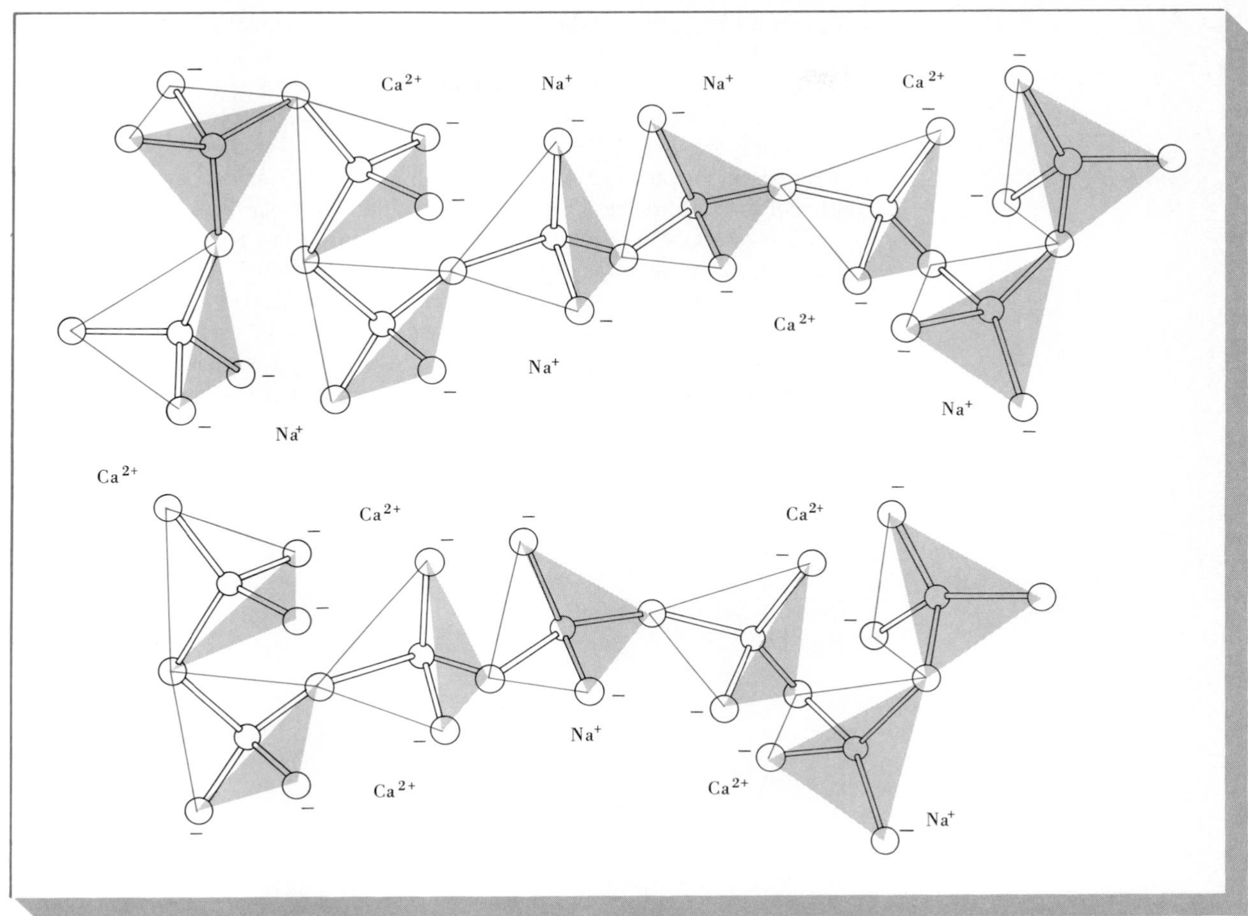

FIGURE 13.4
Glass consists of SiO_3^{2-} chains with cations to balance the negative charge.

A S I D E

COLORED GLASS, GLAZES, AND TRANSITION-METAL COMPOUNDS

Small amounts of compounds of transition metals, elements in which 3*d* subshells are partially filled with electrons, are used to give glass color. In addition, many of the glazes for ceramics are compounds of transition metals. Cobalt compounds give glass a deep-blue color, while iron compounds impart a yellow or orange color to glass. The ions of these transition metals are also colored in aqueous solution where the basis of the color has been studied.

For color, a transition-metal ion with a partially filled *d* subshell must be surrounded by Lewis bases such as water molecules, chloride ions, ammonia, or the oxygen atoms of silicate chains in glass. Changing the base changes the color of the solution. For example, Cu^{2+} forms a lime-green complex with chloride ions, a pale-blue complex with water, and an intense deep-blue complex with ammonia.

Compounds of transition metals are colored because they absorb visible light. Consider an ion of a transition metal bound to six bases. The electrons binding the bases to the metal ions are at the corners an octahedron. In such a complex, common for most of these ions, electrons in 3*d* orbitals differ in energy. The lower-energy *d* orbitals place electrons in regions between electrons from the bases. The higher-energy *d* orbitals place electrons closer to the electron pairs furnished by the bases. In these colored compounds, atoms of transition metals have partially filled *d* subshells. They have one or more electrons in the lower-energy *d* orbitals, and they have one or more vacancies in the higher-energy *d* orbitals. A photon of visible light with the right energy can be absorbed to excite an electron from a lower-energy *d* orbital to an open position in a higher-level *d* orbital.

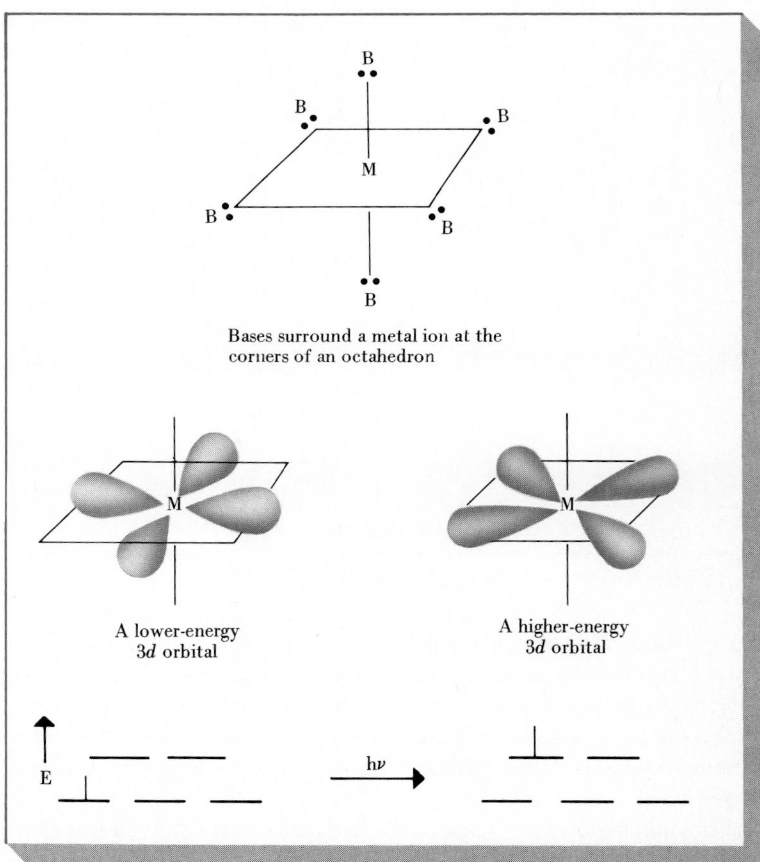

FIGURE 13.5
Transition-metal ions absorb light when an electron is excited from a lower-energy *d* orbital to a higher-energy *d* orbital.

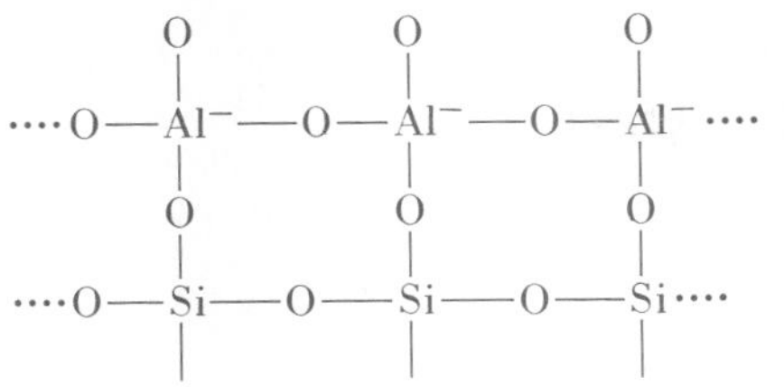

Clays contain large platelike structures composed of aluminum oxide and silicon oxide layers. Because Al nucleus contains one less proton than that of silicon, the substitution of each Al for Si gives a negative charge.

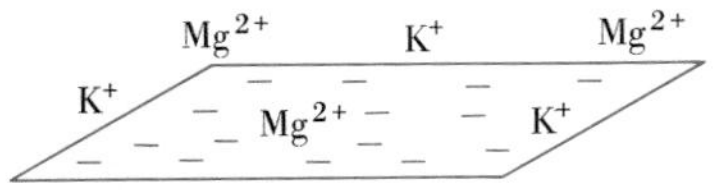

The negative charges on the platelike aluminosilicate anions of clay are balanced by the positive charges of cations.

FIGURE 13.6
Clays consist of large negative aluminosilicate anions with a variety of cations bound to the anions by the attraction of opposite charges.

CLAYS

Clays are *aluminosilicates*. Aluminum oxide, like silicon oxide, forms extended covalent structures. In clays, sheets of aluminum oxide are bonded to sheets of silicon oxide in sandwichlike layers. By substituting aluminum with a +3 kernel charge for silicon with a +4 kernel charge in an SiO_2 lattice or by substituting +2 magnesium for +3 aluminum in an Al_2O_3 lattice, charged anionic plates are formed. As in silicate minerals, cations are required to balance the charge of the polymeric anions.

Clays bind water much more strongly than do silicate minerals. The cations that balance the electric charge of anionic plates are solvated in tightly bound layers of water molecules. Because the water layers in clays can expand and contract, clays readily take up and release water.

A S I D E

CEMENT IS DEHYDRATED ROCK

Cement is made by heating clay with limestone to 1500°C. The clay is dehydrated on heating, and the limestone decomposes. Basic calcium oxide reacts with Lewis acid sites of the clay to cut into the aluminosilicate structure. When water and sand are added to the dry cement, crystals of hydrated aluminosilicate minerals slowly form. The strength of cement increases as crystals grow to form an interlocking network. Sand and gravel are added to the mix to make concrete. Reactions of sand with the basic components of cement help increase the strength at the interface between sand and cement crystals in concrete.

Plaster of paris is also dehydrated rock. When water is mixed with powdered calcium sulfate hemihydrate, $CaSO_4 \cdot \frac{1}{2}H_2O$, to form a slurry, crystals of hydrated calcium sulfate, $CaSO_4 \cdot 2H_2O$, slowly form and the plaster hardens.

$$3H_2O + 2CaSO_4 \cdot \tfrac{1}{2}H_2O \rightarrow 2CaSO_4 \cdot 2H_2O$$

CERAMICS—FROM CLAY TO GLAZE

Potters prepare clay before shaping it. Clays from different sources may be mixed. Ball clay deposited from slow-moving streams has fine particles that contribute plasticity to the mixture. Fire clay, deposited further upstream, has larger particles that contribute strength to ceramics and reduce shrinkage in drying. Potters work clay by adding water and kneading it. Kneading aligns the aluminosilicate plates of clay particles.

Physical changes accompany the drying and firing of clay. When a clay pot is dried, weakly bound water evaporates and some shrinkage occurs. When a pot is fired, the remaining water is lost. (A pot may shatter on heating if a trapped pocket of water turns to steam.) As water leaves and the plates move closer together, forces between the bound cations and the anionic plates become greater. Finally the material is vitrified; no longer can water move between the bound plates. Channels do remain so the fired pot remains porous.

Glazes are used to color and sometimes to seal ceramics. Many glazing compounds are carbonates. On firing, they are converted to oxides that react with the aluminosilicates of clay to give glassy surfaces. Other glazes simply impart a characteristic color to the surface.

The Indians of the American southwest controlled firing conditions to impart orange and black designs to their pottery. They used clays containing iron oxides which gave an orange Fe_2O_3 color when fired under oxidizing conditions. When a reducing atmosphere was introduced, for example by putting green leaves on the fire to give carbon monoxide, the orange Fe_2O_3 was reduced to black Fe_3O_4.

$$CO + 3Fe_2O_3 \rightarrow CO_2 + 2Fe_3O_4$$

FIGURE 13.7
Cements harden as water combines with the anhydrous minerals of cement grains to form a network of interlocking hydrated mineral crystals.

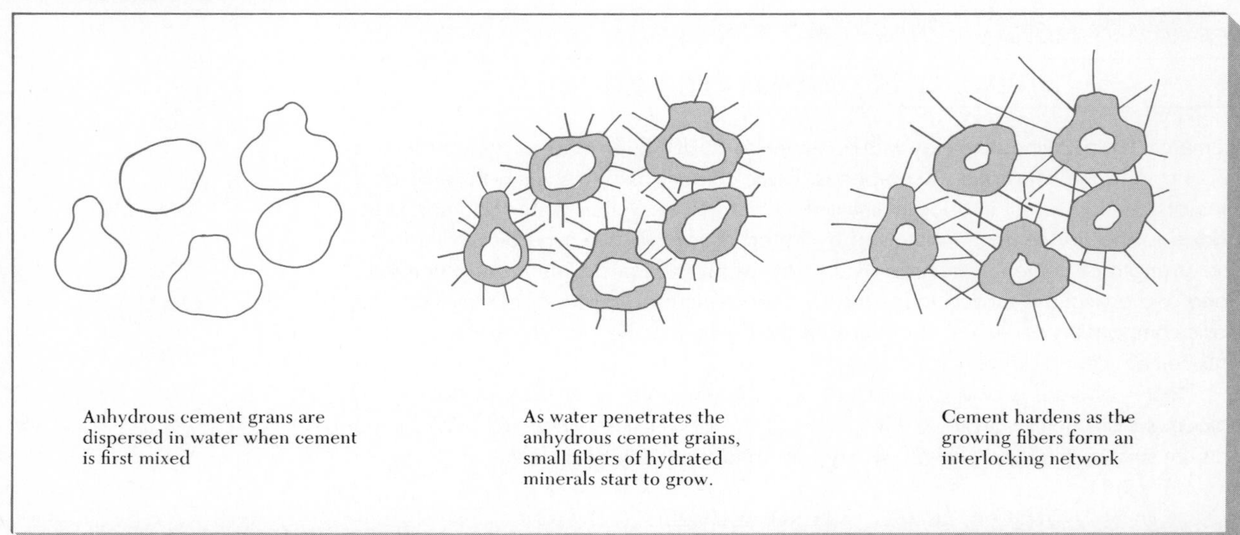

Anhydrous cement grans are dispersed in water when cement is first mixed
As water penetrates the anhydrous cement grains, small fibers of hydrated minerals start to grow.
Cement hardens as the growing fibers form an interlocking network

Indians first fired the pottery under reducing conditions to obtain the black iron oxide. They decorated the pots and covered part of the design with mud. When the pottery was fired under oxidizing conditions, the uncovered areas became orange and the unexposed surfaces remained black.

SOILS RELEASE NUTRIENTS BY ION EXCHANGE

Consider a dilemma concerning potassium, essential for plant growth, that occurs with other nutrients in clay (and other) soils. Since simple potassium compounds are soluble, why hasn't all the potassium in soils been washed into the ocean? On the other hand, if potassium is tightly retained by soils, why is it available to growing plants?

The charge balance about a clay platelet or other charged soil particle must be maintained. Potassium, calcium, and magnesium ions needed by plants are bound to insoluble polymeric anions by strong electrostatic forces and are not free to migrate. Ion-binding sites also occur in organic soils, for organic material decays to produce anions of weak acids that act as cation-binding sites.

Plants take up mineral nutrients by *ion exchange.* They exchange hydronium ions for nutrient cations in the soil. Plants extrude hydronium ions from their roots. One H_3O^+ is extruded for each K^+ taken up, and two H_3O^+ are exchanged for each Mg^{2+}. When a plant dies and decays, the potassium and magnesium ions are released where they can again be captured by soil anions in exchange for H_3O^+.

Clay anions bind cations selectively by charge and size. For example, more highly charged +2 ions are held more tightly than +1 ions. Within a family of the periodic table there is also selectivity. Potassium ions are bound more tightly than sodium ions. (Clay particles washed out to sea selectively bind Mg^{2+} and K^+ from seawater, so soils deposited in river deltas retain high levels of nutrient ions.)

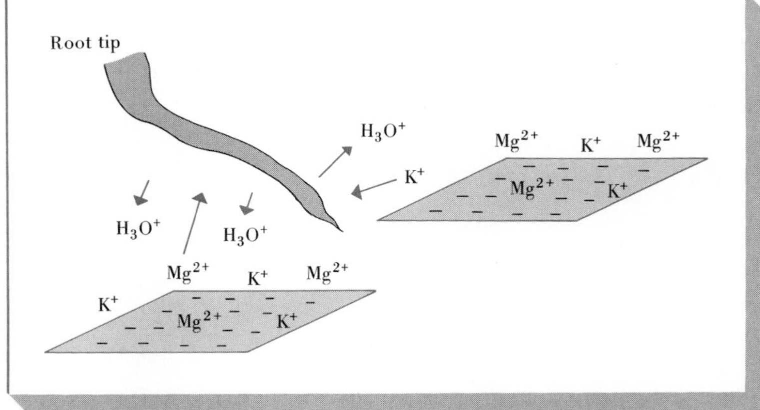

FIGURE 13.8
Plants take up nutrient cations by ion exchange.

The selective binding of potassium ions rather than sodium ions is very important, because as water is used and reused for irrigation the concentration of sodium chloride in the water builds up. The waters in the lower Colorado and Rio Grande rivers are somewhat salty, yet they are still used to irrigate crops. With good drainage, sodium ions do not replace potassium ions extensively. However in poorly drained fields, the accumulation of salt is making once fertile fields unfit for plant growth.

The fertility of a soil depends in part on its ion-exchange capacity. More fertile soils hold greater quantities of nutrients and release them to plants more readily. Some clay soils hold cations increasingly tight as they dry and shrink. In these soils, growth occurs principally in the spring and early summer when the soil retains water from winter rains.

To grow in acid soils, plants must extrude hydronium ions into a more acid environment. To the extent that exchange sites are occupied by hydronium ions, fewer sites are occupied by nutrient ions. Treating a lawn or garden with ground limestone adds needed calcium ions and neutralizes hydronium ions occupying exchange sites. [However, too much lime can cause iron, a nutrient needed in small amounts, to be tied up as insoluble $Fe(OH)_3$ and be unavailable to plants.]

CHEMICAL FERTILIZERS AND NITROGEN FIXATION

In the United States, fewer and fewer farmers produce food for more and more people. This increased production is tied in part to applications of agricultural chemicals. As crops are harvested from fields, needed elements go with them; consequently, plant nutrients need to be replenished. Furthermore, many soils are deficient in one or more nutrients. County agents

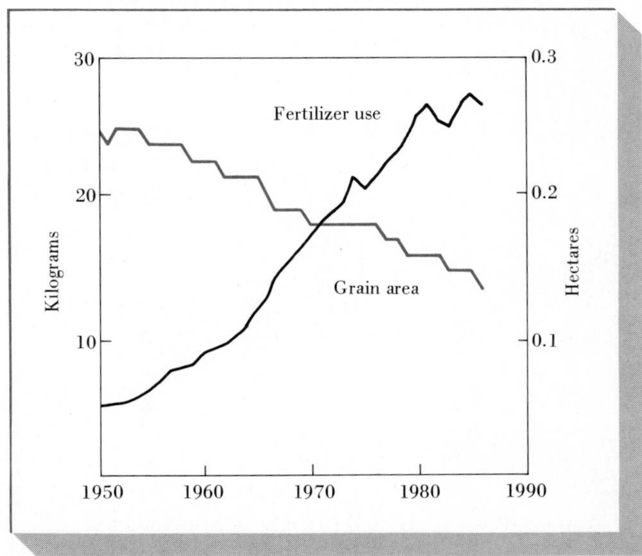

FIGURE 13.9
World fertilizer use and grain area per capita, 1950–86. *(Reproduced from STATE OF THE WORLD, 1987, A Worldwatch Institute Report on Progress Toward a Sustainable Society, Project Editor: Lester R. Brown. By permission of W. W. Norton & Company, Inc. Copyright © 1987 by the Worldwatch Institute.)*

have helped farmers to analyze soils and to determine and correct mineral deficiencies.

In what has been called the "green revolution," new hybrid varieties of rice and wheat have been developed and introduced in many countries. The use of these hybrids has increased both agricultural productivity and the demand for chemical fertilizers; these hybrids require higher levels of nutrients than can be supplied by traditional farming methods using manure for fertilizer.

The most widely used fertilizers contain nitrogen. Nitrogen is essential for making the proteins and nucleic acids of every cell, yet few compounds of nitrogen occur in abundance. (Deposits of sodium and potassium nitrate found in arid regions of Chile are mined for fertilizer.) The incorporation of N_2 into compounds of nitrogen is called *nitrogen fixation*. Bacteria that grow symbiotically in nodules of the roots of legumes fix nitrogen from the air. Rotation of crops by planting legumes such as soybeans and alfalfa one season and crops with high-nitrogen demands such as corn another season has been a time-honored way to maintain soil fertility.

In 1913, Fritz Haber in Germany developed a chemical process for the fixation of nitrogen. In the *Haber process*, nitrogen and hydrogen react over a Fe–FeO catalyst to give ammonia. Even with the catalyst, a temperature of 500°C is required for a satisfactory rate of reaction.

$$N_2 + 3H_2 \rightarrow 2NH_3 + \text{heat}$$

At this high temperature, the exothermic reaction does little to offset the decrease in the entropy of the system; not much ammonia is produced at low pressures. In an application of Le Chatelier's principle, pressures as high as 800 atm are used to increase the yield of ammonia. Upon cooling, ammonia liquefies and is withdrawn. Unreacted nitrogen and hydrogen gas are recycled. By repeatedly passing reactants over the catalyst bed, complete conversion of nitrogen and hydrogen to ammonia is attained.

In the *Ostwald process*, ammonia is oxidized to nitric acid over a platinum catalyst.

$$2NH_3 + 4O_2 \rightarrow 2HNO_3 + 2H_2O$$

Nitric acid is used in the production of ammonium nitrate, NH_4NO_3, a fertilizer rich in nitrogen.

$$HNO_3 + NH_3 \rightarrow NH_4NO_3$$

Nitric acid is also used in the manufacture of explosives, and potassium nitrate is a component of gunpowder. The British navy exercised its superiority during World War I by instituting a blockade to deprive Germany of the needed potassium nitrate that occurs naturally in Chile. The production of munitions and food in Germany depended on the successful application of the Haber and Ostwald processes. (An alternative strategy was used to produce nitrates in the United States during the War of 1812 and in the south during the Civil War. Because nitrogen-fixing bacteria that live on decaying organic matter are abundant in some caves, nitrates were leached from cave dirt.)

FERTILIZERS CONTAINING POTASSIUM AND PHOSPHORUS

The most abundant metal ions in living tissues are sodium, potassium, magnesium, and calcium. Potassium salts deposited from the evaporation of ancient seas are mined as a source of potassium for plants. Limestone and lime from the heating of limestone are used to supplement calcium deficiencies.

Insoluble minerals containing calcium phosphate are mined to provide phosphate for fertilizer. The mineral hydroxyapatite, $Ca_5(PO_4)_3OH$, has the same composition as the hard part of bones and tooth enamel. A harder, less soluble mineral, fluoroapatite, in which fluoride ions have replaced the hydroxy groups of apatite, is often used for phosphate production. (Fluoride-containing toothpastes help increase the hardness of teeth by the same substitution.) In one process used to make the phosphate available to plants, these minerals are treated with sulfuric acid to make superphosphate fertilizer, a mixture of the more soluble compounds $CaSO_4$ and $Ca(H_2PO_4)_2$.

A S I D E

STRONTIUM 90 AND THE TEST BAN TREATY

Public concern for the dangers posed by a single radioactive isotope $^{90}_{38}Sr$ led to the adoption of a limited nuclear test ban treaty. Strontium 90 is one of the radioactive products produced by nuclear fission. It undergoes beta decay with a half-life of 29 years. Atmospheric testing of nuclear weapons introduced radioactive material high into the atmosphere, and the resulting radioactive fallout was scattered over wide areas of the globe.

Since Sr^{2+}, like Mg^{2+} and Ca^{2+} in the same family of the periodic table, binds to clay, strontium 90 in radioactive fallout was retained by soils. By the process of ion exchange it was taken up by grasses. Dairy cows consumed large quantities of contaminated grass, grain, and ensilage, and they concentrated strontium along with calcium in milk.

Radioactive strontium in milk increased the risk of leukemia, a cancer of the blood. It is in the bone marrow that new red blood cells are synthesized. The risk of leukemia was especially great for children since they drank more milk than most adults and they were building new bone tissue. Strontium could accompany calcium in the formation of bones; as a result, the radioactivity of strontium 90 could be concentrated in the bones of growing children.

Linus Pauling, a chemist who had received the Nobel prize for his contributions to bonding theory and to the understanding of molecular architecture, led the drive for

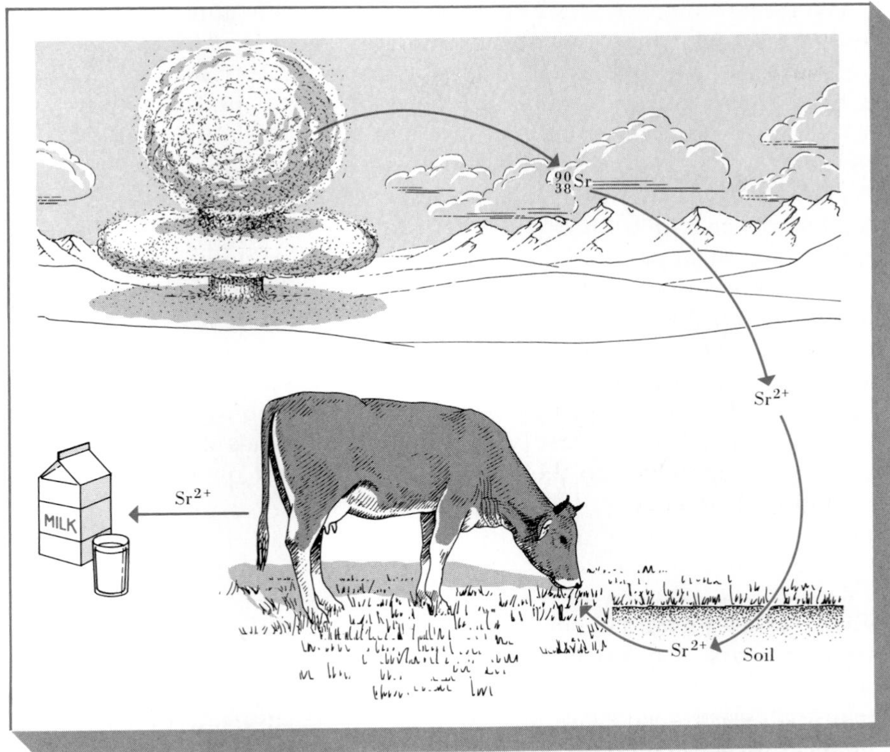

FIGURE 13.10
Concern for leukemia caused by radioactive strontium helped lead to the ban of atmospheric testing of nuclear weapons. Because strontium resembles calcium in its chemistry, it can accompany calcium in the food chain and be deposited in the bones of growing children. Radioactivity in bone tissue can produce leukemia because red blood cells are produced in bone marrow.

the end to atmospheric testing of nuclear weapons. A partial test ban treaty was concluded in 1963. Pauling received the Nobel Peace Prize and became the first chemist since Marie Curie to receive two Nobel prizes.

WATER TREATMENT

There are natural processes for removing biological wastes from waters. Bacteria feed on organic wastes, using them for food and oxidizing them to CO_2 and water. In turn, these bacteria are infected and lysed by viruses called bacteriophages. However, the capacity of these natural processes is limited, and the discharge of large amounts of biological and chemical wastes can overpower the ability of streams to regain purity. As populations have grown, so has the need for people to take a more active role in the treatment of both drinking water and sewage.

Drinking water is treated to kill harmful bacteria and to improve water clarity. Chlorine is added as a germicide in the United States, and ozone, O_3, is used for the same purpose in Europe. Turbid water may be treated with aluminum sulfate and a base. A flocculent precipitate of aluminum hydroxide settles out and carries down suspended clay and other particulates.

Water treatment processes help control the wastes added to streams. In sewage treatment plants, wastes are aerated and held to introduce additional oxygen and to allow time for the organic material to be digested. Many inorganic nutrients remain after sewage treatment. (In some systems, effluents from sewage treatment are sprayed onto fields where the nutrients are retained and contribute to soil fertility.)

The presence of inorganic, nutrient ions, whether from municipal treatment plants or from agricultural runoff, can be a problem for water quality. The plant growth in many lakes is normally limited by deficiencies in one or another nutrient, often nitrogen or phosphorus. As lakes fill with a richer nutrient broth, blooms of algae spread over their surfaces. When these plants die, they sink and decompose. The algae produce oxygen by photosynthesis near the surface, but oxygen is consumed by the decay process. With excessive plant growth and decay, deep waters become oxygen deficient and no longer support quality populations of fish. *Eutrophication* is the name given to the deterioration of water quality by excessive plant growth and decay.

CHEMICAL WASTE DISPOSAL POSES CONTINUING PROBLEMS

Where do we put the chemical wastes from industrial processes? All too often in the past, the answer has been to flush them down the drain. Because natural processes cleanse some wastes, people have hoped that they would cleanse all wastes. This has not proven to be the case, and the disposal of chemical wastes is increasingly being brought to the public attention. Organic compounds in wastes buried in the Love Canal region of Buffalo have risen in soils and volatilized into basements of homes and schools. An accident at a chemical plant near Seveso, Italy, spread dangerous chemicals over many square miles and caused villages to be abandoned. Wastes from manufacturing hexachlorophene, a compound used to control bacterial growth, were mixed with oil and then used to control dust on dirt roads. The persistence of these wastes has caused the town of Times Beach, Missouri, to be abandoned. Clearly not all chemical wastes are removed by natural processes.

The very processes that may help to clean some wastes have made the dimensions of the problems harder to know. Due to their high surface area, soil particles bind some organic and heavy metal ion poisons thereby slowing their spread; health problems from chemical wastes have been discovered years after a dump site may have been used and abandoned. Beginning in the late 1970s, chemical companies have contributed to a Super Fund, to be used for cleaning up abandoned dump sites.

As we better understand the dangers that chemical wastes pose to our health and environment, policies for the effective disposal or destruction of wastes need to be promulgated and enforced. Those who create wastes need to assume the responsibility for their proper disposal. As a society, we need to assume the responsibility for protecting the quality of our chemical environment.

QUESTIONS

1. The mineral wollastonite has the formula $Ca_3Si_3O_9$. What is the charge on the $Si_3O_9{}^{n-}$ anion? Is this anion linear with two oxygen atom bridges between silicon atoms or is it triangular with three bridging oxygen atoms?
2. What structural features of asbestos give rise to the formation of small dustlike particles? What are the advantages and disadvantages of policies that emphasize asbestos removal? Of those that leave asbestos in place but cover it with sealing compounds?
3. Caves are frequently formed in limestone. Decaying organic matter produces acids in groundwater. Write the reaction for dissolving limestone in weak acidic water.
4. Shale is a sedimentary rock formed from compacted clay. Shale is relatively impervious to the flow of groundwater. Why?
5. Stalactites and stalagmites are formations sometimes found in caves. Apply Le Chatelier's principle to show that the loss of CO_2 to the air from water saturated with $CaHCO_3$ leads to an increase in the concentration of $CO_3{}^{2-}$ and, hence, to the deposition of insoluble $CaCO_3$.
6. The properties of glass are dependent on the ratio of SiO_2 to Na_2CO_3. Why would the following changes in the recipe for glass produce unsatisfactory material?
   (a) Large increase in the relative amount of SiO_2
   (b) Large increase in the relative amount of Na_2CO_3
7. Glass is etched by solutions of hydrofluoric acid. Where do H_3O^+ ions attack silicate chains? Where do F^- ions attack?
8. Proposals for the long-term storage of nuclear wastes call for encasing radioactive wastes in glass or ceramic containers prior to burial. It is important that the container material withstand the heat generated by nuclear decay reactions and that it provide an additional barrier to groundwater. Why are glass and ceramic materials superior to metal or concrete for containment?
9. Why is it less attractive economically to collect and recycle glass than it is to collect and recycle aluminum? Consider the energy costs and savings for each.
10. Why is concrete kept wet after pouring?
11. Why does a glaze containing $MgCO_3$ impart a glassy finish to a clay pot?
12. Which of the following compounds would be expected to give color to glass? What do the compounds that color glass have in common?
    (a) Fe_2O_3
    (b) PbO
    (c) $CoCO_3$
    (d) $MnCO_3$
    (e) B_2O_3
13. Like sodium salts, potassium salts are very soluble. Why are potassium ions available to plants? Why are they not found chiefly in the ocean?
14. Why are mineral cations bound more tightly to a dry clay soil than to a water-saturated clay?

15. Can the deposit of acid compounds in watersheds be monitored by measuring the flow and pH of streams carrying runoff from the watersheds? Why or why not?
16. Why do soils differ greatly in their capacity to retain added fertilizer?
17. Why are clay soils on mountain slopes subject to mud slides? Would a single heavy rain or a series of frequent light rains be more apt to set the stage for a mud slide?
18. Clay sediments remain suspended in rivers for a long time. They quickly settle when rivers flow into the ocean. Why would clays settle more rapidly from saltwater?
19. A radioactive isotope of cesium, ${}^{137}_{55}Cs$, is also a by-product of nuclear fission. This isotope has a half-life of 30 years. What nutrient cation does Cs^+ most resemble? Would it be retained by soils and taken up by plants? Does ${}^{137}_{55}Cs$ pose the same health risk as ${}^{90}_{38}Sr$? Why or why not?

CHAPTER

FOURTEEN

AN INTRODUCTION TO ORGANIC CHEMISTRY

A very large majority of the known chemical compounds contain carbon. The branch of chemistry that involves the study of carbon compounds is organic chemistry. This name reflects an incorrect early belief that compounds from living organisms were different from other compounds due to the presence of a vital force. The synthesis of urea, an organic compound first isolated from urine, from inorganic compounds helped bring an end to this vital-force theory in chemistry. The distinction some people make between natural and synthetic vitamins may be a carry-over of the belief in a vital force.

The variety of organic compounds is very great. Thousands of different compounds have be isolated from coal and oil. Volatile compounds serve as trail markers for ants and as sex attractants for boll weevils. Many paints, medicines, insecticides, and textile fibers are products of an organic chemical industry. In the biological world, fats and carbohydrates, nucleic acids, enzymes, and hormones are all carbon-containing compounds.

FUNCTIONAL GROUP ISOMERS—AN ILLUSTRATION IN REASONING

Organic chemists reason by analogy. Compounds are placed in groups with similar structures. From a structure, a chemist infers physical properties and characteristic reactions of a compound. From the physical properties and chemical reactions of a new compound, a chemist infers pieces of structure that are present.

Let us sample the reasoning involved in organic chemistry. Two compounds are known to have the same formula C_2H_6O. Compounds that have the same formula but differ in the way the atoms are connected are called *isomers*. These isomers have very different properties. Under the name given each compound are listed some chemical and physical properties.

	Ethyl Alcohol	Dimethyl Ether
Dissolves in concentrated sulfuric acid	Yes	Yes
Reacts with Na to give H_2 and strong base	Yes	No
Boiling point	78°C	−25°C

Only two possible structures having the formula C_2H_6O can be written. Because each second-row atom has an octet of electrons about it, each carbon atom has four bonds and each oxygen atom has two bonds. Hydrogen atoms are bonded to carbon or to oxygen because hydrogen can accommodate only two electrons in its outer shell. (A hydrogen-hydrogen bond is present only in a hydrogen molecule, H_2.)

$$\begin{array}{ccccccccc} & H & & H & & & \\ & | & & | & & & \\ H- & C & - & C & - & \ddot{\underset{\cdot\cdot}{O}} & -H \\ & | & & | & & & \\ & H & & H & & & \\ & & & \text{A} & & & \end{array} \qquad \begin{array}{ccccccc} & H & & & & H & \\ & | & & & & | & \\ H- & C & - & \ddot{\underset{\cdot\cdot}{O}} & - & C & -H \\ & | & & & & | & \\ & H & & & & H & \\ & & & \text{B} & & & \end{array}$$

To assign these structures to the alcohol and ether, we can reason by an analogy to water. Like water, each reacts as a base with concentrated sulfuric acid. The two compounds behave like water in their reactions with sulfuric acid because each compound has unshared pairs of electrons on its oxygen.

$$H_2SO_4 + H_2O \rightarrow H_3O^+ + HSO_4^-$$

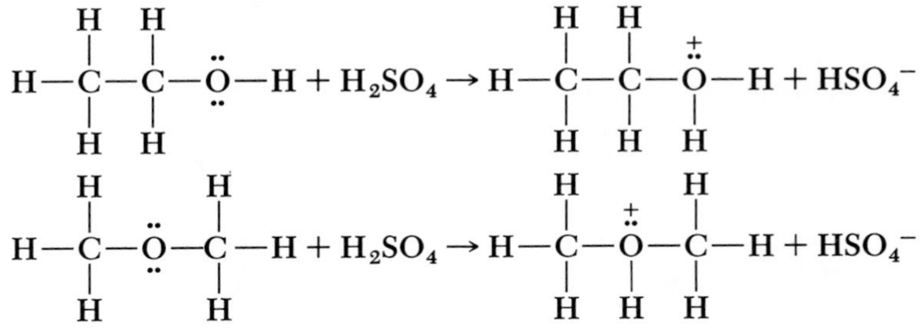

$$\begin{array}{ccccccccccccccc} & H & & H & & & & & H & & H & & & & \\ & | & & | & & & & & | & & | & & + & & \\ H- & C & - & C & - & \ddot{\underset{\cdot\cdot}{O}} & -H + H_2SO_4 \rightarrow & H- & C & - & C & - & \ddot{O} & -H + HSO_4^- \\ & | & & | & & & & & | & & | & & | & & \\ & H & & H & & & & & H & & H & & H & & \end{array}$$

$$\begin{array}{ccccccccccccccc} & H & & & & H & & & H & & & & H & & \\ & | & & & & | & & & | & & + & & | & & \\ H- & C & - & \ddot{\underset{\cdot\cdot}{O}} & - & C & -H + H_2SO_4 \rightarrow & H- & C & - & \ddot{O} & - & C & -H + HSO_4^- \\ & | & & & & | & & & | & & | & & | & & \\ & H & & & & H & & & H & & H & & H & & \end{array}$$

Water is reduced by sodium to form hydrogen gas and a solution of sodium hydroxide. This reduction is a reaction of the —OH group. Only structure A has an —OH group. Like water, it reacts with sodium to give hydrogen gas and a basic solution. On the basis of the reaction with sodium, we can assign structure A to ethyl alcohol and structure B to dimethyl ether.

$$2H_2O + 2Na \rightarrow H_2 + 2Na^+ + 2OH^-$$

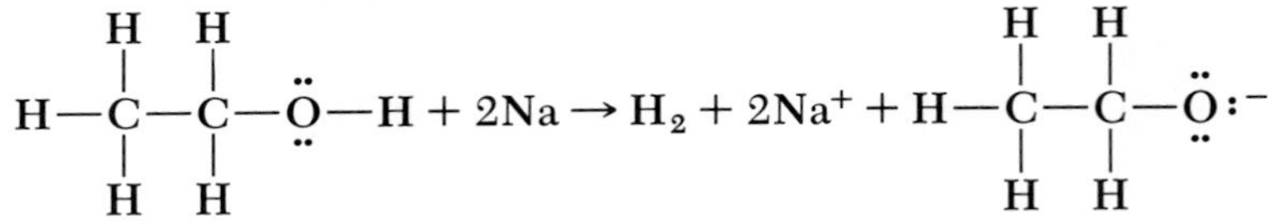

$$\begin{array}{cccccccccccc} & H & & H & & & & & H & & H & \\ & | & & | & & & & & | & & | & \\ H- & C & - & C & - & \ddot{\underset{\cdot\cdot}{O}} & -H + 2Na \rightarrow H_2 + 2Na^+ + & H- & C & - & C & -\ddot{\underset{\cdot\cdot}{O}}:^- \\ & | & & | & & & & & | & & | & \\ & H & & H & & & & & H & & H & \end{array}$$

A check on our assignment of structures is provided by a comparison of boiling points. Ethyl alcohol is expected to have a much higher boiling point than dimethyl ether. Because it has an —OH group like water, the alcohol has strong hydrogen bonds between molecules. With no —OH group, the ether has only weaker forces of attraction between molecules.

With many reagents, the portion of an organic molecule consisting only of C—C and C—H bonds remains unchanged. Carbon-carbon bonds and carbon-hydrogen bonds are notably unreactive. For example, the reactions listed for ethyl alcohol and for dimethyl ether involved only the —O— and the —OH groups. Both the carbon-carbon bond and the carbon-hydrogen

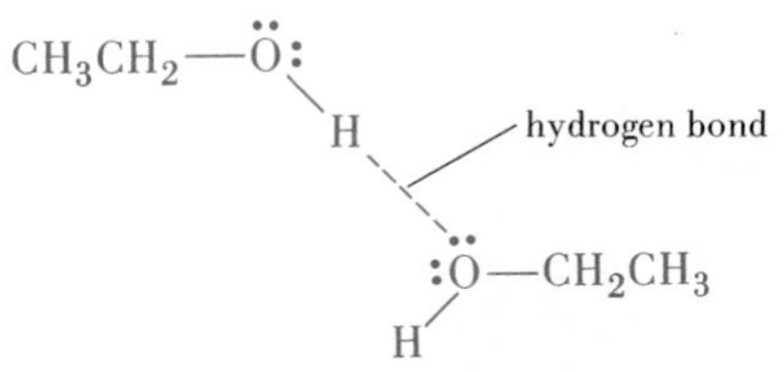

FIGURE 14.1
Ethyl alcohol (ethanol), like water, forms strong intermolecular hydrogen bonds. The partially positive hydrogen of the OH group of one molecule is attracted to a lone pair of electrons on O of another molecule.

bonds were unaffected when the two compounds were subjected to sodium, a powerful reducing agent, and to H_2SO_4, a strong acid.

Ethyl alcohol and dimethyl ether are isomers with different *functional groups*. Ethyl alcohol has the chemistry associated with the alcohol (—OH) functional group, and dimethyl ether has the chemistry associated with the ether (—O—) functional group.

ETHYL ALCOHOL IS OXIDIZED TO PRODUCE A NEW FUNCTIONAL GROUP

Ethyl alcohol, unlike dimethyl ether, reacts with a number of oxidizing agents. Reaction occurs at the site of the alcohol functional group. (The presence of an —OH group on the same carbon atom makes the C—H bonds more susceptible to reaction.) Ethyl alcohol can be readily oxidized to give acetic acid, the acid found in vinegar. Acetic acid is a carboxylic acid; the chemistry of its functional group is discussed in Chapter 15.

$$CH_3CH_2OH \rightarrow CH_3C\begin{matrix} \diagup\!\!\!\!\diagup O \\ \diagdown OH \end{matrix}$$

Abbreviated or condensed structures are used to illustrate the oxidation of ethanol to acetic acid. The number of hydrogen atoms attached to each carbon atom is indicated after the C. A student should be able to draw a Lewis structure if given a condensed structure. The use of condensed structures saves both paper and typesetting expense. In this text, functional groups of compounds will not be written in condensed form, but hydrocarbon portions frequently will be.

ALKANES

Hydrocarbons containing only single bonds between carbon atoms are called *alkanes*. Alkanes react with few other chemicals, and the reactions they do undergo such as combustion tend to be relatively unselective.

Alkanes may be either branched or cyclic. The simplest case of branching is illustrated by the existence of two isomeric alkanes having the formula C_4H_{10}.

$$\underset{\substack{n\text{-butane} \\ \text{bp } -0.5°\text{C}}}{CH_3CH_2CH_2CH_3} \qquad \underset{\substack{\text{Isobutane} \\ \text{bp } -12°\text{C}}}{CH_3\overset{\overset{CH_3}{|}}{C}HCH_3}$$

Butane and isobutane are sometimes called *carbon-skeleton isomers* because the carbon atoms alone show different structures.

Cyclohexane is an example of a cyclic alkane. In a popular convention for writing cyclic compounds, both carbon atoms and hydrogen atoms are omitted. A carbon atom is present at each corner. The number of hydrogen atoms is calculated by subtracting the number of carbon-carbon bonds from four, the total number of bonds to each carbon atom. In cyclohexane, two hydrogens are attached to each carbon atom.

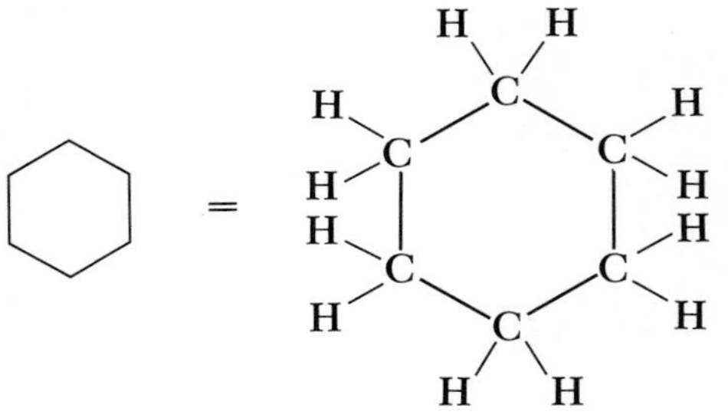

NAMING ORGANIC COMPOUNDS

As science has grown, the concern for information retrieval has also grown. Scientists have developed rules for generating systematic names that are clear and unambiguous. Rules are applied that allow one to go from a structure to a name or from a name to the correct structure. Because it is essential that scientists be able to communicate and that they have access to information in the literature, rules for naming compounds have international recognition.

In systematic nomenclature, organic compounds are named as derivatives of straight-chain (normal) alkanes. (The naming of compounds with rings is often more complex.) The names of the first ten normal alkanes are given in Table 14.1. Students should memorize these names and structures.

Common names continue to be used both for simple compounds and for very complex compounds. For example, common names for simple hydrocarbons extend to the isomers of pentane, C_5H_{12}. Variations in chain branching for these isomers are indicated by the prefixes *n*-, *iso*-, and *neo*-. As carbon chains become longer and more branched, the number and spe-

TABLE 14.1

NORMAL ALKANES WITH ONE TO TEN CARBON ATOMS

Name	Formula
Methane	CH_4
Ethane	CH_3CH_3
Propane	$CH_3CH_2CH_3$
Butane	$CH_3CH_2CH_2CH_3$
Pentane	$CH_3CH_2CH_2CH_2CH_3$
Hexane	$CH_3CH_2CH_2CH_2CH_2CH_3$
Heptane	$CH_3CH_2CH_2CH_2CH_2CH_2CH_3$
Octane	$CH_3CH_2CH_2CH_2CH_2CH_2CH_2CH_3$
Nonane	$CH_3CH_2CH_2CH_2CH_2CH_2CH_2CH_2CH_3$
Decane	$CH_3CH_2CH_2CH_2CH_2CH_2CH_2CH_2CH_2CH_3$

cial meanings of prefixes quickly become intractable. (Systematic names can also have disadvantages. They may be long, awkward, and difficult to say.) For many compounds, the use of common names, like nicknames, persists.

Structure	Common Name	Systematic Name
$CH_3CH_2CH_2CH_2CH_3$	*n*-Pentane	Pentane
$(CH_3)_2CHCH_2CH_3$	Isopentane	2-Methybutane
$CH_3C(CH_3)_2CH_3$	Neopentane	2,2-Dimethylpropane

The names of the isomeric pentanes provide a simple illustration of the naming of branched hydrocarbons. The longest unbranched chain provides the root of the name. Hydrocarbon branches are named by replacing the *-ane* of the corresponding alkane by *-yl.* Each branch replaces an H in the unbranched chain, so it is called a substituent. The position of each alkyl substituent is indicated by a numerical prefix.

Compounds with functional groups are named as derivatives of alkanes. Names of compounds reflect both the functional group present and the parent hydrocarbon skeleton. The longest hydrocarbon chain containing the functional group provides the root that indicates the number of carbon atoms in the chain. The suffix of a name indicates the functional group present. For example, names of alcohols end in *-ol.* Methanol and ethanol are the systematic names of 1- and 2-carbon alcohols.

CH_3OH	CH_3CH_2OH
Methyl alcohol Methanol	Ethyl alcohol Ethanol

When the position of a functional group on a hydrocarbon chain is not obvious, a numeral designates its position. The chain is numbered, beginning from the end that will result in the smaller numeral for the position of the functional group. Using an example of three-carbon alcohols, we note that 1-propanol and 2-propanol are the systematic names for the two alcohols with the formula C_3H_8O. These alcohols are *positional isomers* because they have the same carbon skeleton and the same functional group but they differ in the place of attachment of the —OH group.

$$\begin{array}{ccccccc} & H & & H & & H & \\ & | & & | & & | & \\ H- & C & - & C & - & C & -O-H \\ & | & & | & & | & \\ & H & & H & & H & \end{array}$$

$CH_3CH_2CH_2OH$
n-Propyl alcohol
1-Propanol

$$\begin{array}{ccccccc} & & & H & & & \\ & & & | & & & \\ & H & & O & & H & \\ & | & & | & & | & \\ H- & C & - & C & - & C & -H \\ & | & & | & & | & \\ & H & & H & & H & \end{array}$$

$$\begin{array}{c} OH \\ | \\ CH_3CHCH_3 \end{array}$$

Isopropyl alcohol
2-Propanol

Given a structure, a student should be able to propose a name and, from a systematic name, a structure. The following examples illustrate the application of simple rules of nomenclature to more complex compounds.

$$\begin{array}{llll} & CH_3 & CH_2CH_3 & \\ & | & | & \\ CH_3 & C\;CH_2 & CH\;CH_2CH_2OH & \\ & | & & \\ & CH_3 & & \end{array}$$

Since this compound has an —OH group, it is an alcohol. The longest chain containing the —OH group has six carbons, so it is a hexanol. The hexanol is numbered from that end which gives the lowest number to the functional group. This substituted 1-hexanol has two CH_3— groups attached to C-5 and a CH_3CH_2— group on C-3. The systematic name of the compound is 5,5-dimethyl-3-ethyl-1-hexanol.

Write a structure for 4-ethyl-4-isopropyl-2-octanol. From the suffix *-ol*, the compound is identified as an alcohol. *Oct-* indicates eight. An ethyl substituent has two carbons, and an isopropyl group is a branched 3-carbon substituent.

$$\begin{array}{llll} & OH & CH_2CH_3 & \\ & | & | & \\ CH_3 & CH\;CH_2 & C\;CH_2CH_2CH_2CH_3 & \\ & & | & \\ & CH_3 & CH\;CH_3 & \end{array}$$

ALKENES AND ALKYNES

Alkenes are hydrocarbons that have carbon-carbon double bonds. In the names of alkenes, the suffix *-ene* replaces *-ane* of the corresponding alkane. The hydrocarbon chain is numbered to give the lowest integer to signify the position of the double bond. The simplest alkene has the formula C_2H_4. Ethene, or ethylene as it is commonly called, stimulates the ripening of many fruits.

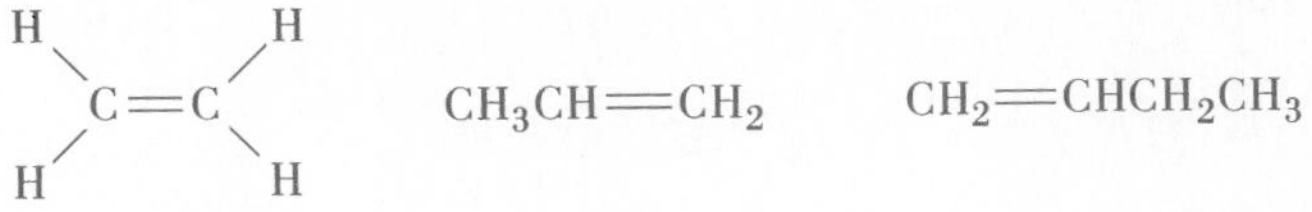

FIGURE 14.2
Alkenes are hydrocarbons that have carbon-carbon double bonds. The structures of three alkenes are illustrated together with systematic and common names (in parentheses).

$$H_2C{=}CH_2$$
Ethene (Ethylene)

$$CH_3CH{=}CH_2$$
Propene (Propylene)

$$CH_2{=}CHCH_2CH_3$$
1-Butene

When each carbon of the double bond is attached to two different groups, isomers exist. For example, there are two isomeric 2-butenes. In the *cis* isomer, the two hydrogens are on the same side of the double bond, and, in the *trans* isomer they are on opposite sides. Isomers differing in the geometry of attachment of groups about double bonds are often called *geometric isomers.* (Geometric isomers do not occur in alkanes because there is rotation about single bonds that does not occur about double bonds.)

Hydrogen can be added to the double bond of an alkene to convert it to the corresponding alkane. Finely divided platinum and palladium are catalysts for this reaction.

$$H_2 + {-}CH{=}CH{-} \rightarrow {-}CH_2CH_2{-}$$

Because double and triple bonds can add hydrogen, they are sometimes called *unsaturated.* In contrast, compounds such as alkanes that have only single bonds are said to be *saturated.*

Double bonds are reactive with a variety of reagents. For example, the electrons of a double bond are available to be shared with an acid or to react with an oxidizing agent. The production of ethyl alcohol and of ethylene glycol illustrates these reactions. Ethyl alcohol is the product of the acid-catalyzed addition of water to the double bond of ethylene.

$$CH_2{=}CH_2 + H_2O \rightarrow CH_3CH_2OH$$

In the production of ethylene glycol, the double bond is first oxidized by oxygen. Then the three-membered ring of the resulting cyclic ether is opened to add water and give a diol in a reaction catalyzed by acid.

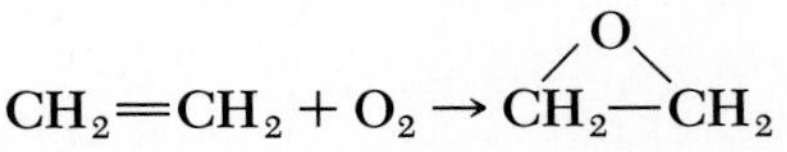

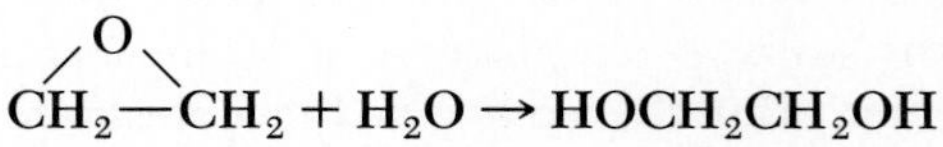

$$CH_2{=}CH_2 + O_2 \rightarrow CH_2{-}CH_2 \text{ (with O bridging both carbons)}$$

$$CH_2{-}CH_2 \text{ (O-bridged)} + H_2O \rightarrow HOCH_2CH_2OH$$

Alkynes are hydrocarbons that have carbon-carbon triple bonds. Like alkenes, alkynes undergo addition reactions.

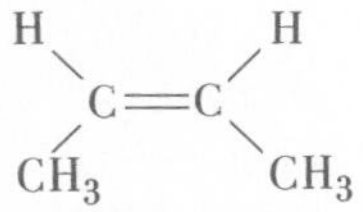

cis-2-Butene: H and H on the same side; CH_3 and CH_3 on the same side of C=C.

cis-2-Butene

trans-2-Butene: H and CH_3 on top; CH_3 and H on bottom of C=C.

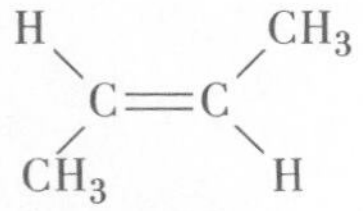

trans-2-Butene

FIGURE 14.3
Geometric isomers are possible when each carbon atom of the double bond is attached to two different groups. In the case of the isomeric 2-butenes, the two hydrogen atoms are on the same side of the double bond in the cis isomer and on opposite sides of the double bond of the trans isomer. (Note that the two 2-butenes are positional isomers of 1-butene shown in Figure 14.2.)

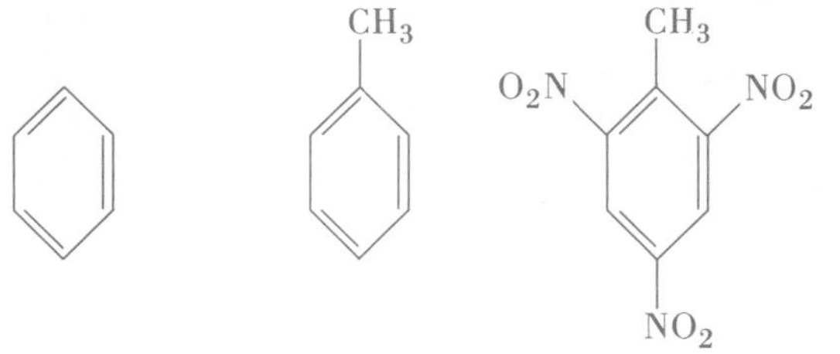

Benzene — Methylbenzene (Toluene) — 2,4,6-Trinitrotoluene — Hydroxybenzene (Phenol)

FIGURE 14.4
Benzene and its derivatives are called aromatic compounds.

$$H{-}C{\equiv}C{-}H$$

Ethyne or acetylene

The simplest alkyne, acetylene, was used as an illuminating gas in the headlights of early automobiles and is still used in lamps carried by cave explorers. The reaction of water with calcium carbide is used to generate acetylene; by controlling the drop rate of water, one controls the formation of gas.

$$CaC_2 + 2H_2O \rightarrow C_2H_2 + Ca(OH)_2$$

Acetylene burns in air with a bright, sooty flame; the light is from glowing carbon particles formed in the flame. Acetylene is burned with pure oxygen to give the hot flame of an oxyacetylene torch used for cutting and welding metal.

AROMATIC HYDROCARBONS

Benzene, C_6H_6, and its derivatives belong to the class of *aromatic hydrocarbons.* Toluene, an aromatic compound used in making the explosive TNT, is methylbenzene. Phenol, a disinfectant also known as carbolic acid, is hydroxybenzene.

Benzene and its derivatives are generally less reactive than alkenes. The explanation for the low reactivity of benzene lies in the stability of the benzene ring. No single Lewis structure using octets of electrons can describe benzene, for all six carbon-carbon bonds are equal in length. Chemists refer to benzene as being resonance stabilized, and they represent it as a hybrid of two (fictional) Lewis structures. The double-headed arrow is used to indicate that the real structure of benzene is intermediate between the two Lewis structures. Compounds that are resonance stabilized are less reactive than would be predicted on the basis of a double bond in a Lewis structure, for their electrons are delocalized.

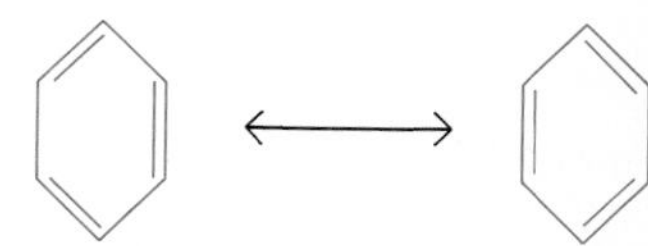

FIGURE 14.5
The real structure of benzene is intermediate between the two structures. Chemists describe benzene as being resonance-stabilized and use the double-headed arrow to indicate resonance rather than change or equilibrium.

Reactions of aromatic hydrocarbons take a different course than reactions of alkenes and alkynes. Acids add to the double bond of alkenes. With

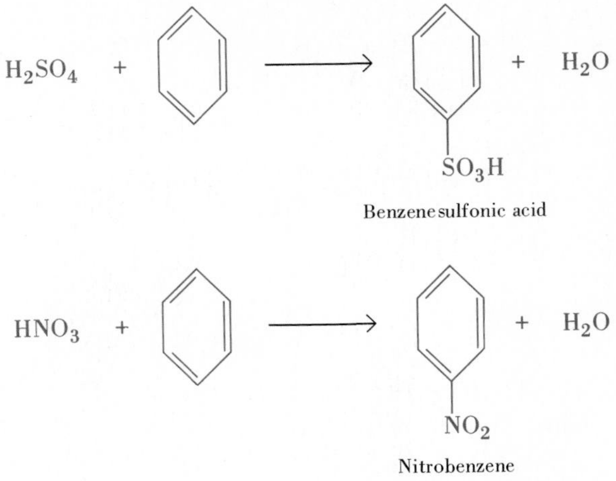

$$H_2SO_4 + C_6H_6 \longrightarrow C_6H_5SO_3H + H_2O$$

Benzenesulfonic acid

$$HNO_3 + C_6H_6 \longrightarrow C_6H_5NO_2 + H_2O$$

Nitrobenzene

FIGURE 14.6
Benzene and other aromatic compounds undergo substitution reactions in which an incoming group replaces a hydrogen atom on the aromatic ring.

benzene, however, the course of reaction is substitution. The preference for substitution rather than addition is related to the stability associated with the benzene ring. Electrons are delocalized in substituted benzenes just as they are in benzene itself.

The production of styrene, a compound used in making plastic materials and synthetic rubber, utilizes both ethylene and benzene. Three reactions are required. In the first, hydrogen chloride adds to ethylene. In the second, a Lewis acid $AlCl_3$ catalyzes the reaction of ethyl chloride with benzene. Finally, hydrogen is catalytically removed from ethylbenzene to give styrene.

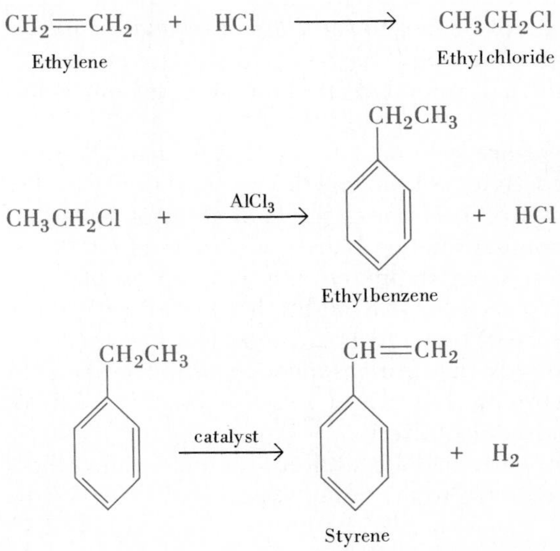

$$\underset{\text{Ethylene}}{CH_2{=}CH_2} + HCl \longrightarrow \underset{\text{Ethyl chloride}}{CH_3CH_2Cl}$$

$$CH_3CH_2Cl + C_6H_6 \xrightarrow{AlCl_3} \underset{\text{Ethylbenzene}}{C_6H_5CH_2CH_3} + HCl$$

$$C_6H_5CH_2CH_3 \xrightarrow{\text{catalyst}} \underset{\text{Styrene}}{C_6H_5CH{=}CH_2} + H_2$$

FIGURE 14.7
The production of styrene utilizes the hydrocarbons ethylene and benzene. Ethylene undergoes an addition reaction, and benzene undergoes a substitution reaction. In a third reaction, hydrogen is catalytically removed from ethyl-benzene to form the final product.

HYDROCARBONS FROM PETROLEUM

Petroleum refining provides the majority of hydrocarbons used in the chemical industry today as well as such products as gasoline and heating oil. Petroleum refining operations are tied to the marketplace. The demands for gasoline, diesel fuel, and heating oil dictate the relative amounts of these products to be produced. During the summer, the demand for gasoline is greatest; in winter, the demand for heating oil is greatest. Refineries represent very substantial capital investments; therefore, operations are de-

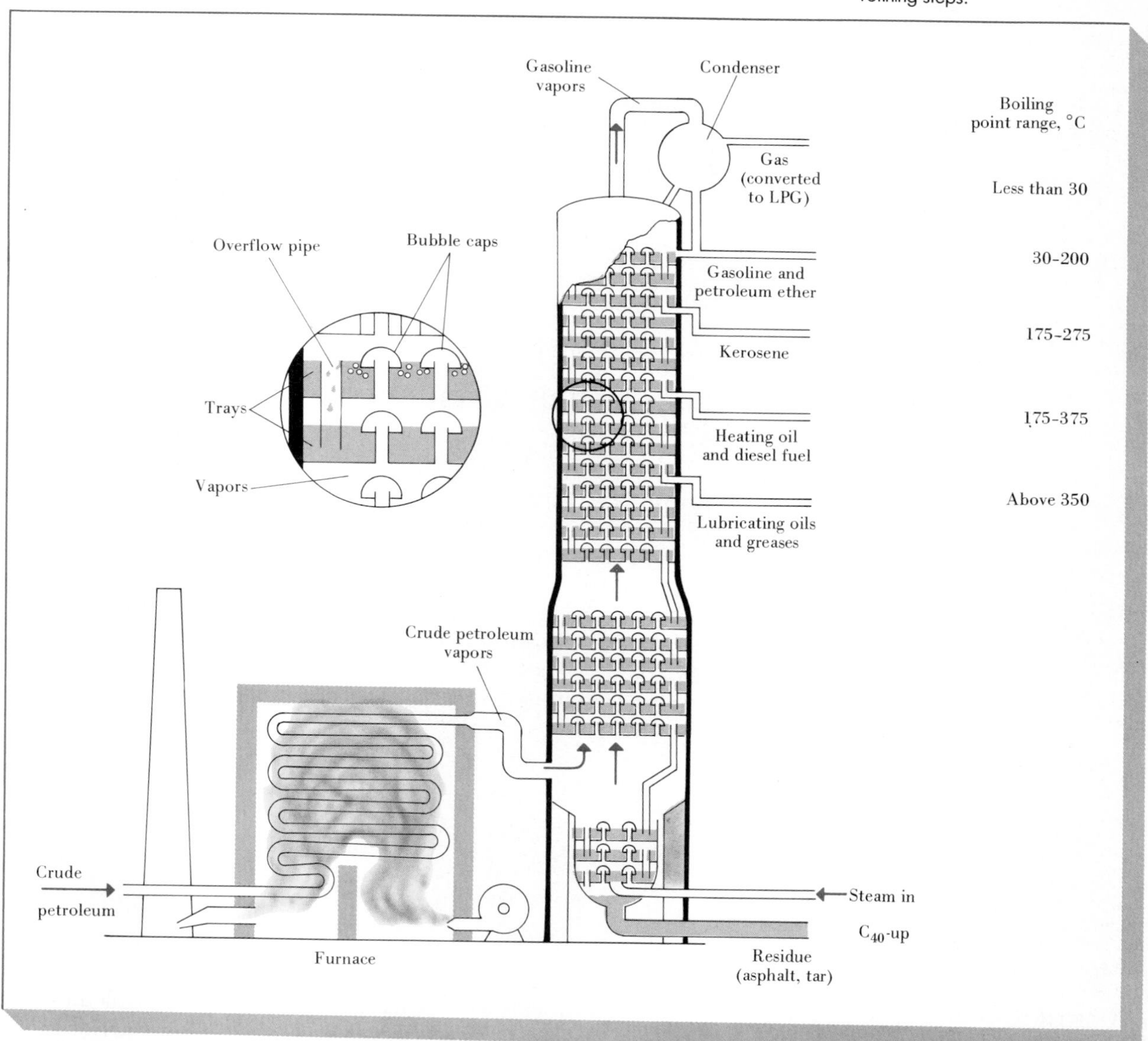

FIGURE 14.8 Distillation is used to separate crude petroleum into fractions that differ in boiling points. Fractions may be used directly or processed in subsequent refining steps.

signed to be continuous, for which there are substantial large-scale benefits.

Several steps are involved in refining crude oil. Nitrogen and sulfur impurities need to be removed. Viscous mixtures of high-molecular-weight compounds need to be converted into lower-molecular-weight, more volatile compounds.

At all stages of refining, there are mixtures to separate. Crude separations are made by distillation. Low-boiling fractions are blended into gasoline. High-boiling fractions are subjected to chemical reactions to convert them to more useful products.

OCTANE RATINGS FOR GASOLINE PERFORMANCE

The octane number of a gasoline is a measure of its performance in internal combustion engines. Higher-octane gasolines burn more smoothly and cause less engine knock. Standard compounds used for evaluating gasoline performance are isooctane (100 octane) and *n*-heptane (0 octane). Octane numbers from 0 to 100 can be measured by comparing the antiknock characteristics of a gasoline blend to a mixture of these standards. A fuel with an octane rating of 87 would give the same engine performance as a mixture of 87% isooctane and 13% *n*-heptane.

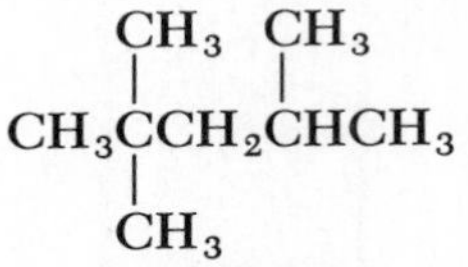

$$CH_3C(CH_3)_2CH_2CH(CH_3)CH_3$$

Isooctane
2,2,4-Trimethylpentane

$$CH_3CH_2CH_2CH_2CH_2CH_2CH_3$$

n-Heptane

Different fuel characteristics are required for good performance in a diesel engine. Diesel fuel is less volatile than gasoline, and the mixture of air and fuel is compressed more and reaches a higher temperature before ignition. A cetane rating scale is used for diesel fuel. Cetane, or hexadecane (value = 100), and 1-methylnaphthalene, an aromatic hydrocarbon (value = 0), serve as standards.

$$CH_3(CH_2)_{14}CH_3$$

Cetane (hexadecane)

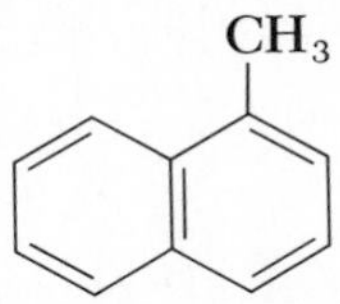

1-Methylnaphthalene

CHEMICAL REACTIONS IN PETROLEUM REFINING

Catalytic cracking is a process by which high-molecular-weight alkanes are broken or cracked to give smaller fragments. High-boiling fractions (bp 250–500°C) are heated to about 500°C with aluminosilicate catalysts. The product stream from this process is rich in lower-molecular-weight, branched compounds. Many of the compounds formed by catalytic cracking have carbon-carbon double bonds.

In another refining process, low-molecular-weight alkanes and alkenes from catalytic cracking are combined using a sulfuric acid catalyst to give branched chain alkanes. In this step, compounds too volatile for use in gasoline are converted to high-octane products.

Catalytic reforming converts cyclic alkanes and some linear alkanes into aromatic hydrocarbons. Aromatic compounds have high octane ratings in gasoline. A platinum catalyst is used in this process. Several years ago, a major gasoline producer implied a high value for its gasoline by saying, accurately, that it was made with "platformate" (i.e., a platinum catalyst). Some of the hydrogen that is a by-product of reforming reactions is used to remove sulfur compounds from petroleum. Much of it is used in the manufacture of ammonia for fertilizer.

The petrochemical industry converts hydrocarbons produced from petroleum into a wide variety of high-volume organic compounds. Some are

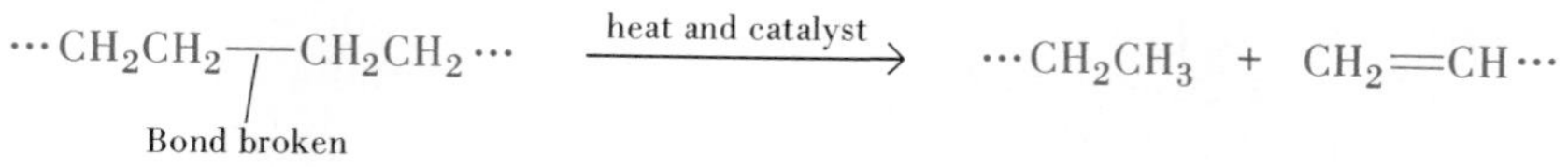

$$\cdots CH_2CH_2 - CH_2CH_2 \cdots \xrightarrow{\text{heat and catalyst}} \cdots CH_2CH_3 + CH_2{=}CH \cdots$$

Bond broken

FIGURE 14.9
In catalytic cracking reactions, high-molecular-weight alkanes are converted mixtures of lower-molecular-weight alkanes and alkenes.

$$CH_3CH(CH_3)CH_3 + CH_2{=}C(CH_3)CH_3 \xrightarrow{H_2SO_4} CH_3CH(CH_3)CH_2C(CH_3)_2CH_3$$

Isobutane (Methylpropane) + Isobutene (Methylpropene) → "Isooctane" (2,2,4-Trimethylpentane)

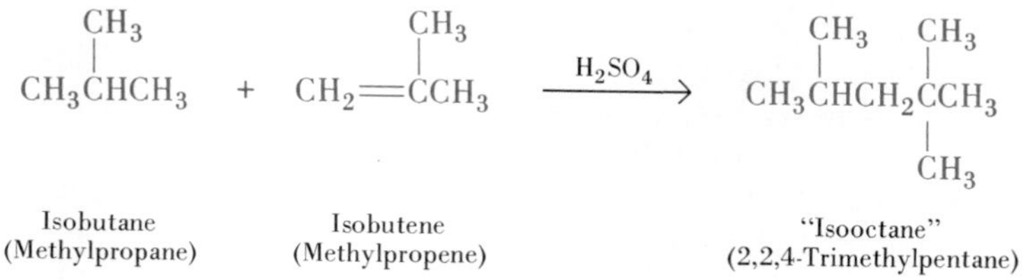

FIGURE 14.10
Small alkanes and alkenes from catalytic cracking are combined to give branched-chain alkanes.

$$\text{Cyclohexane} \xrightarrow{\text{catalyst}} \text{Benzene} + 3\,H_2$$

$$CH_3CH_2CH_2CH_2CH_2CH_2CH_3 \xrightarrow{\text{catalyst}} C_6H_5CH_3 + 4\,H_2$$

Heptane → Toluene

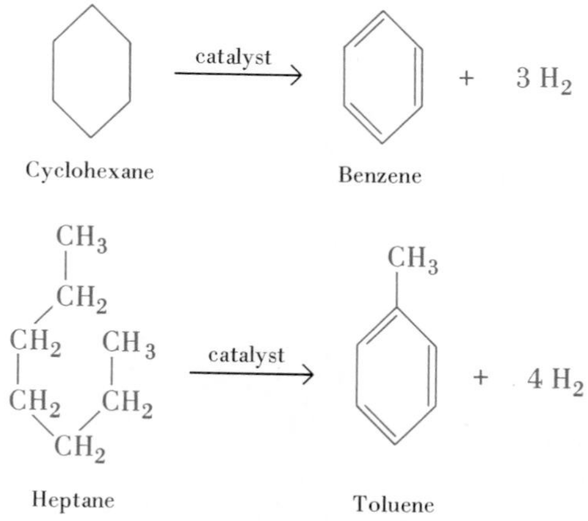

FIGURE 14.11
Aromatic hydrocarbons are formed from cycloalkanes and from some linear alkanes through catalytic reforming steps.

used as solvents and others are used in the formation of plastics and other useful materials. Quantities in production are large to take advantage of large-scale economies. Alkenes and benzene derivatives obtained from petroleum are feedstocks for the petrochemical industry. Low-cost chemicals are used whenever possible. Sulfuric acid, oxygen, water, chlorine, and sodium hydroxide are used in large quantities. As the use of chemicals produced from petroleum has grown, oil refining companies have moved into the production of chemicals, and large chemical companies have acquired oil producing companies.

HALOGENATED HYDROCARBONS— BENEFITS AND RISKS

Few compounds containing only the carbon, hydrogen, and halogen atoms are found in nature. Those that occur are synthesized by plants living in the sea where halide ion concentrations are high. A far greater variety and volume of halogenated hydrocarbons have been placed in the environment by people. These manufactured compounds have been used as refrigerants, as dry cleaning solvents, as electrical insulators, as insecticides, and for the manufacture of phonograph records and floor coverings.

The use of many halogenated hydrocarbons has come under criticism. Microorganisms that degrade organic compounds found in nature lack enzymes designed to react with these manufactured compounds. While some halogenated hydrocarbons react readily with water or with oxygen and are converted into compounds that can be degraded by microorganisms, many are very resistant to reaction. The resistance to degradation frequently increases when two or more halogen atoms are attached to the same carbon atom. Most halogenated hydrocarbons have little solubility in water; they do not form hydrogen bonds with water. Hence they cannot be readily disposed of by simple dilution. After enjoying the benefits of applications of these compounds, we are becoming aware of the environmental costs associated with their continued use.

Three examples of halogenated hydrocarbons that illustrate both the range of utility of these compounds and their associated risks are chloroflu-

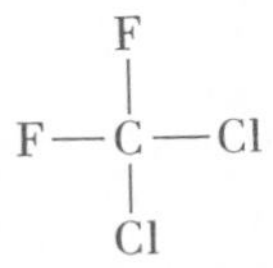

CF_2Cl_2 (F—C(F)(Cl)—Cl)

Dichlorodifluoromethane
Freon-22

$BrCH_2CH_2Br$

Ethylene bromide
1,2-Dibromoethane

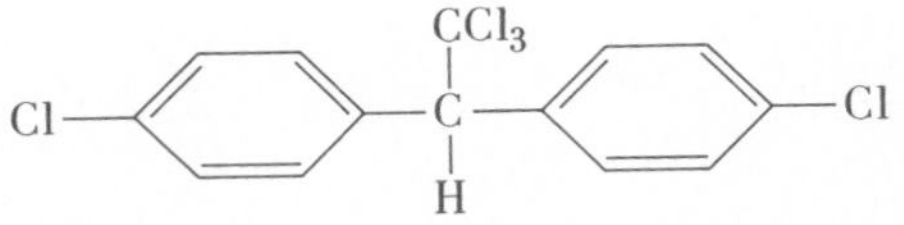

Cl—C_6H_4—C(H)(CCl_3)—C_6H_4—Cl

DDT

FIGURE 14.12
Manufactured halogenated hydrocarbons have served in important applications, but their continued use is in question. The compounds shown have been used as a refrigerant, a fumigant, and an insecticide.

Formation of ozone:

$$O_2 \xrightarrow[\text{ultraviolet}]{h\nu} 2\,O$$

$$O + O_2 \longrightarrow O_3$$

Destruction of ozone:

$$CCl_2F_2 \xrightarrow[\text{ultraviolet}]{h\nu} CClF_2 + Cl$$

$$Cl + O_3 \longrightarrow ClO + O_2$$

$$ClO + O \longrightarrow Cl + O_2$$

FIGURE 14.13
Chlorofluorocarbons pose a threat to the ozone layer of the upper atmosphere. The process is initiated by high-energy ultraviolet light and takes place by a chain mechanism.

orocarbons, ethylene bromide, and the chlorinated insecticide DDT. Chlorofluorocarbon gases are easily condensed, nontoxic, noncorrosive, nonflammable gases used as heat-transfer materials in refrigerators and air conditioners. Ethylene bromide (1,2-dibromoethane) is a fungicide used to prevent molds from growing on grains in storage. The introduction of the insecticide DDT has done much to reduce the incidence of the insect-borne diseases, plague and malaria, and has led to increased yields in western forests and southern cotton fields.

Chlorofluorocarbons pose a threat to the ozone layer of the upper atmosphere. Remarkably stable compounds, they react with neither water nor oxygen. Instead, they diffuse slowly to the upper atmosphere where they catalyze reactions that convert ozone, a form of the element oxygen that has the molecular formula O_3, back to the more normal form of oxygen, O_2. This process is initiated by high-energy ultraviolet light. The reaction takes place by a chain mechanism; a small amount of a chlorofluorocarbon can initiate the destruction of a large amount of ozone. The destruction of the ozone layer is dangerous, for ozone absorbs the harmful ultraviolet radiation coming from the sun. Scientists believe that depletion of the ozone layer will result in a greatly increased incidence of skin cancer.

Ethylene bromide is a volatile liquid that has been used as an additive in lead gasolines and as a fumigant in grain elevators. However, it causes mutations in microorganisms and cancers in test animals. The use of ethylene bromide to protect grains in storage is now banned.

Finally, DDT has been found to cause environmental damage far from where it is used as an insecticide. A relatively nonpolar molecule, DDT is more soluble in fatty tissue than in water. Plants low in the food chain have very low concentrations of DDT in their tissues, but concentrations of DDT rise as one ascends the food chain from plants to plant-eating fish and then to fish-eating birds. Predators such as sea gulls are especially at risk, for accumulations of DDT interfere with their reproductive cycle.

Are there solutions to the problems posed by the wide use of chlorinated hydrocarbons? If there are solutions, will the correct choices be made? The answers are not yet known. The growth in the production of chlorofluorocarbons has been reduced, and hydrocarbons have replaced chlorofluorocarbons as propellants in aerosol cans. Less DDT is used now than in previous decades, but there remain important applications for which DDT is still the insecticide of choice. Compromises between difficult options may be important. If sprayed in the right place at the right time, small amounts of DDT may be highly effective, yet cause little environmental damage.

ASIDE

IS ALL-NATURAL BETTER?

Many people assume that foods grown without the use of pesticides and processed without chemical preservatives are inherently safer to eat as well as more flavorful than foods grown and processed with chemicals. Furthermore, these people have been joined by a much larger group of people in efforts to ban the use of compounds such as DDT and ethylene bromide. For these people, the risks of cancer and of environmental damages associated with the use of these chemicals outweigh their benefits of increased food production and decreased losses on storage.

Rules governing the use of pesticides differ widely. Different rules apply to pesticides in use before 1978 and those introduced later. Rules for applications to crops and for applications to processed foods also differ. Even traces of carcinogenic substances are banned from processed foods under the terms of the Delaney amendment to the Pure Food and Drug Act. Yet, since the passage of this amendment scientists have increased their ability to detect trace amounts of chemicals more than a thousandfold.

The actual situation is complex, and short-sighted decisions may be made using only a partial understanding of the complexities involved. During the 1980s, Bruce Ames and his associates studied the mutagenic properties of large numbers of compounds, both natural and synthetic. Some of the more powerful chemical mutagens are produced by plants themselves and by molds that grow on plants. Levels of these chemicals may rise when the plant is diseased or attacked by insects. Efforts to develop plants that are more resistant to insects can serve to increase the quantities of some very dangerous, naturally occurring chemicals. Ames argues that we cannot have a risk-free chemical environment and that the benefits offered by small amounts of synthetic chemicals, even dangerous ones, may far exceed the risks.

QUESTIONS

1. Write the products of the reaction of $CH_3CH_2CH_2OH$ with each of the following.
   (a) H_2SO_4
   (b) Na
2. Which of the following functional group isomers has the higher boiling point? Why?

$CH_3CH_2OCH_2CH_3$	$CH_3CH_2CH_2CH_2OH$
Diethyl ether	*n*-Butyl alcohol

3. Draw the three isomeric ethers having the formula $C_4H_{10}O$.
4. Write structures for a compound fitting each of the following descriptions.
   (a) Branched alkane
   (b) 3-Carbon alkene
   (c) Alkylbenzene
   (d) 3-Carbon ether
   (e) Chlorinated hydrocarbon
5. $$\begin{array}{rcl} & \mathrm{OH} & \\ & | & \\ \mathrm{CH_3}— & \mathrm{C} & —\mathrm{CH_2CH_2CH_3} \\ & | & \\ & \mathrm{H} & \end{array}$$

   Each of the following compounds is an isomer of the 2-pentanol shown above. Match the correct isomeric relationship to each compound.

   (a) $CH_3CH_2OCH_2CH_2CH_3$ (a′) Functional group isomer

   (b) $\begin{array}{rcl} & \mathrm{OH} & \\ & | & \\ \mathrm{CH_3} & \mathrm{C} & \mathrm{CH_2CH_3} \\ & | & \\ & \mathrm{CH_3} & \end{array}$ (b′) Positional isomer

   (c) $\begin{array}{rcl} & \mathrm{OH} & \\ & | & \\ \mathrm{CH_3CH_2} & \mathrm{C}\mathrm{H} & \mathrm{CH_2CH_3} \end{array}$ (c′) Carbon-skeleton isomer
6. Name the following compounds.
   (a) $CH_3CH_2OCH_2CH_3$
   (b) $CH_3CH_2CH_2CH_2CH_2OH$
   (c) $\begin{array}{rcl} & \mathrm{OH} & \\ & | & \\ \mathrm{CH_3} & \mathrm{C}\mathrm{H} & \mathrm{CH_3} \\ & | & \\ & \mathrm{CH_3} & \end{array}$
   (d) $CH_3CH{=}CH_2$
7. Write structures for the following compounds.
   (a) Diisopropyl ether
   (b) 3-Pentanol
   (c) 2,3-Dimethylbutane
   (d) 1-Butene
8. Why are alcohols more soluble in water than are hydrocarbons?
9. Draw cis and trans isomers of 2-pentene, $CH_3CH{=}CHCH_2CH_3$.
10. In which of the following conversions is ethylene oxidized? Reduced? Neither oxidized nor reduced?
    (a) Ethene to ethane
    (b) Ethene to ethanol
    (c) Ethene to ethylene glycol (1,2-dihydroxyethane)
11. Draw a complete Lewis structure for cyclohexene. What is the formula of cyclohexene?

12. Draw resonance structures of ortho-xylene (1,2-dimethylbenzene). Why is the existence of a single isomer of this compound evidence for resonance? (If the Lewis structures existed, would you expect to see isomers?)
13. Give reactions to account for the following observations.
    (a) When bromine is added to an alkene, cyclohexene, the bromine color fades.
    (b) When bromine is added to an aromatic hydrocarbon, benzene, the bromine color persists. If, however, a Lewis acid is present, the bromine color fades and the odor of hydrogen bromide can be noted.
14. Why is the boiling point of ethylene glycol used in antifreeze higher than that of either ethyl alcohol or water?
15. In most petroleums, the carbon chains are long and predominantly linear. Why is it the case that the higher the octane number of gasoline produced by a refinery, the fewer gallons produced per barrel of petroleum?
16. Why are many of the plants for manufacturing ammonia located in Texas and Louisiana? (Why is it less expensive and less dangerous to transport ammonia for fertilizer than it is to transport hydrogen?)
17. Nitrogen oxides from the engines of supersonic transport planes were predicted to deplete the ozone layer. This concern was a part of the decision not to develop these planes. Why are nitrogen oxides produced by fires and automobiles not as great a threat to the ozone layer?
18. Draw a Lewis structure for ozone. Is ozone a linear molecule? Or is it bent?
19. Draw structures for the products of each of the following reactions.

    (a)  [benzene ring] $+ HNO_3 \rightarrow$

    (b) $CH_2{=}CHCH_3 + H_2O \rightarrow$ H_2SO_4 catalyst
20. The structure of cholesterol is shown. What functional groups are present in this molecule? Why is this molecule more soluble in fats than in water?

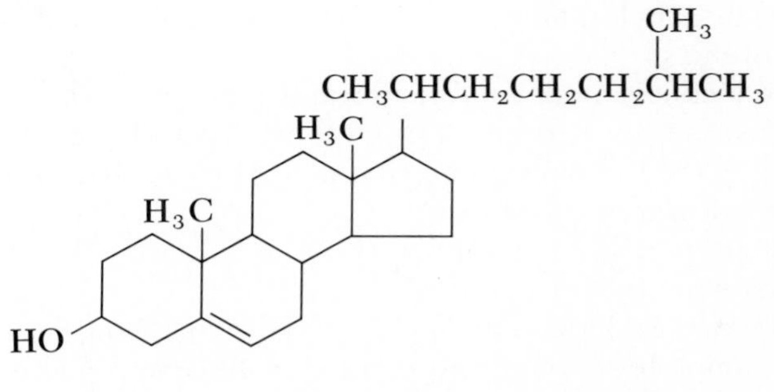

CH_3

$CH_3CHCH_2CH_2CH_2CHCH_3$

H_3C

H_3C

HO

CHAPTER

FIFTEEN

THE VARIETY OF ORGANIC COMPOUNDS

The majority of organic compounds contain oxygen, nitrogen, or both of these elements. These compounds exhibit a far greater range of physical and chemical properties than do compounds containing only carbon and hydrogen. In this chapter, a variety of organic functional groups containing oxygen and/or nitrogen are introduced.

The organic chemist uses these compounds to synthesize textile fibers, perfumes, food additives, and medicines. The body uses similar compounds as building blocks for making proteins, fats, and membranes. The chapter closes with a look at strategies employed by chemists for making complex compounds from simple starting materials.

EXTENDING THE VARIETY OF ISOMERS

The number of possible isomers with a given formula increases rapidly as the number of carbon atoms increases. For example, there are four alcohols having the formula $C_4H_{10}O$. The structures of these alcohols, their boiling points, and the two names for each are given in Table 15.1.

These 4-carbon alcohols show different carbon skeletons as well as different points of attachment of the —OH group. The first two alcohols are positional isomers that are derivatives of the straight-chain hydrocarbon, *n*-butane. The other two alcohols are also positional isomers, but they have the same carbon skeleton as the branched hydrocarbon isobutane.

Compounds with the same functional group show both similarities and differences. For example, the 4-carbon alcohols in Table 15.1 are all hydrogen-bonded, and they all react in a similar way with acids and bases. However, one of these alcohols, 2-methyl-2-propanol, fails to react with some reagents that oxidize the other butyl alcohols.

TABLE 15.1

FOUR CARBON ALCOHOLS

Alcohol	bp, °C	Common Name	Systematic Name
$CH_3CH_2CH_2CH_2OH$	118	*n*-Butyl alcohol	1-Butanol
$CH_3CH(OH)CH_2CH_3$	100	*sec*-Butyl alcohol	2-Butanol
$CH_3CH(CH_3)CH_2OH$	108	Isobutyl alcohol	2-Methyl-1-propanol
$CH_3C(OH)(CH_3)CH_3$	82	*t*-Butyl alcohol	2-Methyl-2-propanol

CH_3CH_2OH	$HOCH_2CH_2OH$	$HOCH_2CH(OH)CH_2OH$
Ethyl alcohol (Ethanol)	Ethylene glycol (1,2-Ethanediol)	Glycerol, glycerin (1,2,3-Propanetriol)

FIGURE 15.1 Three widely used alcohols are ethanol, ethylene glycol, and glycerol.

USES OF ALCOHOLS

A wide variety of alcohols are readily available either from fermentation processes or as products of the petrochemical industry. Many are used for solvents or as intermediates in the chemical industry. Three simple alcohols of major importance are ethanol, ethylene glycol, and glycerol.

Ethyl alcohol is used as a solvent and as a feedstock for other chemical reactions. (For human consumption, alcohol is required by law to be produced by fermentation.) In Brazil, an extensive effort is underway to produce by fermentation ethanol to be used in gasoline. A greater use of alcohol in fuels would reduce the amount of oil that Brazil must import and would ease its balance of payments difficulties.

Beginning in the mid 1980s, blends of methanol or ethanol and gasoline came on the market. To blend these low-molecular-weight alcohols and the hydrocarbons in gasoline, it was necessary to find a way to prevent water from collecting in the fuel mixture. (Water is far more soluble in alcohols than in hydrocarbons due to hydrogen bonding.) Small amounts of additives are used to stabilize the alcohol-gasoline mixtures.

Ethylene glycol (1,2-ethanediol) is used both in making textile fibers and as a coolant for automobile engines. Like water, it has a high heat capacity and can transfer heat from the engine to the radiator where it is exchanged with surrounding air. With two —OH groups, it is a strong-hydrogen-bonded liquid with both a high boiling point and a low freezing point, properties desirable for a compound that must remain a liquid in a running engine on a hot day and in an idle car on a freezing night.

Glycerol or glycerin (1,2,3-propanetriol) is a by-product of manufacturing soap. It is used as a moisturizer in cosmetics, as a starting material for the manufacture of dynamite, and in making protective coatings for appliances.

ETHERS

Ethers are much less widely used than alcohols. Two ethers of some importance are diethyl ether and methyl *t*-butyl ether. Diethyl ether was an important and early general anesthetic in surgery. The widespread use of this compound did much to reduce pain and to extend the range of surgery, but its use required extreme care due to the highly explosive nature of ether vapor. Diethyl ether has been replaced as an anesthetic by less flammable compounds to reduce the danger of fire and explosion. The production of methyl *t*-butyl ether is growing rapidly because it is one of the compounds being used to replace tetraethyl lead as a gasoline additive.

FIGURE 15.2
Two ethers of importance are diethyl ether and methyl isobutyl ether. Diethyl ether was used as an early general anesthetic in surgery, and methyl *t*-butyl ether is used as a gasoline additive.

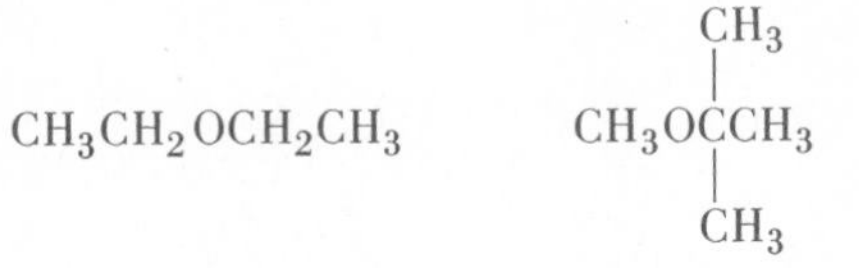

$CH_3CH_2OCH_2CH_3$

Diethyl ether

$$CH_3OC(CH_3)_2CH_3$$

Methyl *t*-butyl ether

ALDEHYDES AND KETONES

Aldehydes and *ketones* are compounds that have a carbon-oxygen double bond. Formaldehyde is the simplest aldehyde, and acetone is the simplest ketone. In a ketone, the C=O or *carbonyl group* is flanked by carbon atoms. In an aldehyde, one or two hydrogens are attached to the carbonyl group. This difference is important, for aldehydes are readily oxidized to carboxylic acids while ketones are more resistant to oxidation.

Aldehydes and ketones are named by adding a suffix to the root of the name of the parent hydrocarbon. The characteristic suffix in the name of an aldehyde is *-al*; in ketones, it is *-one.*

Bases add to the polar carbon-oxygen double bond of aldehydes and ketones as they do with carbon dioxide. The extent of the reaction varies. A solution of formaldehyde in water is almost completely hydrated while very little of the adduct of water and acetone is present in aqueous acetone. This Lewis acid-base chemistry of aldehydes and ketones, important in the structure of sugars and other carbohydrates, is developed further in Chapter 17.

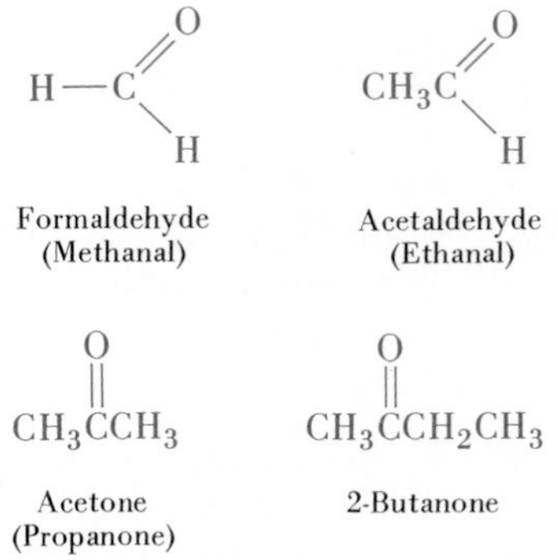

$H{-}CHO$

Formaldehyde
(Methanal)

CH_3CHO

Acetaldehyde
(Ethanal)

CH_3COCH_3

Acetone
(Propanone)

$CH_3COCH_2CH_3$

2-Butanone

FIGURE 15.3
Aldehydes and ketones are compounds that have a carbon-oxygen double bond called a carbonyl group. In an aldehyde, the carbonyl-carbon is attached to one or two hydrogens. In a ketone, the carbonyl-carbon is attached to two carbons.

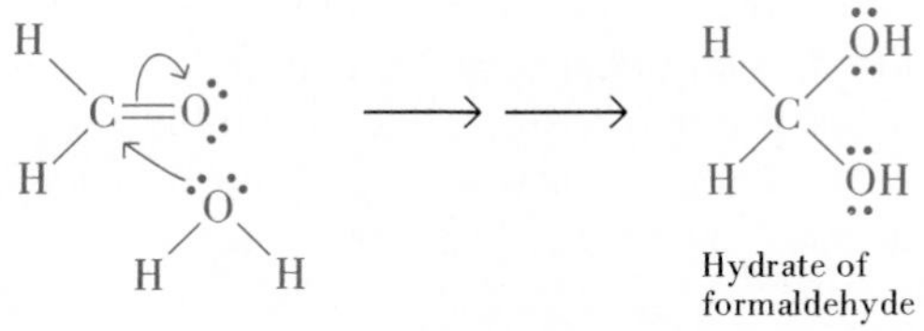

$$H_2C{=}O + H_2O \longrightarrow \longrightarrow H_2C(OH)_2$$

Hydrate of formaldehyde

FIGURE 15.4
Bases add to the carbonyl-carbon atom of aldehydes and ketones. This Lewis acid-base chemistry is important in the chemistry of sugars and other carbohydrates.

CARBOXYLIC ACIDS

Carboxylic acids are among the earliest known and the most important organic compounds. Among the organic acids are formic acid or methanoic acid, formerly isolated from ants, and acetic or ethanoic acid found in vinegar. Common names of carboxylic acids remain in use. Systematic names end in the suffix *-oic acid*. The names and structures of some carboxylic acids are listed in Table 15.2.

Acid-base chemistry facilitates the separation and purification of carboxylic acids. They can be separated from mixtures of organic compounds by extraction into sodium hydroxide solution, for the weak carboxylic acids

$HC(=O)OH$ — Formic acid (Methanoic acid)

$CH_3C(=O)OH$ — Acetic acid (Ethanoic acid)

FIGURE 15.5
Carboxylic acids are among the most important organic compounds. Among the organic acids are formic acids found in ants and acetic acid found in vinegar.

TABLE 15.2

STRUCTURES OF SOME CARBOXYLIC ACIDS

Acid	Common Name	Systematic Name
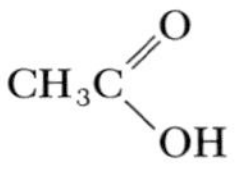 $CH_3C(=O)OH$	Acetic acid	Ethanoic acid
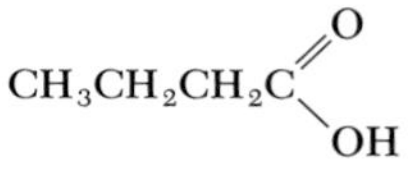 $CH_3CH_2CH_2C(=O)OH$	Butyric acid	Butanoic acid
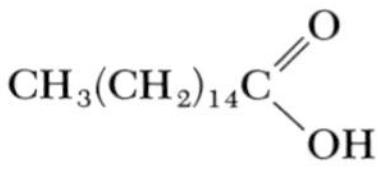 $CH_3(CH_2)_{14}C(=O)OH$	Palmitic acid	Hexadecanoic acid
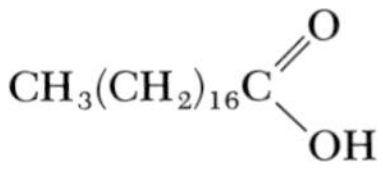 $CH_3(CH_2)_{16}C(=O)OH$	Stearic acid	Octadecanoic acid
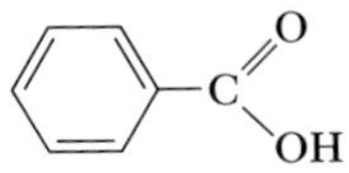 $C_6H_5{-}C(=O)OH$	Benzoic acid	Benzoic acid
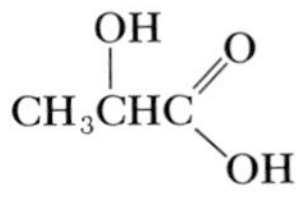 $CH_3CH(OH)C(=O)OH$	Lactic acid	
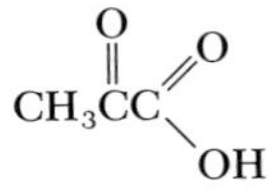 $CH_3C(=O)C(=O)OH$	Pyruvic acid	
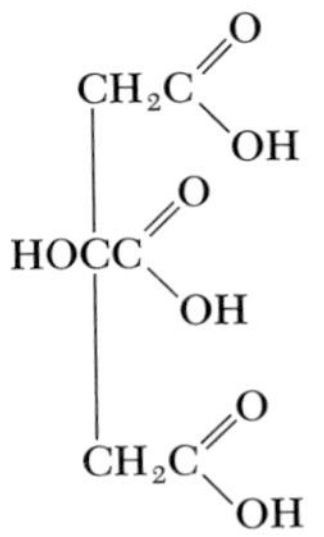 $HOC(CH_2C(=O)OH)_2C(=O)OH$	Citric acid	

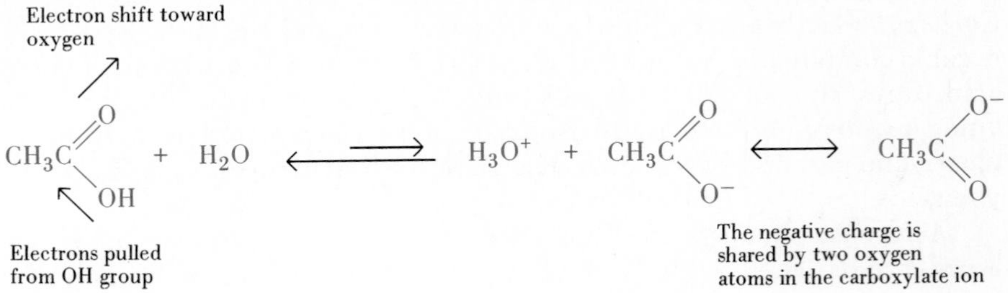

Electron shift toward oxygen

$$CH_3C(=O)OH + H_2O \rightleftharpoons H_3O^+ + CH_3C(=O)O^- \longleftrightarrow CH_3C(O^-)=O$$

Electrons pulled from OH group

The negative charge is shared by two oxygen atoms in the carboxylate ion

FIGURE 15.6
Carboxylic acids are more acidic than water or ethanol.

are converted entirely to sodium salts by reaction with strong base. Sodium carboxylates are far more soluble in water than the free acids. Upon acidification with a strong acid, the carboxylate ion accepts a proton to reform the carboxylic acid which is then recovered from the aqueous solution.

Why are carboxylic acids more acidic than water or alcohols? All have an —OH group, yet the —OH of a carboxylic acid is a better proton donor than the —OH of either an alcohol or water. Both the polarity of the carbon-oxygen double bond of the acid and resonance in the conjugate base contribute to the acidity. The polar C=O group pulls electron density from the —OH group making it more positive and better able to donate a proton than the —OH of an alcohol. In the carboxylate ion, the negative charge is shared equally by the two oxygen atoms. Because the negative charge is more diffuse, the carboxylate ion is a weaker base than hydroxide ion.

STEREOISOMERS

The discovery of compounds that seemed to have the same structure but had different properties posed an important puzzle for chemists. When lactic acid produced by the fermentation of milk was compared to lactic acid isolated from muscles, the samples had different properties.

Two young chemists, van't Hoff in the Netherlands and Le Bel in France, independently ascribed these differences in properties to differences in geometry. In 1874, more than thirty-five years before Rutherford proposed a nuclear structure for the atom, they postulated that the four bonds to a carbon atom are directed toward the corners of a tetrahedron. A tetrahedral shape for carbon accounted for the existence of mirror image isomers because the carbon atoms bonded to four different groups are asymmetric. The lactic acid from muscle consisted of a single isomer and that from milk was shown to be a mixture of equal amounts of mirror image isomers.

Molecules that differ only in geometry are called *stereoisomers*. Stereoisomers that differ only in the attachment of four groups about an asymmetric carbon atom are mirror image compounds called *enantiomers*. Enantiomers can be thought of as right- and left-handed molecules. Consider the C—H bond of the central carbon atom of lactic acid to be a steering column and the other three bonds to be spokes of a steering wheel. In the two

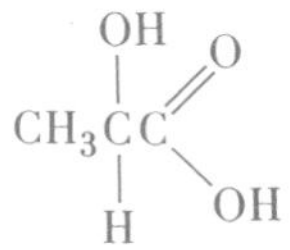

OH
|
CH_3CC(=O)OH
|
H

Lactic acid has an asymmetric carbon atom at C-2

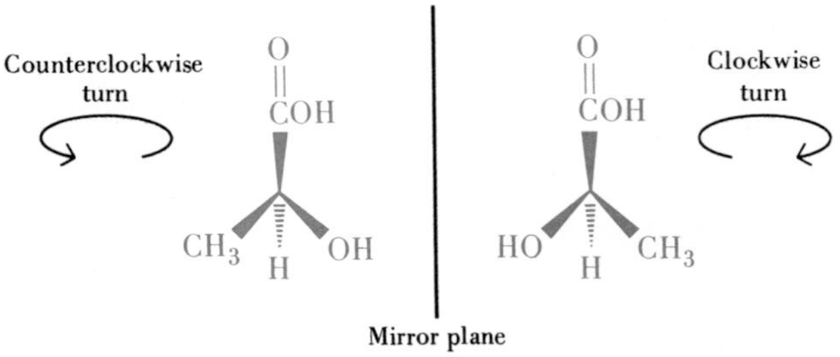

Counterclockwise turn | COOH, CH_3, H, OH || Mirror plane || COOH, HO, H, CH_3 | Clockwise turn

FIGURE 15.7
Enantiomers are mirror image isomers. The relationship is like that between left and right hands.

enantiomers, the three groups correspond to opposite rotations of the wheel just as the fingers of right and left hands curve in opposite directions. These molecules are *chiral*; they have handedness. The relation between the isomers labeled D and L is that of left and right hands. They are nonidentical mirror images.

Enantiomers can be distinguished by their interaction with polarized light. Most physical properties such as melting point, solubility, and density are identical, but enantiomers rotate plane-polarized light in opposite directions. By passing light through a polarizing lens (similar to those used to reduce glare in sunglasses), then through a solution of the compound being tested, and finally through a second polarizing lens, the rotation of polarized light can be studied. If a sample is chiral, the second lens will need to be turned or rotated to restore maximum brightness. Enantiomers rotate light to an equal extent but in opposite directions.

ASIDE

CHIRAL COMPOUNDS, ODOR, AND TASTE

Molecules carry chemical messages to our sense receptors. These receptors are proteins, and, as is the case with enzymes, a signal molecule fits these receptors as a key fits a lock. Since proteins are chiral, signal molecules having right- and left-handed shapes may not fit the same receptor protein, just as a right hand does not fit a left glove.

Stereoisomers of the cyclic ketone, carvone, are isolated from caraway seeds and dill, and from spearmint. Carvones contribute to the taste and odor of each of these spices. The carvone isolated from dill and caraway is the enantiomer of that found in spearmint. The different odors of the carvones from caraway and spearmint illustrate that receptor sites can distinguish between stereoisomers.

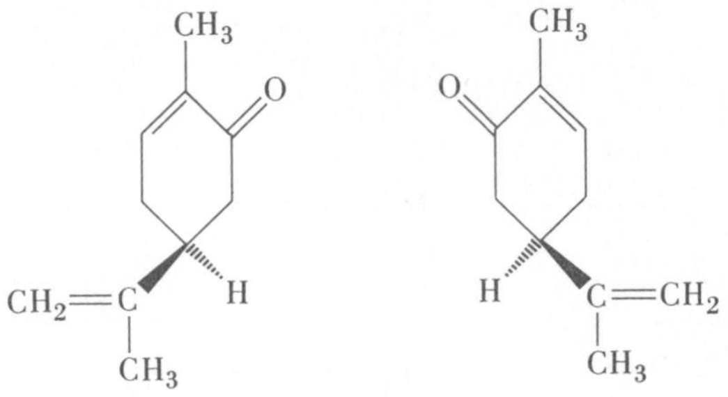

FIGURE 15.8
Enantiomeric carvones from spearmint and caraway differ in taste and odor.

ESTERS

Carboxylic acids react with alcohols to form products known as *esters*. In the reaction, a molecule of water is also formed. Reactions in which two molecules combine by the elimination of a small molecule such as water are called condensation reactions.

$$CH_3\overset{\overset{\displaystyle O}{\|}}{C}OH + HOCH_2CH_2CH_3 \rightarrow CH_3\overset{\overset{\displaystyle O}{\|}}{C}OCH_2CH_2CH_3 + H_2O$$

Acetic acid + *n*-propyl alcohol → *n*-propyl acetate + water
Ethanoic acid + 1-propanol → 1-propyl ethanoate + water

Esters are named as derivatives of carboxylic acids. The alcohol portion of an ester is named first but with the suffix *-yl*. Then the acid portion is named with the ending *-oic acid* changed to *-oate* in the ester.

Esters have lower boiling points than alcohols or acids of the same molecular weight because they are not hydrogen-bonded. Volatile, low-molecular-weight esters contribute to the sweet odor of many fruits.

SOAPS AND DETERGENTS

When the pioneers made soap using wood ashes (or lye) and animal fats, they were not aware that they were doing chemistry on esters. The wood

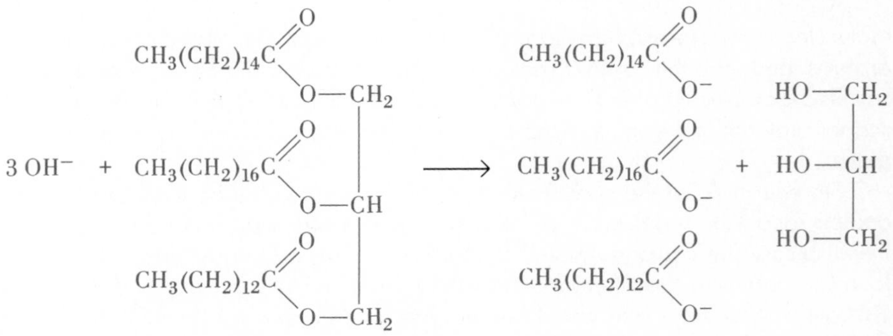

FIGURE 15.9
Fats and esters of long-chain carboxylic acids and glycerol. They are cleaved by reaction with hydroxide ion to form soaps and glycerol.

ashes were a source of the base potassium carbonate. Fats are esters of carboxylic acids and the triol, glycerol. The hydrolysis (reaction with water) of an ester in base, the key reaction in making soap, is called *saponification*.

Soaps are sodium or potassium salts of long-chain carboxylic acids. The acids obtained from animal fats have long, saturated hydrocarbon chains containing an even number of carbon atoms, frequently 14, 16, or 18. Soaps have a *hydrophilic* (water-loving) ionic head, $—CO_2^-$, and a *hydrophobic* (water-hating) tail, $—(CH_2)_nCH_3$.

A tiny sliver of soap placed on water moves about as a monolayer of soap spreads over the water. The hydrophilic head of the soap particles are dissolved in the water, and the hydrophobic tails float on the water surface like a film of oil. If the water is agitated spherical micelles form. In the interior of a micelle the hydrocarbon tails of the soap molecules are dissolved in one another. The ionic carboxylate heads on the outer surface of the micelle cause it to be soluble in water.

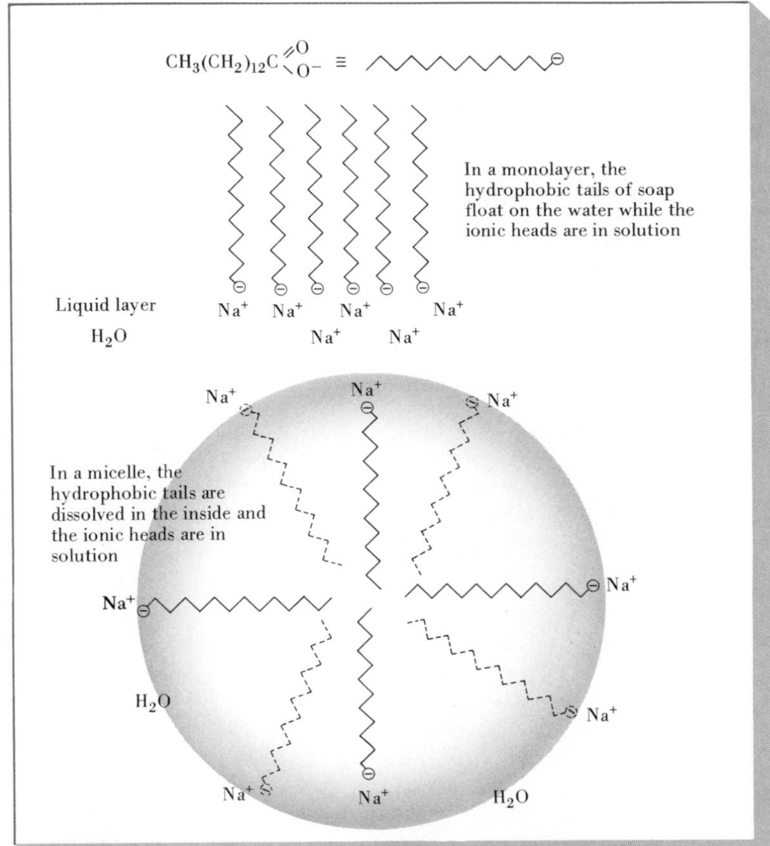

FIGURE 15.10
Soaps have a hydrophylic head and a hydrophobic tail. The hydrophylic head of a soap dissolves in water, but the hydrophobic tail does not. Soaps can form monolayers or micelles when put into water.

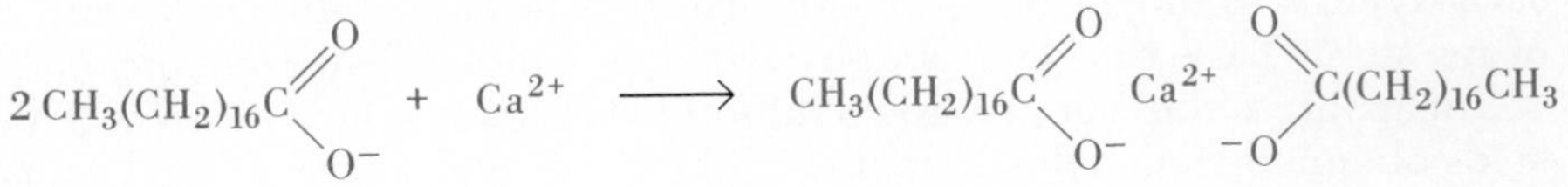

$$2\,CH_3(CH_2)_{16}C(=O)O^- + Ca^{2+} \longrightarrow CH_3(CH_2)_{16}C(=O)O^- \; Ca^{2+} \; {}^-OC(=O)(CH_2)_{16}CH_3$$

FIGURE 15.11
Soap scums form in hard water (water containing dissolved calcium or magnesium salts) when the +2 cations combine with carboxylate ions to form insoluble compounds.

Soaps clean by dissolving grease in the interior of micelles. Although oils and grease are insoluble in water, they dissolve in the hydrophobic interior of micelles and can be carried away in water.

Synthetic detergents perform better than soaps in hard water. Hard water contains dissolved calcium and magnesium salts. Soaps precipitate with these +2 ions forming films or scum. Detergents do not precipitate with Mg^{2+} or Ca^{2+}, so they retain their cleaning power in hard water.

Synthetic detergents can be made from animal or vegetable lipids. To be effective, a detergent molecule must have both a hydrophilic head and hydrophobic tail. Fatty acids can be reduced to produce long-chain alco-

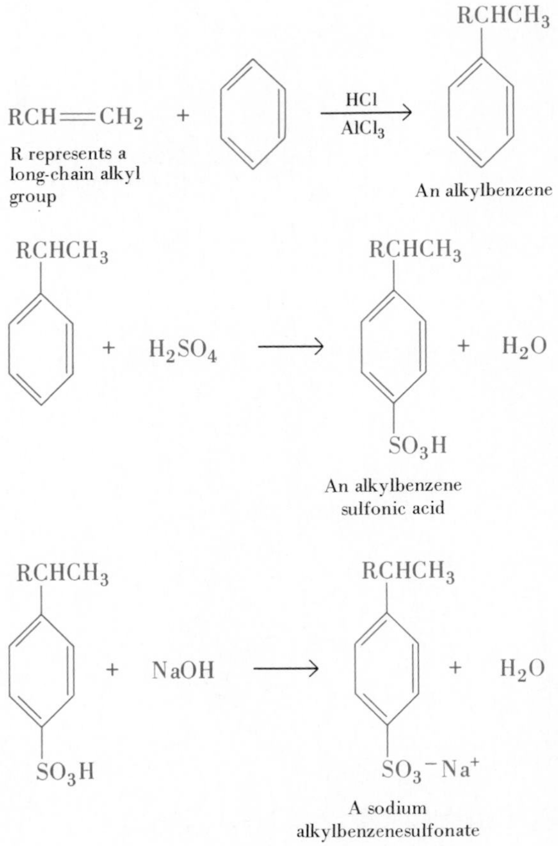

$$RCH{=}CH_2 + C_6H_6 \xrightarrow[AlCl_3]{HCl} RCHCH_3{-}C_6H_5$$

R represents a long-chain alkyl group

An alkylbenzene

$$RCHCH_3{-}C_6H_5 + H_2SO_4 \longrightarrow RCHCH_3{-}C_6H_4{-}SO_3H + H_2O$$

An alkylbenzene sulfonic acid

$$RCHCH_3{-}C_6H_4{-}SO_3H + NaOH \longrightarrow RCHCH_3{-}C_6H_4{-}SO_3^-Na^+ + H_2O$$

A sodium alkylbenzenesulfonate

FIGURE 15.12
The most widely used detergents are made from alkenes, benzene, sulfuric acid, and sodium hydroxide in a series of reactions.

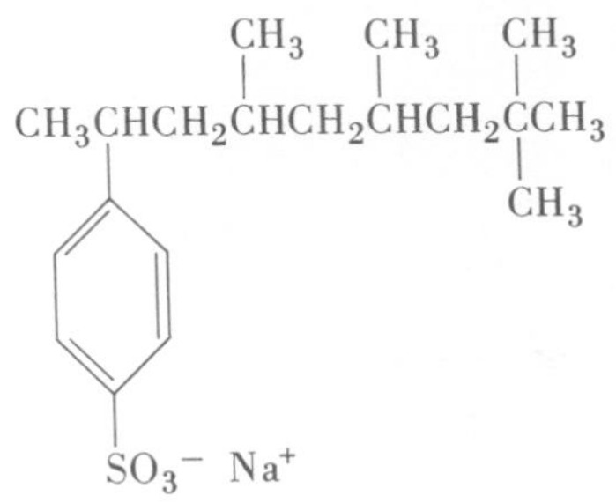

```
        CH3   CH3   CH3
         |     |     |
CH3CHCH2CHCH2CHCH2CCH3
    |                |
    |               CH3
  (benzene ring)
    |
  SO3-  Na+
```

FIGURE 15.13
Detergents used in the 1950s had highly branched side chains that were not broken down by microorganisms. The failure to degrade detergents led to foam formation in sewer effluents and in streams. Detergents now in use have more linear side chains and are more readily degraded.

hols. (These alcohols are also obtained by the saponification of some vegetable oils.) The alcohols are converted to detergents first by reaction with sulfuric acid and then with sodium hydroxide. Lauryl sulfate is an ester of sulfuric acid prepared by a condensation reaction of lauryl alcohol with sulfuric acid. Sodium, potassium, and ammonium salts of lauryl sulfate are mild detergents frequently used in shampoos.

$$\underset{\text{Lauryl alcohol}}{CH_3(CH_2)_{10}CH_2OH} + H_2SO_4 \rightarrow \underset{\text{Lauryl sulfate}}{CH_3(CH_2)_{10}CH_2OSO_3H} + H_2O$$

$$CH_3(CH_2)_{10}CH_2OSO_3H + NaOH \rightarrow \underset{\text{Sodium lauryl sulfate}}{CH_3(CH_2)_{10}CH_2OSO_3^-Na^+} + H_2O$$

Sodium alkylbenzene sulfonates have been widely used, high-volume detergents. These detergents are produced using sulfuric acid, sodium hydroxide, and hydrocarbons from petrochemical sources. The reactions of benzene are used to link a hydrophobic alkyl group and a sulfonic acid group. In the final reaction, sodium hydroxide is used to convert the acid end of the molecule to a hydrophilic salt.

Synthetic detergents entering streams need to be degraded by microorganisms, or downstream water quality suffers. In water, lauryl sulfate slowly hydrolyzes to the alcohol which is oxidized to a fatty acid and degraded by microorganisms. Microorganisms can oxidize long-chain hydrocarbons, but this degradative action is blocked by chain branching found in early alkylbenzene sulfonates. During the 1950s, the appearance of detergent foam in streams pointed to the failure to degrade or dispose of synthetic organic detergents. Now manufacturers produce detergents with more nearly linear alkyl groups that are degraded by microorganisms.

A S I D E

CELL MEMBRANES ARE AMPHIPHILIC

The lipids found in cell membranes are amphiphilic. Like soap and detergent molecules, membrane molecules have hydrophilic heads and hydrophobic tails. Like soaps and detergents, membrane lipids can form monolayers and micelles.

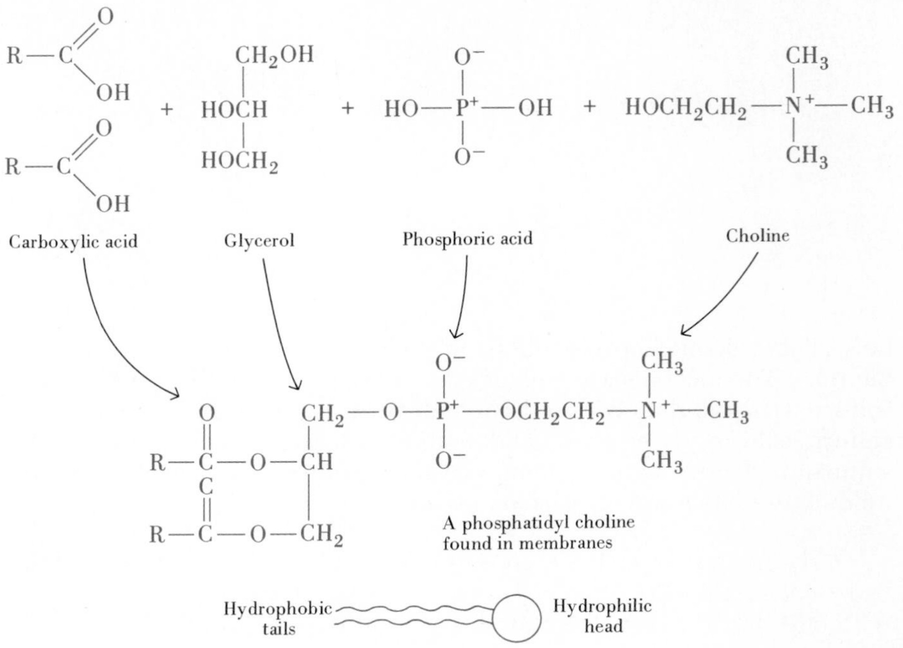

$$2\ R{-}C(=O){-}OH \;+\; \begin{matrix} CH_2OH \\ | \\ HOCH \\ | \\ HOCH_2 \end{matrix} \;+\; HO{-}\overset{\displaystyle O^-}{\underset{\displaystyle O^-}{P^+}}{-}OH \;+\; HOCH_2CH_2{-}\overset{\displaystyle CH_3}{\underset{\displaystyle CH_3}{N^+}}{-}CH_3$$

Carboxylic acid   Glycerol   Phosphoric acid   Choline

$$\begin{matrix} & & CH_2{-}O{-}\overset{\displaystyle O^-}{\underset{\displaystyle O^-}{P^+}}{-}OCH_2CH_2{-}\overset{\displaystyle CH_3}{\underset{\displaystyle CH_3}{N^+}}{-}CH_3 \\ & & | \\ R{-}\overset{\displaystyle O}{\overset{\|}{C}}{-}O{-}CH \\ \overset{\|}{C} & & | \\ R{-}C{-}O{-}CH_2 \end{matrix}$$

A phosphatidyl choline found in membranes

Hydrophobic tails   Hydrophilic head

FIGURE 15.14
Esters of phosphotidic acid are found in membranes. These esters have nonpolar hydrophobic tails and ionic hydrophilic heads. In water, these compounds readily form bilayers.

Many membrane molecules are esters of phosphotidic acid. In this compound, as in fats, two hydrophobic fatty acids are esterified with glycerol. However, the third —OH group of glycerol is esterified to phosphoric acid. The phosphate portion of the ester bears a negative charge and is linked to an ionic or polar residue that serves to increase the size of the hydrophilic head.

Chemical tools for the investigation of membrane lipids include enzymes isolated from snake venoms. These enzymes catalyze the cleavage of membrane lipids at selected sites. By breaking down membrane lipids in a snakebite victim, these enzymes assist the spread of venom.

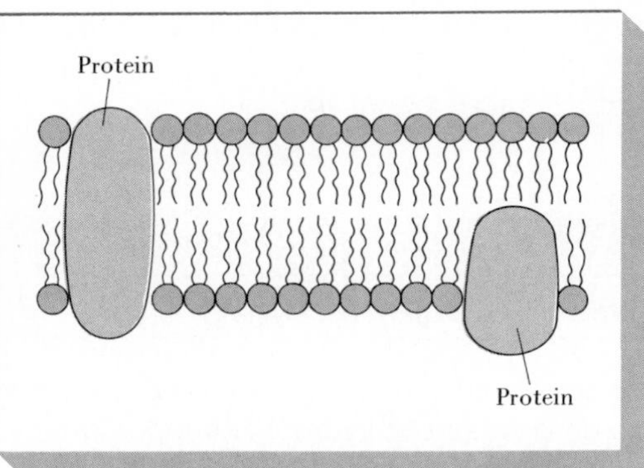

FIGURE 15.15
According to the fluid mosaic model, cell membranes are composed of lipid bilayers and membrane-bound proteins. Hydrophilic heads of the lipids extend into the aqueous layer, and hydrophobic tails form a barrier to the passage of ions and polar molecules. Proteins in the membrane serve as channels and pumps to pass polar species across the membrane.

Membranes consist of lipid bilayers in which various proteins are embedded. Hydrophilic heads of the lipids extend into the aqueous layer on either side of the membrane barrier. The hydrophobic tails form a barrier to the passage of ions and polar molecules. Proteins in the membranes serve as channels and pumps to pass polar species across the membrane. This description of membrane structure is known as the fluid mosaic model.

VEGETABLE OILS ARE UNSATURATED

The fatty acids of plants differ from those found in animals. Lipids need to be flexible or liquid under physiological conditions. Vegetable oils remain liquid at temperatures at which fats from warm-blooded animals are solids. Most fatty acid residues in animal fats are saturated; those obtained from vegetables contain double bonds or unsaturation.

Stearic acid, a C-18 straight-chain acid, melts at 70°C. In contrast, oleic acid with one double bond melts at 14°C. The double bonds in naturally occurring unsaturated fatty acids have the cis configuration. The two hydrogens are on one side of the rigid double bond and the two alkyl groups are on the other. The cis double bond imparts a bend to the tail of the acid. Because they do not stack well in lattices, unsaturated lipids remain liquid at temperatures where their saturated counterparts are solids.

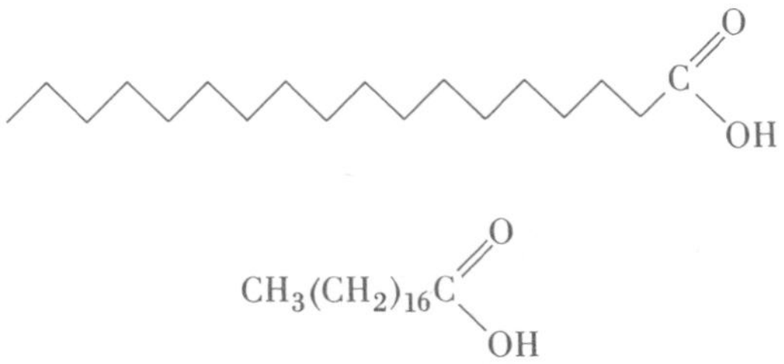

$CH_3(CH_2)_{16}C(=O)OH$

Stearic acid; mp 70

H H

$CH_3(CH_2)_7CH{=}CH(CH_2)_7C(=O)OH$

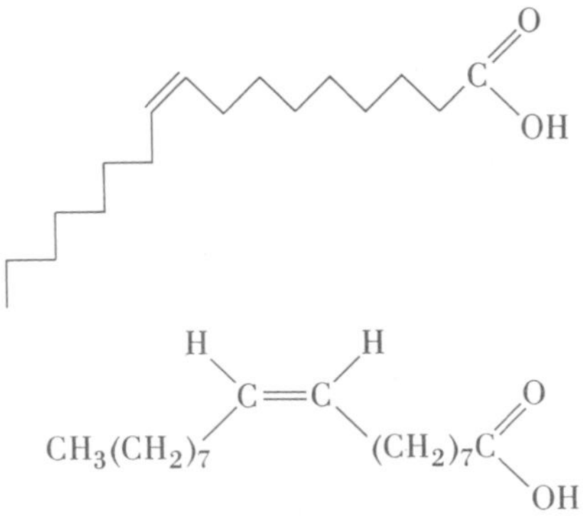

Oleic acid; mp 14

FIGURE 15.16
Stearic acid, with a saturated straight chain, melts at 70°C. In contrast, oleic acid, with one double bond, melts at 14°C. Unsaturated lipids do not stack well in lattices and remain liquid at temperatures where their saturated counterparts are solids.

Margarines are made from vegetable oils by hydrogenating the double bonds of some of the unsaturated fatty acid esters. Adding hydrogen to double bonds gives a higher-melting product. Since we have a dietary requirement for some polyunsaturated fatty acids, the oils used to make many softer margarines are only partially reduced.

A S I D E

ANTIOXIDANTS ARE USED TO PRESERVE FOOD

Compounds with carbon-carbon double bonds are subject to air oxidation. The reaction takes place by a chain process involving intermediates with odd numbers of electrons. The oxidation of alkenes can be used to advantage. Compounds with double bonds are used in oil-based paints. Upon oxidation, these compounds form resistant coats that protect painted surfaces. The drying of an oil-based paint is an oxidation process.

In foods, this same oxidation contributes to spoiling. Upon oxidation, unsaturated fats and oils are converted to carboxylic acids that give food an undesirable flavor and odor. To protect foods, antioxidants are sometimes added. Two such compounds, abbreviated BHA and BHT, react with the odd electron species that are intermediates in air oxidation. A BHA or BHT molecule donates a hydrogen atom from its OH group to the reactive intermediate, rendering it unreactive. The BHA or BHT then has an odd electron, but the bulky butyl groups hinder further reaction. The chain reactions of oxidation leading to spoilage are stopped, and the food is protected.

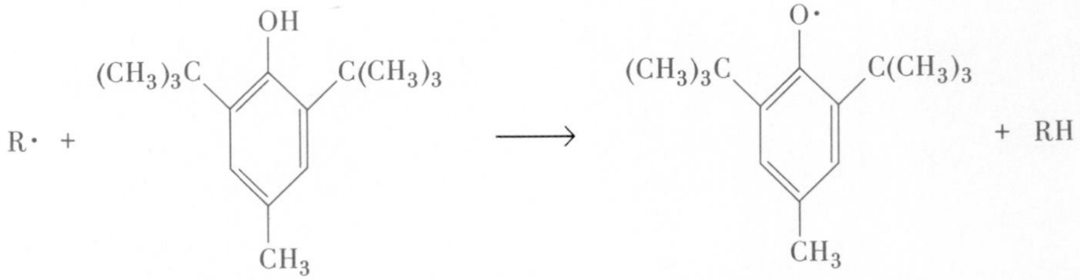

OH
$(CH_3)_3C$ $C(CH_3)_3$
R· + ⟶ + RH
CH_3
Butylated hydroxytoluene, BHT

O·
$(CH_3)_3C$ $C(CH_3)_3$
CH_3
An unreactive free radical

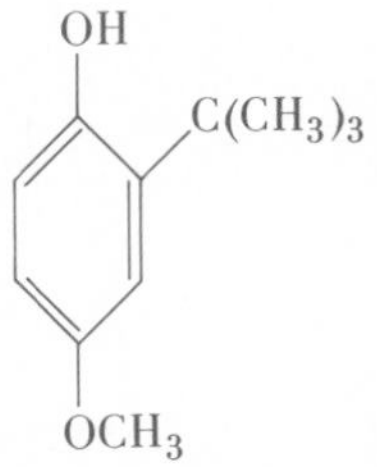

OH
$C(CH_3)_3$
OCH_3
Butylated hydroxyanisole, BHA

FIGURE 15.17
Antioxidants, such as BHT and BHA, are used as food preservatives. These compounds interrupt free-radical chain reactions by donating a hydrogen atom to a reactive radical and forming a relatively unreactive intermediate. The chain reactions leading to food spoilage are interrupted and the food is protected.

Inorganic antioxidants include sodium nitrite and sodium bisulfite. Sodium nitrite is used as a preservative in bacon, and sodium bisulfite is often used in restaurants to help keep cut vegetables and fruit from discoloring. There is concern that all food additives be listed on labels, because sodium nitrite may act as a carcinogen and some people have intense allergic reactions to sodium bisulfite.

AMINES AND AMIDES

Amines are organic bases. Just as alcohols can be considered to be derivatives of water, amines can be considered to be derivatives of ammonia. Amines may have one, two, or three alkyl groups substituted for hydrogens on ammonia. The basic character of amines depends upon an unshared pair of electrons on nitrogen.

$$CH_3CH_2NH_2 \qquad CH_3\overset{\displaystyle CH_3 \atop |}{N}CH_3$$

Ethyl amine — Trimethyl amine

Alkaloids are physiologically active organic amines isolated from plants. For example, caffeine, nicotine, quinine, cocaine, and morphine are all alkaloids.

The acid-base chemistry of alkaloids is simple, so the control of the chemical processing of alkaloids used in illicit drugs is very difficult. For example, cocaine is obtained from the coca plant by first extracting the dried leaves with kerosene and then reacting the base with strong acid to separate it from the kerosene and from the neutral compounds. When the acid salt of cocaine is later reacted with baking soda ($NaHCO_3$), the free base is formed by the removal of an acidic proton.

Amides are neutral compounds that can be considered products of a condensation reaction of an amine or ammonia with a carboxylic acid. Amides have structures analogous to those of esters in which an amine moiety has replaced the alcohol moiety of the ester. The importance of amide functional groups to proteins and to nylon is discussed in following chapters.

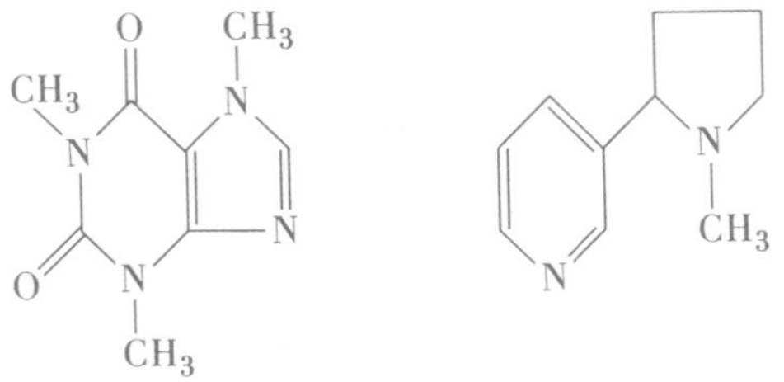

Caffeine — Nicotine

FIGURE 15.18
Alkaloids are organic bases isolated from plants. Caffeine is found in coffee and tea, and nicotine is found in tobacco. Many alkaloids have potent physiological effects.

FIGURE 15.19
Amides are neutral compounds that can be formed by the condensation reaction of a carboxylic acid with ammonia or an amine.

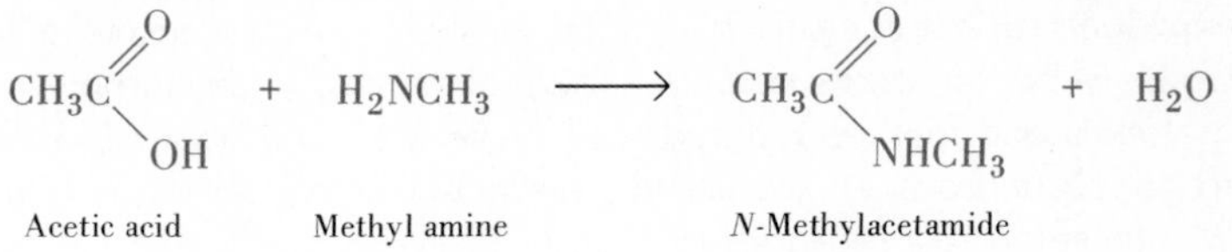

$$\underset{\text{Acetic acid}}{CH_3C(=O)OH} + \underset{\text{Methyl amine}}{H_2NCH_3} \longrightarrow \underset{N\text{-Methylacetamide}}{CH_3C(=O)NHCH_3} + H_2O$$

SYNTHESIS OF CARBON-CARBON BONDS

The synthesis of organic compounds involves both the construction of a carbon skeleton and the elaboration of functional groups. There are many reactions that convert one functional group to another, but there are far fewer reactions that combine small molecules to form larger molecules. Reactions that form new carbon-carbon bonds are important in the synthesis of compounds for use in medicine and in industry.

One major approach to the synthesis of new carbon-carbon bonds takes advantage of the two major types of chemical reactivities found in esters. Substitution reactions occur at the carbon atom bonded to two oxygens. For example, esters are converted to free acids by reaction with water and to amides by reaction with amines. A second reactivity is at the α carbon, the carbon adjacent to that bearing oxygens. Strong bases can remove a proton from the α carbon to generate a basic anion.

These two reactivities are combined in a reaction leading to the formation of a new carbon-carbon bond. The anion generated by removing a hydrogen from the α carbon of one ester molecule can attack the Lewis acid site of a second ester molecule to substitute for the alcohol moiety. Because they have reactivity at two sites, esters can undergo a condensation reaction to form new carbon-carbon bonds.

FIGURE 15.20
Esters can undergo condensation reactions to form new carbon-carbon bonds.

$$CH_3CH_2O^- + CH_3C(=O)OCH_2CH_3 \longrightarrow CH_3CH_2OH + {}^-CH_2C(=O)OCH_2CH_3$$

Electrons are pulled from the methyl group toward oxygen making the C-H bond of the methyl group more acidic

$$CH_3C(=O)OCH_2CH_3 + {}^-CH_2C(=O)OCH_2CH_3 \longrightarrow CH_3C(=O)CH_2C(=O)OCH_2CH_3 + {}^-OCH_2CH_3$$

A new carbon-carbon is formed in a Lewis acid base reaction

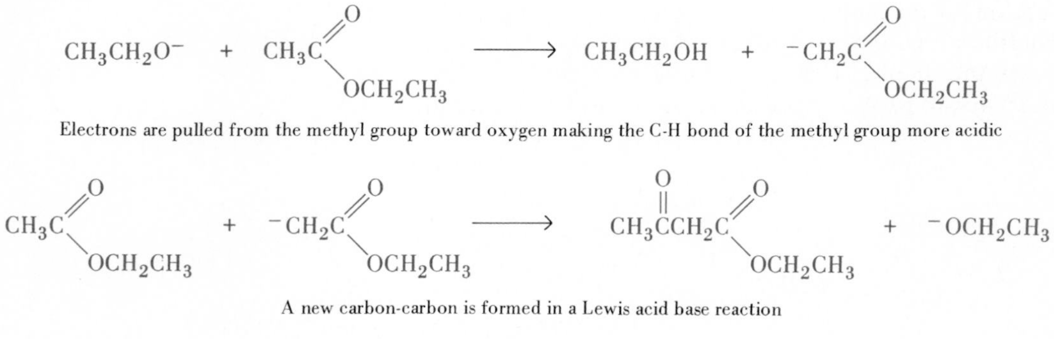

FIGURE 15.21
Aldehydes react under basic conditions to form new carbon-carbon bonds. In a second reaction, water may be eliminated to form a carbon-carbon double bond.

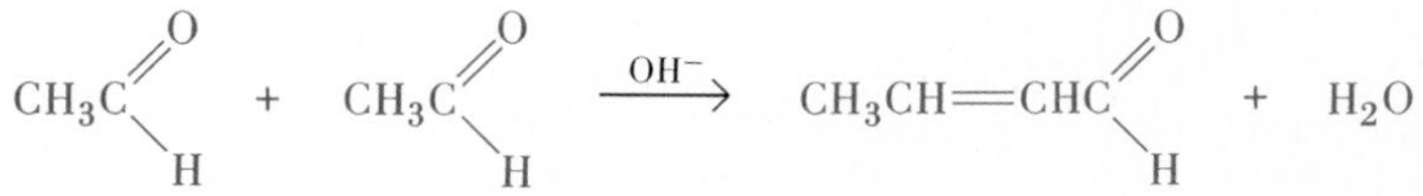

$$CH_3C(=O)H + CH_3C(=O)H \xrightarrow{OH^-} CH_3CH{=}CHC(=O)H + H_2O$$

Aldehydes and ketones undergo similar condensation reactions to form new carbon-carbon bonds. In these compounds, as in esters, the carbonyl group makes a hydrogen atom on an α carbon more acidic. The carbonyl group of a second molecule provides a Lewis acid site for attack.

A S I D E

THE SYNTHESIS GAME

The synthesis of complex molecules found in nature has been a challenge to organic chemists for 100 years. Initially, the synthesis of a compound by a series of rational reactions was the final step in the proof of structure. For many decades, structures were determined by breaking complex compounds into simpler pieces using chemical reactions. Just as a paragraph in a foreign language might be interpreted by interpreting individual sentences and by breaking sentences into phrases and words, so were complex molecules examined by isolating and analyzing fragments. Just as knowing the words in a sentence or the sentences in a paragraph does not always lead to the correct translation, knowing the fragments of a compound does not always lead to a correct structure. The successful synthesis of a proposed structure was a proof for the assignment of structure for a compound.

Chemical synthesis which began as a tool for structure proof took on a life of its own. The synthesis of complex compounds found in nature is an important part of organic chemistry. Nature provides a very large variety of target molecules. Plants and animals have been waging complex chemical warfare for millions of years. Further, many of the target molecules are physiologically active, offering potential for use in medicine. Apart from any potential benefits, synthesis is also a game in which young chemists might demonstrate their promise and veteran chemists can exhibit their versatility.

The synthesis game can be played with flair and elegance. Some compounds are relatively easy to synthesize; others present special problems. For example, one compound may be particularly sensitive to all but the mildest chemical conditions and another may have a stereochemistry that is difficult to achieve. Or a chemist might accept the challenge of trying to imitate a biochemical pathway for synthesis. A chemist must solve the problem presented by a particular structure. Often a new

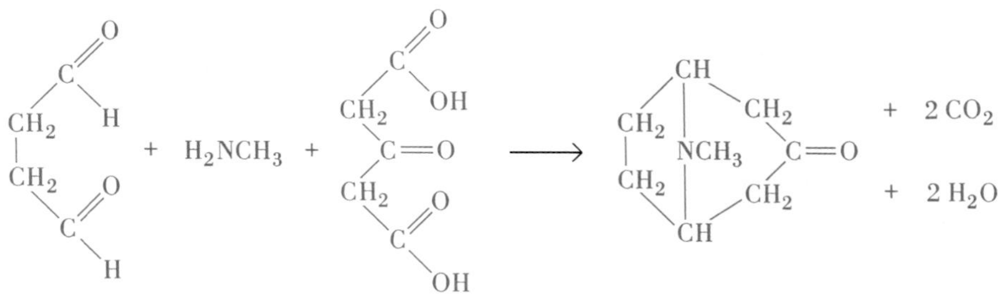

$$OHC{-}CH_2{-}CH_2{-}CHO \; + \; H_2NCH_3 \; + \; HOOC{-}CH_2{-}C(=O){-}CH_2{-}COOH \longrightarrow \text{tropinone} \; + \; 2\,CO_2 \; + \; 2\,H_2O$$

FIGURE 15.22
The synthesis of the alkaloid tropinone is an early example of elegance in synthesis.

reagent or a new strategy for solving a general chemical problem is developed as a part of the solution to a synthetic problem. Just as an athlete can win fame by his or her performance in a championship game, a chemist can win fame by synthesizing a "championship" molecule.

Syntheses can also be considered chemical works of art. Chemists use the word "elegant" to describe a particularly beautiful piece of chemistry. The synthesis of the alkaloid tropinone by Sir Robert Robinson in 1917 set an early standard for elegance in synthesis. His synthesis was biomimetic, for it pointed to the way plants make a variety of physiologically active alkaloids. It was both simple and sophisticated. It was simple because only three compounds were mixed and the desired compound formed. It was sophisticated, for it showed a profound insight into the structure of tropinone and how this alkaloid is formed in nature.

QUESTIONS

1. One of the largest plants manufacturing methyl *t*-butyl ether is located in Saudi Arabia. Why would the plant be there? What is the use for this product? Where would it be used?
2. Just as water reacts as a base to add to the carbon-oxygen double bond of formaldehyde, alcohols also add to this Lewis acid. Draw the structure of the product formed when methanol adds to formaldehyde.
3. Acetic acid, $C_2H_4O_2$, has the same density in the vapor state as molecules having twice the molecular weight. What forces would exist in acetic acid dimers? Draw a structure for a dimer of acetic acid in the vapor state.
4. Account for the observation that trifluoroacetic acid, CF_3CO_2H, is a stronger acid than acetic acid.
5. Write structures for a compound fitting each of the following descriptions.
   (a) 2-Carbon aldehyde
   (b) 4-Carbon ketone
   (c) 3-Carbon carboxylic acid
   (d) 2-Carbon amine
   (e) 2-Carbon amide
6. Write structures for the following compounds.
   (a) Ethanal
   (b) 2-Pentanone
   (c) Propanoic acid
   (d) Methyl butanoate
   (e) Dimethyl amine
7. One of the butyl alcohols consists of a pair of enantiomers. Which butyl alcohol has a chiral center? Sketch the enantiomers of this butyl alcohol.
8. Soap is ineffective in an acidic solution. Why?
9. Why would the sodium salt of a short chain (4- to 8-carbon) fail to be a satisfactory soap? Why is it necessary to add an alkyl group to benzene in making sodium alkylbenzene sulfonate detergents?

10. Draw the structure of a detergent micelle.
11. Why is the lipid bilayer found in membranes impermeable to the passage of ions?
12. What functional group is present in each of the following compounds?

(a) $CH_3CH_2CH_2C(=O)OH$

(b) $CH_3CH_2C(=O)H$

(c) $CH_3CH_2\overset{O}{\overset{\|}{C}}CH_2CH_3$

(d) $H\overset{O}{\overset{\|}{C}}OCH_2CH_3$

(e) $CH_3CH_2CH_2NH_2$

(f) $CH_3C(=O)NH_2$

13. Give a name for each of the compounds in problem 12.
14. Read the labels on a soft and a hard margarine. Why do the margarines differ in softening temperature? What chemical reaction has been used to modify vegetable oils for use in margarines?
15. Just as the —OH group in acetic acid is more acidic than water, the $—NH_2$ group in acetamide is less basic than ammonia. How does the C=O group make the nitrogen in the amide less basic?
16. Draw the structures of the products of each of the following reactions.

(a) $CH_3C(=O)OH + CH_3OH \rightarrow$

(b) $CH_3CH_2CH_2CH_2\overset{O}{\overset{\|}{C}}OCH_3 + OH^- \rightarrow$

17. In the following pairs of compounds, one isomer has a single functional group and the other isomer has two functional groups. Name the functional groups present in each compound.

(a) $CH_3—\overset{O}{\overset{\|}{C}}—CH_2OH$ and $CH_3CH_2—\overset{O}{\overset{\|}{C}}—OH$

(b) $NH_2CH_2CH_2C(=O)H$ and $CH_3CH_2C(=O)NH_2$

18. Architectural designs are solutions to design problems involving use, cost, and space. In what ways are architecture and chemical synthesis similar? In what ways may each be elegant?

CHAPTER

SIXTEEN

POLYMERS

Polymers are composed of very large molecules with repeating subunits. Polymer science is a development of the twentieth century. The growth of polymer science has been so rapid that now more than one-half of the chemists working in industry are involved with polymer-related problems. The explosion of knowledge about polymers and the application of that knowledge to technology have occurred in both the areas of synthetic polymers and naturally occurring polymers such as proteins and nucleic acids.

Polymer chemistry is directed toward the production of materials that can meet the needs generated by a wide variety of applications. Whether it be a new textile fiber, a strong, lightweight composite for use in transportation, a human growth hormone, or human insulin, the goal is to meet a defined need with a well-designed product.

In this chapter and in the following one, we will sample the variety of synthetic and natural polymers. We will look at relationships between the properties of these polymers and their chemical structures.

BIG MOLECULES ARE DIFFERENT

Polymer science was slow to develop. The traditional experimental methods and logic used to study organic compounds were inadequate to deal with polymer problems. Organic chemists had developed experimental methods to analyze compounds composed of small- or medium-sized molecules, but polymer molecules, which may consist of hundreds or thousands of atoms, presented new problems. They are hard to dissolve and cannot be purified by distillation or recrystallization. When polymers formed accidentally in a reaction mixture, they were often ignored or discarded.

To explore the unique properties of polymers, scientists needed to develop new experimental methods and new theories to interpret experimental results. A polymer is different because it is composed of large molecules. (Indeed, as late as 1920, some scientists were reluctant to believe that polymers were molecules.) To understand polymers, scientists needed to analyze contributions to physical and chemical properties made by size alone as well as those made by functional groups. The methods developed to analyze the size of macromolecules included light scattering, osmotic pressure measurements, and ultracentrifugation.

Light is scattered by small particles like dust. Short-wavelength blue light is scattered more than long-wavelength red light. Widely scattered blue light from the sun reaches us from many angles, so the sky looks blue during the day. At dawn and at sunset, the red light which has traveled a more nearly straight path through the atmosphere colors the sky. Very large molecules are also large enough to scatter light. Because solutions of polymers contain molecules large enough to scatter light, they appear opalescent blue. (This can be seen by looking at a dilute solution of powdered milk which contains protein macromolecules.) From measurements of light scattering, scientists can determine the size of polymer molecules.

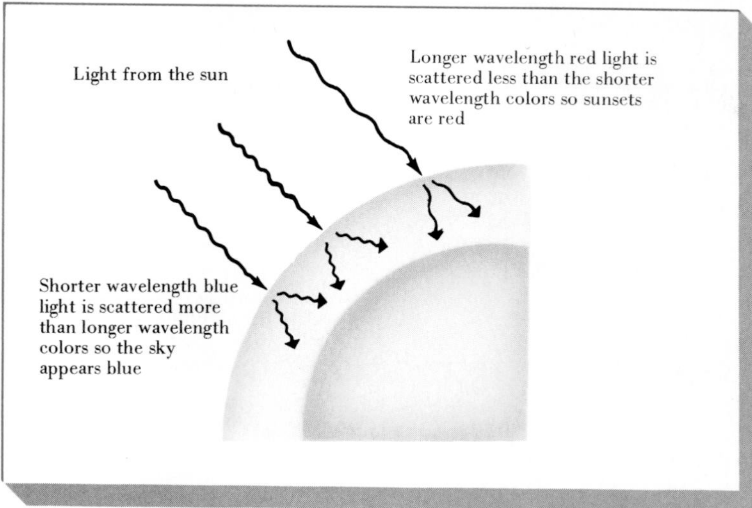

FIGURE 16.1
Dust in the atmosphere scatters light, causing the daytime sky to look blue and sunsets to look red. Like tiny dust particles, large polymer molecules scatter light. Light-scattering experiments are used to determine the size of polymer molecules.

Osmosis occurs across membranes. Trees draw in water at the roots, and osmotic pressure across cell membranes lifts water to the leaves. When a polymer solution is separated from excess solvent by a membrane that is permeable to the solvent but not the polymer, solvent will pass into the solution and dilute the polymer. The pressure driving solvent molecules through the membrane is called osmotic pressure. Scientists use osmotic pressure measurements to analyze the concentrations of very dilute polymer solutions.

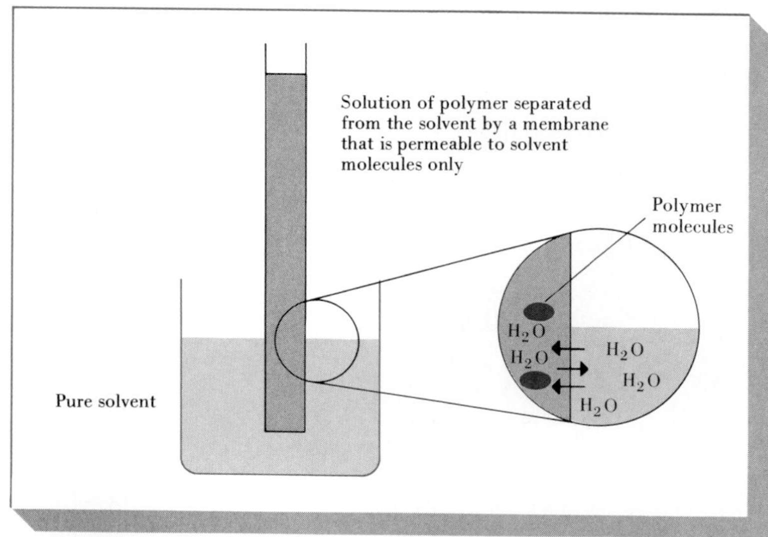

FIGURE 16.2
Osmotic pressure exists across a semipermeable membrane, as solvent molecules flow from the solvent side of the membrane to dilute the solution of polymer. Osmotic pressure measurements are used to measure the concentration of polymer solutions.

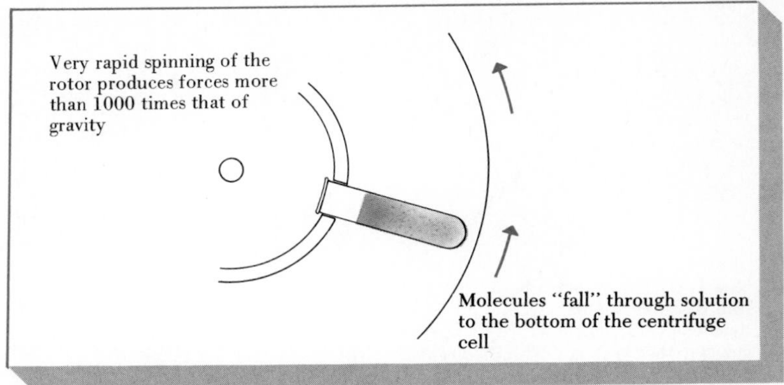
Very rapid spinning of the rotor produces forces more than 1000 times that of gravity
Molecules "fall" through solution to the bottom of the centrifuge cell

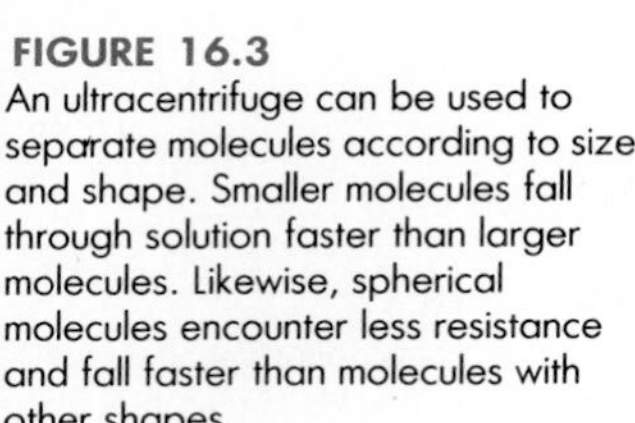

FIGURE 16.3
An ultracentrifuge can be used to separate molecules according to size and shape. Smaller molecules fall through solution faster than larger molecules. Likewise, spherical molecules encounter less resistance and fall faster than molecules with other shapes.

Scientists separate polymers using ultracentrifuges. In ultracentrifuges, samples are spun thousands of times per minute. Spinning provides forces thousands of times greater than that of gravity and permits the separation of molecules based on differences in size, shape, and density.

Population growth created increased demands for natural products that could not be met. The first synthetic polymers, prepared by modifying naturally occurring cellulose fibers, were used as substitutes for ivory and silk. (The structures of these polymers, nitrocellulose and cellulose acetate, are discussed in Chapter 17.) With the spread of automobiles, the petroleum industry grew to be capable of providing low-cost hydrocarbon starting materials for the synthesis of new polymers. Shortages associated with World War II hastened the development of nylon and synthetic rubber.

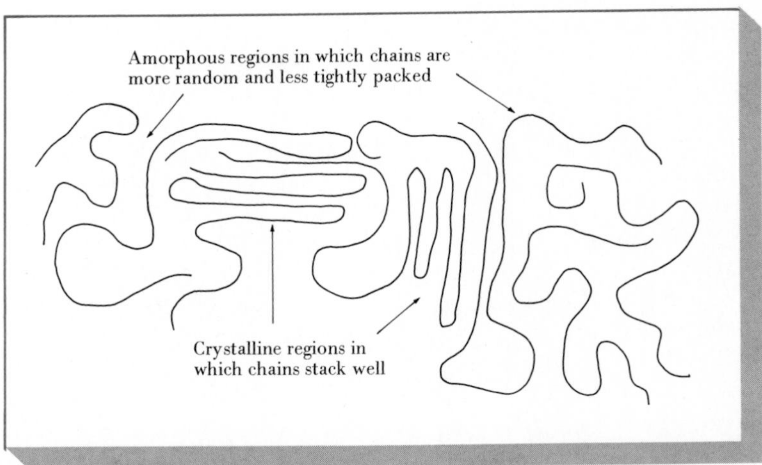
Amorphous regions in which chains are more random and less tightly packed
Crystalline regions in which chains stack well

FIGURE 16.4
The properties of many polymers reflect a structure in which there is a mixture of ordered crystalline regions and more random amorphous regions.

TABLE 16.1

A SAMPLE OF ADDITION POLYMERS

Monomer	Polymer	Uses
$CH_2{=}CH_2$ Ethylene	$(-CH_2CH_2-)_n$ Polyethylene	Low-cost containers for food, toys, films
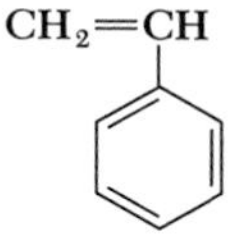 $CH_2{=}CH-C_6H_5$ Styrene	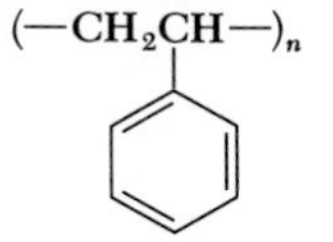 $(-CH_2CH(C_6H_5)-)_n$ Polystyrene	Synthetic rubber, packaging, appliance parts, toys
$CH_2{=}CHCl$ Vinyl chloride	$(-CH_2CHCl-)_n$ Polyvinyl chloride (PVC)	Pipes, floor tiles, adhesives, phonograph records
$CH_2{=}CH-C{\equiv}N$ Acrylonitrile	$(-CH_2CH(C{\equiv}N)-)_n$ Polyacrylonitrile	Textile fiber, rugs
$CH_2{=}CH-O-C({=}O)CH_3$ Vinyl acetate	$(-CH_2CH(O-C({=}O)CH_3)-)_n$ Polyvinyl acetate	Latex paint
$CH_2{=}CH-C({=}O)OCH_3$ Methyl methacrylate	$(-CH_2CH(C({=}O)OCH_3)-)_n$ Polymethyl methacrylate	Glass substitute, safety glass
$CF_2{=}CF_2$ Tetrafluoroethylene	$(-CF_2CF_2-)_n$ Polytetrafluoroethylene	Bearings, nonstick frying pans

THERMOPLASTIC POLYMERS

Many, but not all, synthetic polymers are thermoplastic materials. For example, both polyethylene, used in milk containers, and polystyrene, used in coffee cups, are *thermoplastic* polymers. At elevated temperatures, thermoplastics are viscous liquids that can be molded. Upon cooling, they harden and retain the shape of the mold. Note that glass is thermoplastic and clay is plastic according to this definition. Note also that most so-called plastic objects are not plastic at the temperatures at which they are used.

The properties of many thermoplastic polymers reflect a structure that is a mixture of crystalline and amorphous regions. In crystalline regions, chains stack in an ordered fashion. Crystalline regions contribute greater strength and dimensional stability to a polymer. In amorphous regions, chains do not stack well and the arrangement of adjacent chains is more random. Amorphous regions contribute greater toughness and flexibility to a polymer. Many polymers are translucent or opaque because crystalline regions and amorphous regions differ in density and light is scattered as it passes from region to region.

Because intermolecular forces in polymers are relatively weak, crystalline regions melt at moderate temperatures allowing the polymers to be shaped in molds. On cooling rapidly, some crystalline regions reform, but much of the polymer remains amorphous.

Control of molecular weight for thermoplastic polymers is important. Polymers with very high molecular weights are difficult to fabricate, for they are exceedingly viscous even at elevated temperatures. Polymers with too low molecular weights lack strength and toughness.

ADDITION POLYMERS

The simplest and most widely used synthetic polymer is polyethylene, which is prepared from the *monomer* ethylene or ethene. *(Monomers are the subunits from which polymers are made.)* Polyethylene and its derivatives prepared from modified ethylene monomers are called *addition polymers.* Since polyethylene is prepared from low-cost ethylene, variations on the polyethylene structure need to supply added desirable properties to justify the increased costs of monomer feedstocks.

A brief sampling of addition polymers reveals a great diversity of properties and applications. Polymethyl methacrylate has bulky side groups along the polymer chain. It stacks very poorly and is entirely glassy or amorphous. It is a tough transparent polymer used in paints and safety glass. Polyvinyl chloride also lacks crystallinity. Due to the large chlorine atoms on every other carbon atom, forces between chains are stronger than in most other addition polymers. Polyvinyl chloride is both oil and water resistant; PVC is used widely in pipes, hoses, and phonograph records where strength, flexibility, and toughness are desired. In contrast, the polymer chains in Teflon, the polymer of tetrafluoroethylene, are very straight and stiff. In the extended chains, relatively negative fluorine atoms are at their maximum distance from one another. The surface of Teflon consists largely of long linear chains. This surface is remarkably smooth and friction-free.

FORMATION OF ADDITION POLYMERS

Addition polymers can be produced by free-radical reactions. A small amount of an initiator is mixed with a large quantity of monomer. Initiator molecules have weak bonds that break on heating to give radicals, species with odd numbers of electrons. Radicals induce polymerization by adding rapidly to double bonds. The product of each addition is itself a new radical. Additions of monomers continue, generating long chains. Termination reactions end the polymer growth. Two growing chains may join, or one growing end may abstract a hydrogen from another.

The bulky methyl and ester groups on every other carbon atom give rise to twists and turns along the chain

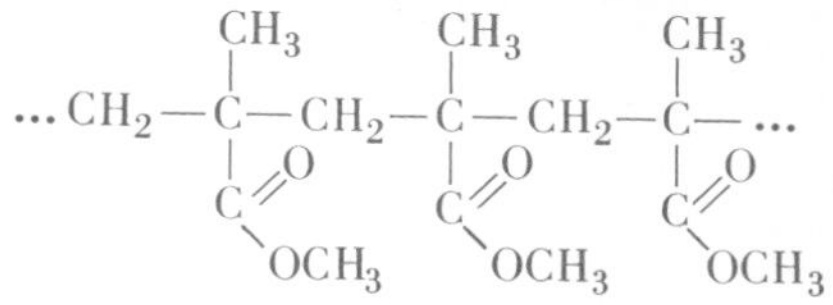

$$\begin{array}{c}
\quad CH_3 \qquad\quad CH_3 \qquad\quad CH_3 \\
\cdots CH_2-\underset{\substack{|\\ C\overset{\diagup\!\!\diagup O}{}\\ \diagdown\\ OCH_3}}{\overset{|}{C}}-CH_2-\underset{\substack{|\\ C\overset{\diagup\!\!\diagup O}{}\\ \diagdown\\ OCH_3}}{\overset{|}{C}}-CH_2-\underset{\substack{|\\ C\overset{\diagup\!\!\diagup O}{}\\ \diagdown\\ OCH_3}}{\overset{|}{C}}-\cdots
\end{array}$$

Polymethyl methacrylate

Chlorine atoms, with more electrons than second row elements, give rise to stronger forces of attraction between adjacent chains than are present in most addition polymers

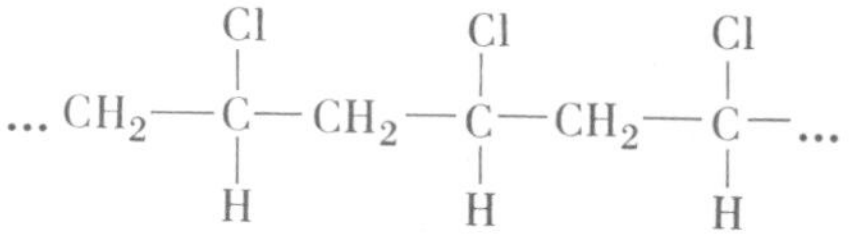

$$\cdots CH_2-\underset{\substack{|\\H}}{\overset{\substack{Cl\\|}}{C}}-CH_2-\underset{\substack{|\\H}}{\overset{\substack{Cl\\|}}{C}}-CH_2-\underset{\substack{|\\H}}{\overset{\substack{Cl\\|}}{C}}-\cdots$$

Polyvinyl chloride (PVC)

Forces of repulsion between partially negative fluorine atoms cause the chains in Teflon to be relatively straight and stiff

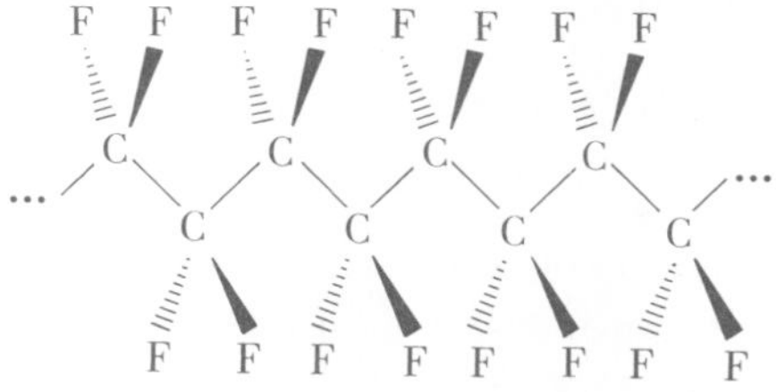

$$\begin{array}{c}
F\ \ F\quad F\ \ F\quad F\ \ F\quad F\ \ F \\
C \qquad C \qquad C \qquad C \\
\cdots \diagup \quad \diagdown \diagup \quad \diagdown \diagup \quad \diagdown \diagup \quad \diagdown \diagup \cdots \\
C \qquad C \qquad C \qquad C \\
F\ \ F\quad F\ \ F\quad F\ \ F\quad F\ \ F
\end{array}$$

Polytetrafluoroethylene (Teflon)

FIGURE 16.5 The properties of addition polymers depend in part on the substituents along the chain and on how these substituents affect the interaction of neighboring chains.

Note that the ends of a polymer chain may differ from the bulk of the polymer. End-group reactivity can present special problems to the polymer chemist just as end-group twists or catches may present special problems in opening a zipper or keeping it closed.

A knowledge of the mechanism of polymer formation enables scientists to better control the properties of polymers. For example, a lower range of molecular weights results from the use of a higher concentration of initiator because more numerous and shorter chains are formed.

PLASTICIZERS

Additives called *plasticizers* are included in many polymer formulations. Plasticizers are added to increase softness and pliability of some plastics and to increase the useful lifetime of others. For example, esters of phthalic acid, a diacid, with branched-chain alcohols are used as plasticizers. Because compounds used as plasticizers do not stack well and are not very

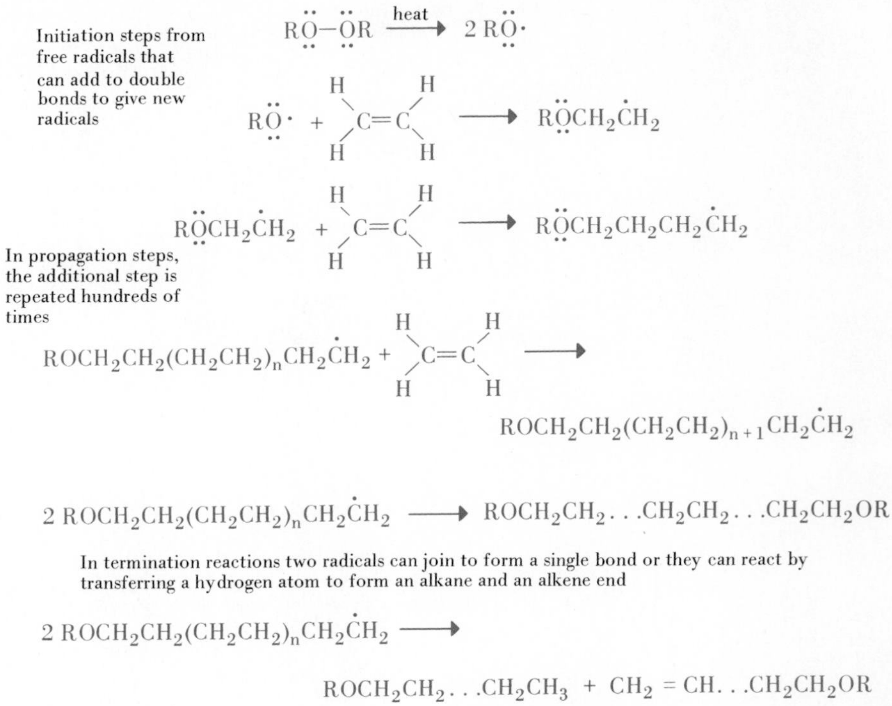

Initiation steps from free radicals that can add to double bonds to give new radicals

$$R\ddot{O}-\ddot{O}R \xrightarrow{\text{heat}} 2\,R\ddot{O}\cdot$$

$$R\ddot{O}\cdot + H_2C{=}CH_2 \longrightarrow R\ddot{O}CH_2\dot{C}H_2$$

In propagation steps, the additional step is repeated hundreds of times

$$R\ddot{O}CH_2\dot{C}H_2 + H_2C{=}CH_2 \longrightarrow R\ddot{O}CH_2CH_2CH_2\dot{C}H_2$$

$$ROCH_2CH_2(CH_2CH_2)_nCH_2\dot{C}H_2 + H_2C{=}CH_2 \longrightarrow ROCH_2CH_2(CH_2CH_2)_{n+1}CH_2\dot{C}H_2$$

$$2\,ROCH_2CH_2(CH_2CH_2)_nCH_2\dot{C}H_2 \longrightarrow ROCH_2CH_2...CH_2CH_2...CH_2CH_2OR$$

In termination reactions two radicals can join to form a single bond or they can react by transferring a hydrogen atom to form an alkane and an alkene end

$$2\,ROCH_2CH_2(CH_2CH_2)_nCH_2\dot{C}H_2 \longrightarrow ROCH_2CH_2...CH_2CH_3 + CH_2=CH...CH_2CH_2OR$$

FIGURE 16.6
Addition polymers can be formed by a free-radical mechanism. First, free radicals are generated to initiate chain growth. Then, growing radical chains repeatedly add to double bonds, forming new radicals in each addition step. Finally, two radicals can react to terminate the growth process.

volatile, they dissolve in the polymer to enable it to remain amorphous over time. It is the escape of plasticizers which causes plastic dashboards and seat covers to become brittle and crack. (The haze on the windshield of a car parked in the sun on a hot summer day may be from the evaporation of plasticizers from the dashboard.)

FIGURE 16.7
Diesters of phthalic acid and branched-chain alcohols are used as plasticizers to increase the softness and flexibility of polymers such as PVC.

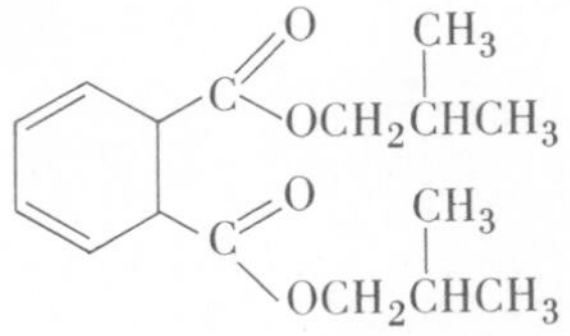

$$C_6H_4[C(=O)OCH_2CH(CH_3)CH_3]_2$$

Diisobutyl phthalate

A S I D E

NEW CATALYSTS AND IMPROVED PROPERTIES FOR ADDITION POLYMERS

The free-radical process for forming addition polymers has certain drawbacks. First, monomers such as propylene cannot be used due to a side reaction. A growing radical chain can abstract a hydrogen atom from a propylene molecule to terminate one chain and form a new radical. The resonance-stabilized radical formed by removing a hydrogen atom from propylene is too unreactive to undergo chain growth. Second, a growing chain itself can undergo hydrogen-abstraction reactions. The softening temperature of polyethylene prepared by a free-radical process is lowered by the presence of short side chains resulting from hydrogen-abstraction reactions followed by growth from new radical sites.

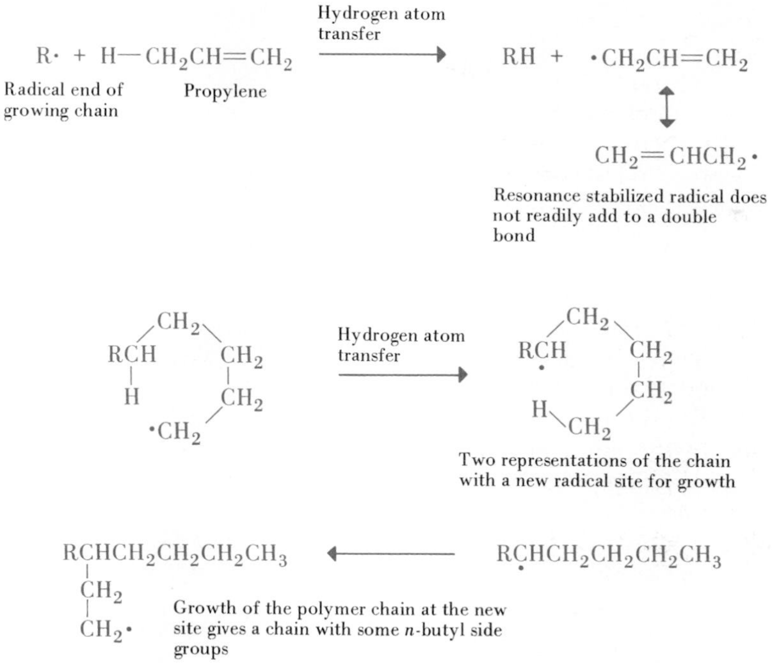

$$R\cdot + H{-}CH_2CH{=}CH_2 \xrightarrow{\text{Hydrogen atom transfer}} RH + \cdot CH_2CH{=}CH_2 \leftrightarrow CH_2{=}CHCH_2\cdot$$

Radical end of growing chain

Propylene

Resonance stabilized radical does not readily add to a double bond

Hydrogen atom transfer

Two representations of the chain with a new radical site for growth

$$RCHCH_2CH_2CH_2CH_3 \text{ (with } CH_2CH_2\cdot \text{ on C)} \longleftarrow R\dot{C}HCH_2CH_2CH_2CH_3$$

Growth of the polymer chain at the new site gives a chain with some *n*-butyl side groups

FIGURE 16.8
There are problems with free-radical polymerizations. Polypropylene cannot be prepared since propylene can donate a hydrogen atom to form an unreactive radical, which terminates the chain. Polyethylene prepared by a free-radical process has side chains that cause the polymer to be too low melting for uses that involve sterilization temperatures.

The use of Lewis acid catalysts has permitted the development of a new generation of addition polymers designed to meet applications requiring more demanding specifications. Linear polyethylene made using these catalysts is stronger and has a higher density than the branched polymer made by free-radical processes. Because it can be heated to 100°C without softening, it can be sterilized. It is used as a low-cost material for food packaging. Propylene can be polymerized; polypropylene has

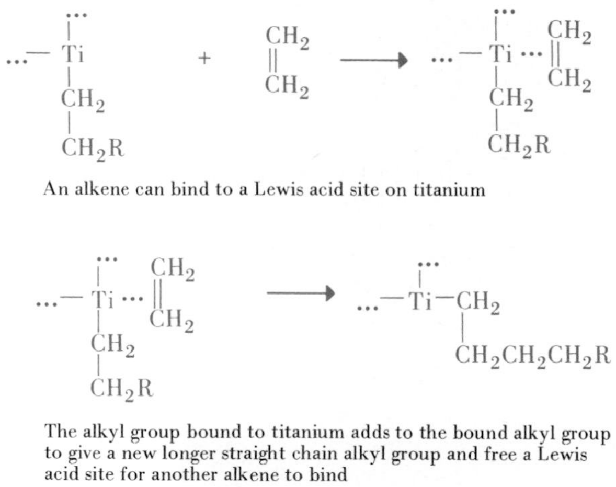

$$\cdots{-}Ti(CH_2CH_2R) + CH_2{=}CH_2 \longrightarrow \cdots{-}Ti(CH_2CH_2R)\cdots(CH_2{=}CH_2)$$

An alkene can bind to a Lewis acid site on titanium

$$\cdots{-}Ti(CH_2CH_2R)\cdots(CH_2{=}CH_2) \longrightarrow \cdots{-}Ti{-}CH_2CH_2CH_2CH_2R$$

The alkyl group bound to titanium adds to the bound alkyl group to give a new longer straight chain alkyl group and free a Lewis acid site for another alkene to bind

FIGURE 16.9
New Lewis acid catalysts are made to make high-density polyethylene, polypropylene, and other addition polymers with improved properties.

found application in outdoor carpeting and as a mildew- and stain-resistant furniture fabric.

The use of Lewis acid catalysts permits much greater control of polymer geometry because both monomers and the growing chain are bound to the same metal atom. As each monomer is inserted into the chain, a new position on the metal becomes available to first coordinate and then deliver a monomer unit to the growing chain.

RUBBER, AN ELASTIC POLYMER

Natural rubber was a laboratory curiosity. When the latex from the rubber tree was pressed and dried, the resulting sticky polymer would both bounce and yield. If struck, it behaved elastically. When pressed over a longer period of time, it deformed permanently.

The discovery of vulcanization of rubber by Charles Goodyear in 1839 led to the applications we know today. Vulcanization is the process of cross-linking polymer chains in rubber. Green rubber is mixed with a small amount of sulfur, shaped into a product, and then heated. For example, a tire is constructed from casing and tread, and then it undergoes precise shaping in a mold which impresses a tread design. Vulcanization occurs on heating.

Rubber is poly-*cis*-isoprene, an unsaturated hydrocarbon polymer. Sulfur reacts at a few double-bond sites to form cross-links or interchain bonds. The more highly cross-linked the rubber, the harder it is. Cross-links are far closer together in rubber used for tire treads than in rubber used in household gloves.

The polymer chains in rubber are extremely flexible. Weak forces between chain segments and the cis geometry of the double bond result in chains that turn and twist but do not form ordered crystalline regions. When stretched, rubber becomes more ordered. Chains are extended in the direction of stretching, but, because of the cross-links, they cannot slide past one another. When the stretched rubber is released, it retracts. Random motions along the polymer backbone cause chains to shorten and cross-links to draw closer together.

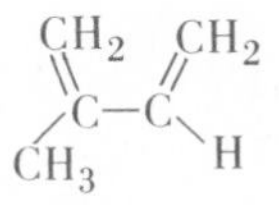

$$\begin{array}{c} CH_2 \quad CH_2 \\ \| \quad\ \ \ // \\ C{-}C \\ / \qquad \ \backslash \\ CH_3 \qquad\quad H \end{array}$$

Isoprene
2-methyl-1,3-butadiene

Isoprene is formed when rubber is decomposed by heating.

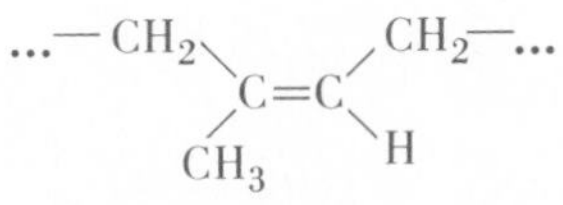

$$\cdots{-}CH_2\diagdown \qquad \diagup CH_2{-}\cdots \\ C{=}C \\ \diagup \qquad \diagdown \\ CH_3 \qquad\qquad H$$

Rubber is *poly-cis*-isoprene

FIGURE 16.10
Rubber is a hydrocarbon polymer with a 5-carbon alkene as the repeating unit.

$$\begin{array}{c} \ldots-CH_2(CH_3)C{=}CH-CH_2-\ldots \\ \ldots-CH_2(CH_3)C{=}CH-CH_2-\ldots \end{array} + S \xrightarrow{\text{heat}} \begin{array}{c} \ldots-CH_2-C(CH_3){=}C(H)-CH_2-\ldots \\ | \\ S \\ | \\ S \\ | \\ \ldots-CH_2-C(CH_3){=}C(H)-CH_2-\ldots \end{array}$$

FIGURE 16.11
In the vulcanization process, green rubber is heated with sulfur. In reactions that involve the double bonds of rubber, some —SS— bridges form to cross-link the chains.

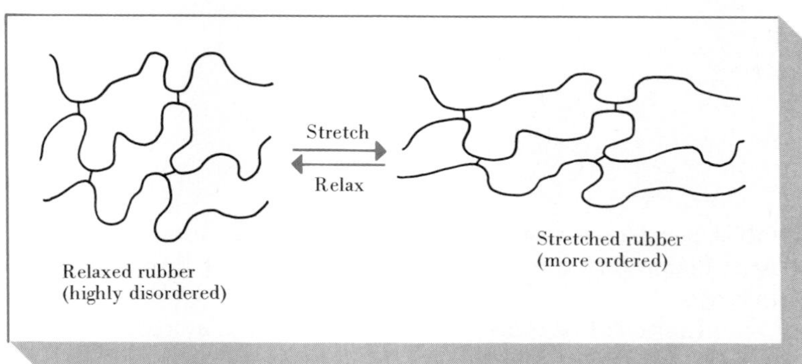

FIGURE 16.12
When rubber is stretched, the chain segments between cross-links become straighter and more ordered. When the tension is removed, the chains return to a more random orientation.

[Since increased randomness (higher entropy) is favored more at higher temperatures, the restoring forces in rubber increase with increasing temperature. A hot golf ball can be driven farther than a cold golf ball. If rubber objects are cooled sufficiently, they become leathery. The rate of relaxation of extended chains slows and the rubber loses its elasticity.]

Ozone, a component of air pollution, attacks rubber. Ozone adds to the double bonds of rubber, resulting in their cleavage. Increased ozone con-

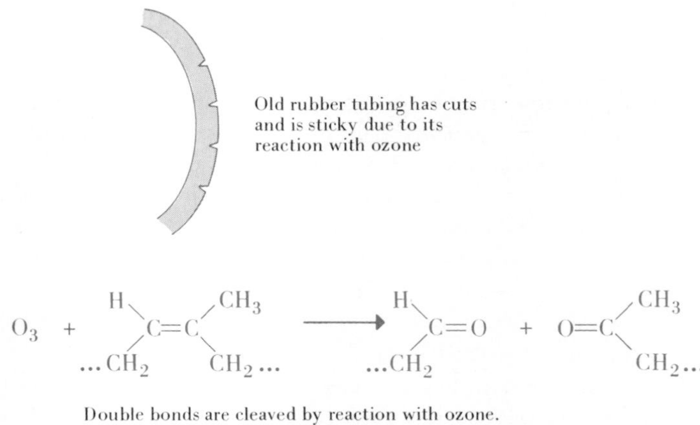

Old rubber tubing has cuts
and is sticky due to its
reaction with ozone
$O_3 + \ldots CH_2(H)C{=}C(CH_3)CH_2\ldots \longrightarrow \ldots CH_2(H)C{=}O + O{=}C(CH_3)CH_2\ldots$
Double bonds are cleaved by reaction with ozone.

FIGURE 16.13
Ozone, a component of air pollution, shortens the useful lifetime of rubber.

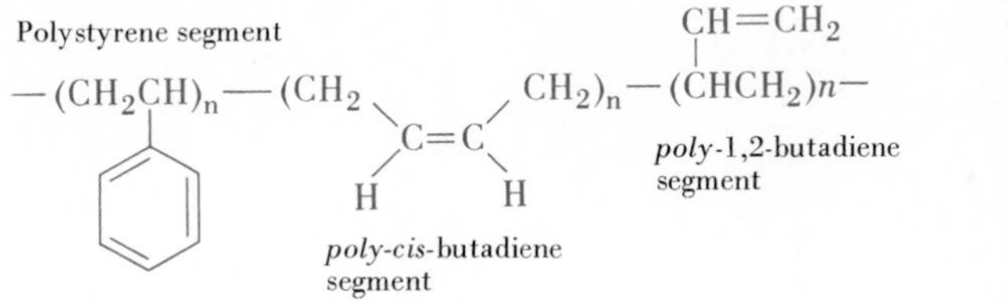

Polystyrene segment

$$-(CH_2CH)_n-(CH_2-CH=CH-CH_2)_n-(CHCH_2)_n-$$

(phenyl group on the polystyrene CH; H and H on the C=C; $CH{=}CH_2$ on the CHCH$_2$)

poly-cis-butadiene segment

poly-1,2-butadiene segment

Polymers of styrene and butadiene were used as synthetic rubber. These early synthetic rubbers did not have a uniform structure and did not perform as well as did natural rubber or later synthetics.

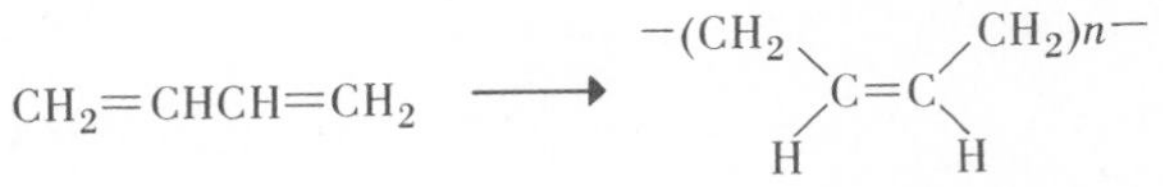

FIGURE 16.14
Synthetic rubbers are made from 1,3-butadiene and styrene and from 1,3-butadiene, alone. By using Lewis acid catalysts, a high-quality synthetic polymer with the same cis geometry as natural rubber can be prepared.

$$CH_2{=}CHCH{=}CH_2 \longrightarrow -(CH_2-CH{=}CH-CH_2)_n-$$

poly-cis-butadiene

centrations are associated with photochemical smog. An indirect economic cost of air pollution is a decreased useful life for hoses, windshield wipers, and tires.

Synthetic rubbers are used more widely than natural rubber. The shift has been dramatic. Before World War II, natural rubber accounted for more than 95 percent of all use. Today, synthetics comprise 80 percent of rubber in use. An early synthetic rubber was a copolymer of styrene and butadiene. However, early synthetic rubbers tended to build up more heat when flexing than did natural rubbers; excessive heat destroys cross-links and weakens rubber. Natural rubber remained a preferred product for demanding uses such as tire casings, while synthetic rubbers were used in treads.

With the introduction of Lewis acid catalysts in 1955, improved stereochemical control of polymerization became possible. Since then, poly-*cis*-butadiene and poly-*cis*-isoprene have been synthesized. These polymers are superior to earlier synthetic rubbers and are widely used.

CONDENSATION POLYMERS

Polyamide and polyester polymers (discussed below) are sometimes referred to as *condensation polymers.* Recall that an amide group can be prepared by a condensation reaction in which a carboxylic acid combines with an amine to split out a small molecule (H_2O), and that an ester can be prepared by the condensation reaction of a carboxylic acid with an alcohol. By using monomers with a functional group at each end, condensation reactions can be used to prepare long, chainlike molecules having many functional groups incorporated into the polymer backbone.

Nylons constitute a family of synthetic polyamides. (Proteins, naturally occurring polyamides, are discussed in Chapter 17.) Nylons may be prepared by condensation reactions of diamines with dicarboxylic acids as

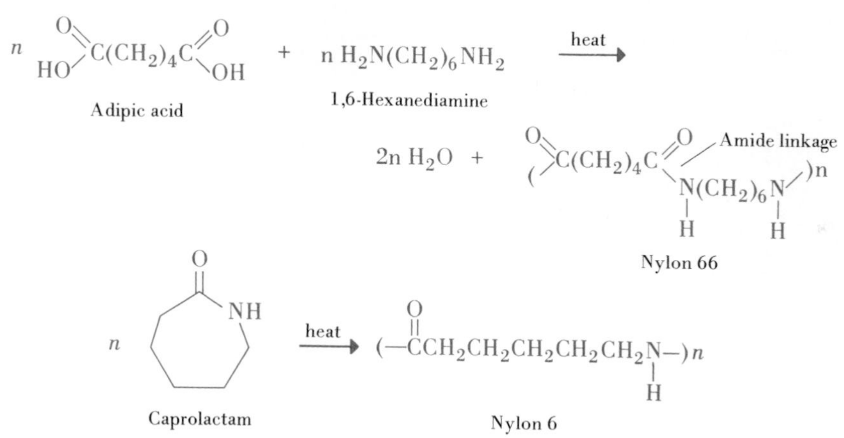

$$n\ \underset{\text{Adipic acid}}{HO-\overset{O}{\overset{\|}{C}}(CH_2)_4\overset{O}{\overset{\|}{C}}-OH} + \underset{\text{1,6-Hexanediamine}}{n\ H_2N(CH_2)_6NH_2} \xrightarrow{\text{heat}}$$

$$2n\ H_2O + \underset{\text{Nylon 66}}{(\overset{O}{\overset{\|}{C}}(CH_2)_4\overset{O}{\overset{\|}{C}}-\underset{H}{\underset{|}{N}}(CH_2)_6\underset{H}{\underset{|}{N}})_n}$$

(Amide linkage: the $-\overset{O}{\overset{\|}{C}}-\underset{H}{\underset{|}{N}}-$ bond)

$$n\ \underset{\text{Caprolactam}}{\text{(cyclic } \overset{O}{\overset{\|}{C}}(CH_2)_5NH\text{)}} \xrightarrow{\text{heat}} \underset{\text{Nylon 6}}{(-\overset{O}{\overset{\|}{C}}CH_2CH_2CH_2CH_2CH_2\underset{H}{\underset{|}{N}}-)_n}$$

FIGURE 16.15
Synthetic polyamide polymers are called nylons. Nylons may be prepared from a diacid and a diamine as shown for nylon 66, or they may be prepared from cyclic amides as shown for nylon 6.

illustrated for the reaction of nylon 66. The structures of two nylons, nylon 6 and nylon 66, are shown.

One important use of nylon polymers is in the production of textiles. Synthetic textile fibers are oriented, highly crystalline polymers with molecular weights near 10,000. The melted polymer is extruded through small holes and then immediately stretched. Polymer chains become oriented along the axis of the fiber. Cross-sectional strength arises from the covalent bonds of many parallel chains. Longitudinal strength, resistance to rupture upon stretching, is dependent on strong interchain forces.

The hydrocarbon portions of the nylon chain contribute flexibility to the polymer; the amide groups provide strong interchain hydrogen bonds. By varying the length of the hydrocarbon segments between amide groups, polymer scientists prepare nylons with different properties. Shorter chains result in a harder, more crystalline, rigid product. For example, nylon used in a sweater may be 15–35% crystalline, but that used in fishing line is more than 90% crystalline.

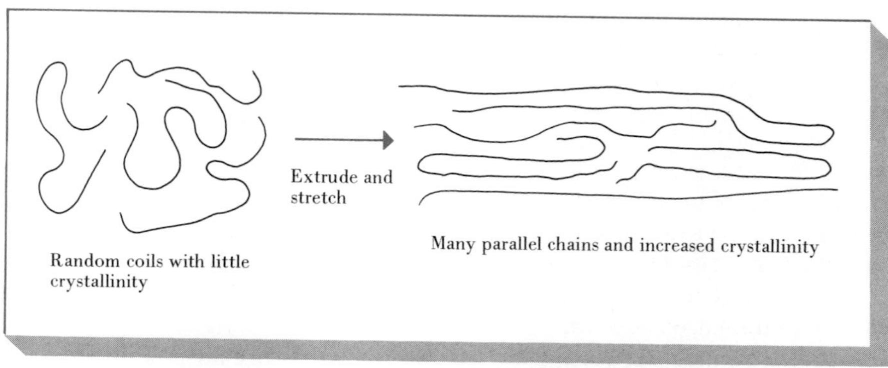

FIGURE 16.16
Textile fibers are prepared by squeezing the molten polymer through small dies and then quickly stretching and cooling the emerging thread. Because of the stretching, many of the polymer strands are aligned parallel to the length of the fiber. A textile fiber has greater crystallinity and strength than the bulk polymer with its more random orientation of chains.

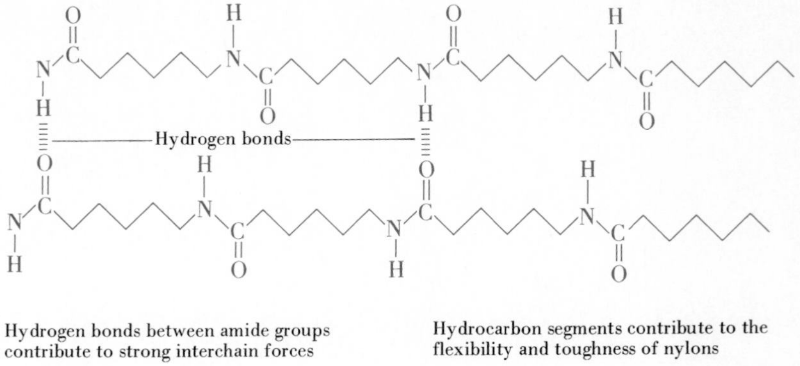

Hydrogen bonds

Hydrogen bonds between amide groups contribute to strong interchain forces

Hydrocarbon segments contribute to the flexibility and toughness of nylons

FIGURE 16.17
The combination of strong interchain hydrogen bonds and flexible hydrocarbon segments gives nylon 6 strength and flexibility for use in textile fibers.

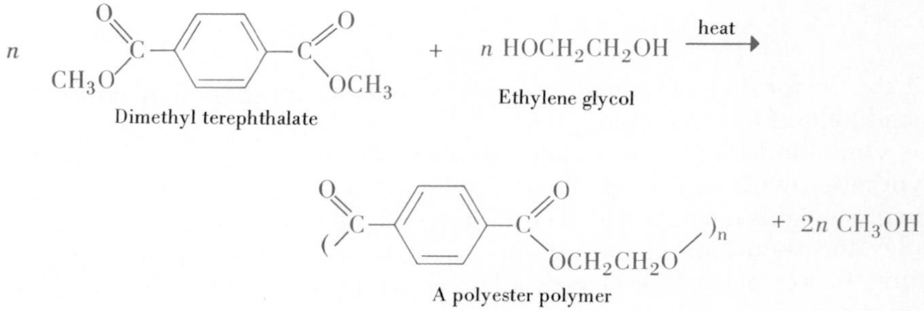

n Dimethyl terephthalate + $n\ HOCH_2CH_2OH$ (Ethylene glycol) $\xrightarrow{\text{heat}}$ A polyester polymer $+\ 2n\ CH_3OH$

FIGURE 16.18
Polyester textile fibers are prepared by reacting the diester of an aromatic dicarboxylic acid with a short-chain diol.

Polyester fibers are made by the condensation polymerization of a diester, dimethyl terephthalate, and a diol, ethylene glycol. Interchain forces in polyesters are potentially weaker than those in polyamides. Although the ester groups are polar, there are no hydrogen bonds present. To provide satisfactory performance in a fiber, the hydrocarbon portions of the chains between functional groups are shorter and more rigid than in nylons. A 2-carbon diol is used, and the diacid contains a rigid, planar benzene ring. Chains stack better in a polyester than in a polyamide.

In the early 1980s, aromatic amide polymers were introduced in tire reinforcing cords and in other applications requiring high-strength, lightweight polymers. Both the rigidity of aromatic rings and the attraction of hydrogen bonds contribute to high strength of aromatic amide polymers.

THERMOSETTING POLYMERS

Most polymers used for high-temperature applications or for applications requiring high strength and durability are thermosetting resins. *Thermosetting resins are rigid, highly cross-linked, three-dimensional arrays that cannot be remelted and reshaped.* The final cross-linking steps occur in the mold or in a final heat treatment.

The structures of two widely used thermosetting resins are illustrated. Phenol formaldehyde resins, known as Bakelite resins, were among the

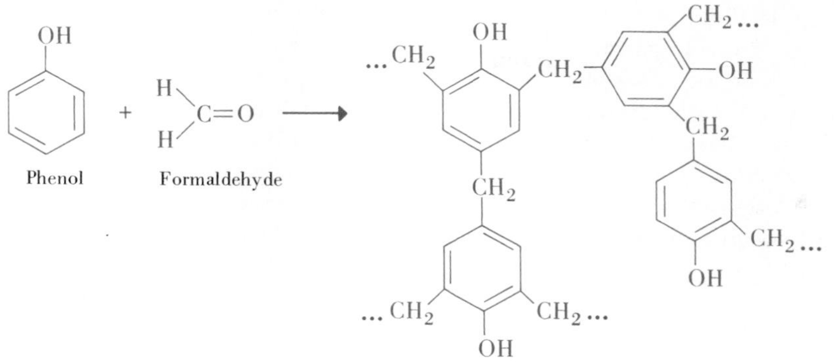

A phenol-formaldehyde polymer (Bakelite)

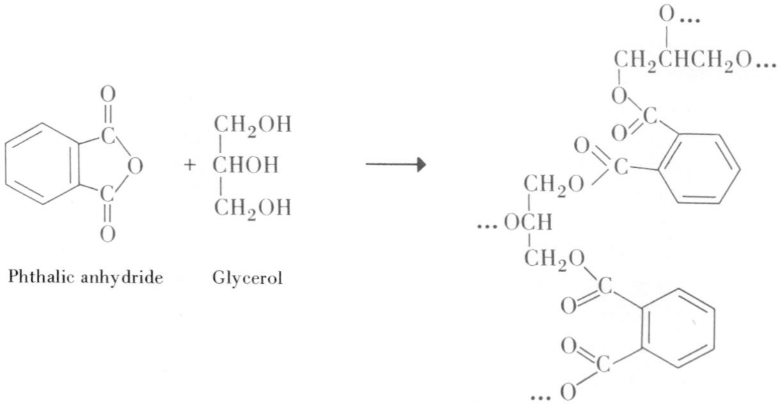

A 3-dimensional polyester

FIGURE 16.19
Thermosetting polymers have three-dimensional structures. Once a thermosetting resin is fully formed, it cannot be melted and reshaped. Phenol formaldehyde polymers and polyesters made using the triol glycerol are among the most widely used thermosetting polymers.

most widely used thermosetting resins. Polyesters, similar to but different from those used in textile fibers, are another important group of thermosetting resins. By substituting the triol glycerol for ethylene glycol, the diol used in textile fibers, a three-dimensional polyester can be prepared. A polyester prepared from glycerol and phthalic acid is used in baked-enamel finishes of appliances.

ASIDE

NEW COMPOSITE MATERIALS MEET DEMANDING USES

For uses ranging from fishing rods to sports cars, the use of polymeric materials is growing rapidly. Materials are combined to give a *composite* that has properties superior to either material alone. For example, new fishing rods are light, strong, and supple. Fibers of a borosilicate glass (low thermal expansion) or graphite run the length of each rod segment, and a three-dimensional polymer matrix binds them together.

Because composite materials can be molded into complex shapes and because they are highly resistant to corrosion, they are finding new uses in automotive manufacture. The stylish front end of a sports car may be a composite while the straighter side panels are sheet metal. In contrast to the example of the fishing rod, the reinforcing fibers may be woven to give strength in two dimensions. Laminated layers also give increased strength. Thermosetting polyesters are the most widely used matrix materials.

POLYMERS FOR ION EXCHANGE

Synthetic ion exchange resins are used in water softening and purification systems as well as in chemical separation applications. Ion-exchange resins can be considered to be modified polystyrene resins. By including a very small amount of divinylbenzene with the styrene monomer, a cross-linked polymer is formed.

The benzene rings in the polymer are chemically modified to introduce charged substituents and to convert the polymer into either a giant anion or a giant cation. Beads of ion-exchange polymers swell with water. If the polymer is anionic, then the water contains dissolved cations attracted to the charges on the polymer. If the polymer is a giant cation, there are bound anions in solution. These dissolved ions cannot wash out due to the strong attraction of opposite charges, but they can be exchanged for other cations.

Hard water contains dissolved calcium or magnesium compounds. When hard water is heated, it leaves deposits that can narrow or close pipes and cause boilers used to produce steam to lose efficiency. (Some calcium and magnesium compounds are less soluble in hot water than in cold.) When hard water containing dissolved calcium or magnesium salts passes through a cation-exchange resin loaded with sodium ions, calcium and magnesium ions are retained and sodium ions elute. The resulting soft water is suitable for use in boilers or for washing but, because of its high sodium content, it is not desirable for drinking.

Ion-free or deionized water can be prepared by using first a cation-exchange resin and then an anion-exchange resin. Mineral cations such as Na^+

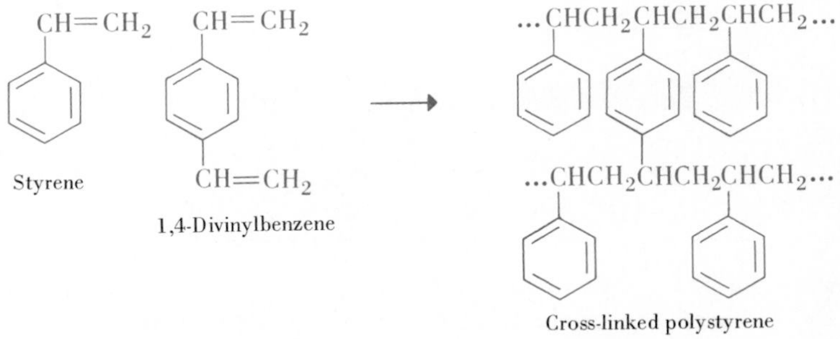

$CH{=}CH_2$ (Styrene) + $CH{=}CH_2$ / $CH{=}CH_2$ (1,4-Divinylbenzene) $\longrightarrow$ $\ldots CHCH_2CHCH_2CHCH_2 \ldots$ / $\ldots CHCH_2CHCH_2CHCH_2 \ldots$ (Cross-linked polystyrene)

Styrene

1,4-Divinylbenzene

Cross-linked polystyrene

FIGURE 16.20 Three-dimensional, cross-linked polystyrene can be prepared by the polymerization of styrene to which a small amount of divinyl benzene has been added. Each of the alkene groups of divinylbenzene becomes part of a different polystyrene chain. To produce a polymer with closer cross-links, more divinyl benzene is added.

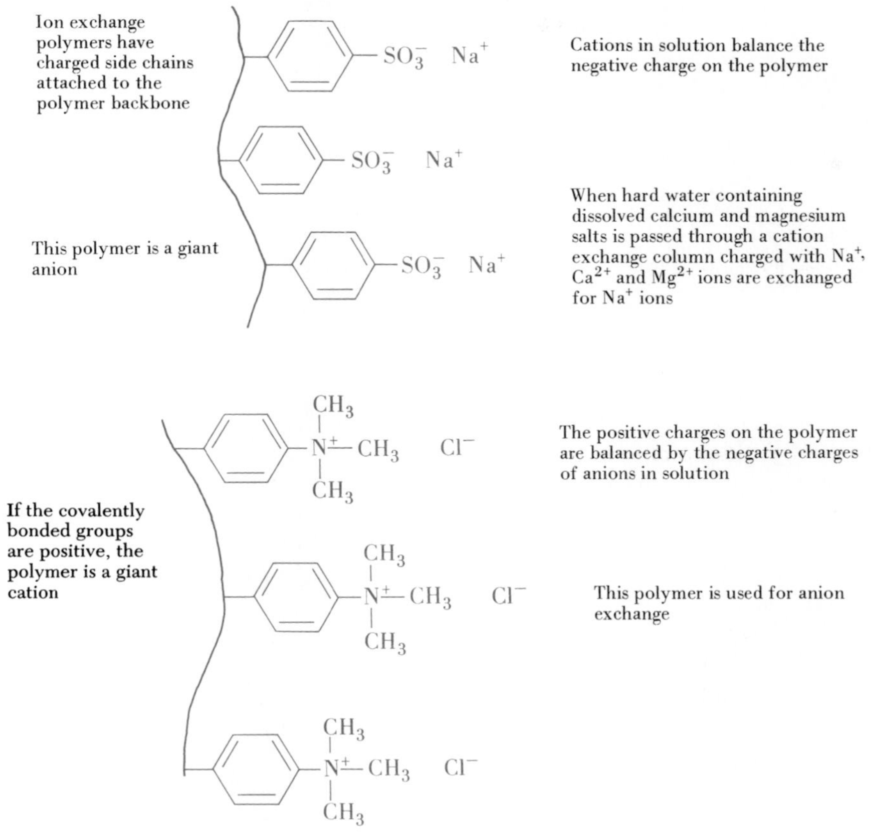

Ion exchange polymers have charged side chains attached to the polymer backbone

SO_3^- Na^+

Cations in solution balance the negative charge on the polymer

SO_3^- Na^+

When hard water containing dissolved calcium and magnesium salts is passed through a cation exchange column charged with Na^+, Ca^{2+} and Mg^{2+} ions are exchanged for Na^+ ions

This polymer is a giant anion

SO_3^- Na^+

CH_3 N^+ CH_3 CH_3 Cl^-

The positive charges on the polymer are balanced by the negative charges of anions in solution

If the covalently bonded groups are positive, the polymer is a giant cation

CH_3 N^+ CH_3 CH_3 Cl^-

This polymer is used for anion exchange

CH_3 N^+ CH_3 CH_3 Cl^-

FIGURE 16.21
Chemists chemically modify polymers to produce ion-exchange resins for use in chemical separations.

and Ca^{2+} are exchanged for hydronium ions, and mineral anions such as Cl^- and SO_4^{2-} are exchanged for hydroxide ions. Hydronium ions and hydroxide ions combine to form water.

A S I D E

SYNTHETIC POLYMERS TO SEPARATE ENZYMES

A final illustration of polymer design features polymers used in separations of proteins. Proteins may be sorted according to electric charge or according to size. Both methods are used in many enzyme preparations. These separations are done using polymer matrices, either beads packed in a column or porous gels. If a protein is an enzyme or a hormone, a suitable assay can be used to monitor separations and purification.

Protein separations based on charge can be done by ion exchange. If a mixture of proteins dissolved in an acid buffer is put onto a cation-exchange column, those

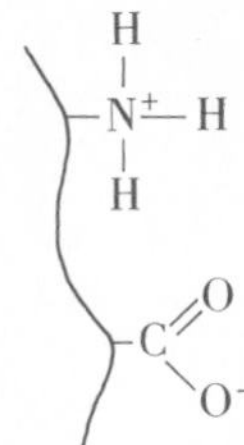

At a low pH, most proteins have a positive charge
As some acidic groups donate protons, the net charge on a protein decreases
At still higher pHs, proteins may have a net negative charge

FIGURE 16.22
Most proteins have acidic and basic groups. The net charge on a protein depends on the pH and on the numbers and strengths of these acids and bases.

proteins having a positive charge are retained while neutral or negatively charged proteins elute with the solvent. When the pH is raised, acidic groups on proteins donate protons, making the charge on bound proteins less positive. Hence, the proteins are less tightly bound to the column. By using a series of buffers with increasing pHs, one can separate proteins differing in numbers and strengths of acidic groups.

Frequently, the proteins eluting together from an ion-exchange column are concentrated and then separated according to size using a second column packed with synthetic cross-linked polyacrylamide beads. Depending on the relative amount of cross-linking reagent used, the relative size of the holes in the polymer network may be small, medium, or large.

If the pore size of the beads is medium, then small molecules and small proteins pass through the beads. (A fish net designed to catch tuna does not catch minnows.) Because very large proteins do not enter the beads, their movement is restricted to channels between polymer beads and they are eluted from the column much sooner than small molecules that pass through the beads. Because intermediate-sized proteins pass through some pores but not through others, these proteins can be separated by size with the larger proteins eluted earlier.

Scientists use molecular exclusion polymers (polymers with a range of pore sizes) to separate proteins with different molecular weights. Proteins of known molecular weight are used to calibrate these resins, and the molecular weight or size of an unknown protein can be estimated from its position of elution. At this point we have come full circle. At the beginning of the chapter we emphasized that the inability to determine the sizes of large molecules delayed the study of polymers. We close this chapter by noting that molecular exclusion resins are now synthesized to separate polymers on the basis of molecular size.

QUESTIONS

1. In an early approach to determine the molecular weight of hemoglobin, a weighed sample of the protein was burned and the ash was analyzed for iron. Hemoglobin contains 0.34% iron by weight. If there is one iron per hemoglobin molecule, what would be the molecular

weight of the hemoglobin? The molecular weight of hemoglobin is 64,500. How many iron atoms are present in each hemoglobin?

2. The monomer $CH_2{=}CHCN$ is used to make Orlon, a polymer used in textile fibers. Draw the structure of a segment of this polymer.
3. The polymer $—(CH_2CCl_2)_n—$ is used to manufacture films for wrapping food. Draw a Lewis structure of the monomer used to make this addition polymer.
4. Polyethylene glycol, $—(CH_2CH_2O)_n—$, is soluble in water. Why is this polymer water soluble while polyethylene is not?
5. Which of the following properties of polymers is most dependent on the presence of both crystalline and amorphous regions in a polymer?
   (a) Dimensional stability
   (b) Flexibility
   (c) Toughness
6. Common foam plastic coffee cups are made of polystyrene. How do the benzene rings contribute to the high-temperature strength of polystyrene compared to polyethylene?
7. Why is linear or high-density polyethylene stronger than low-density polyethylene that has alkyl side groups?
8. In the free-radical polymerization of ethylene, would an increase in the concentration of initiator used result in a lower-molecular-weight or a higher-molecular-weight polymer? Explain briefly.
9. A compound with the structure $C_4H_9N{=}NC_4H_9$ is used to initiate the formation of some addition polymers. When heated, this compound decomposes to give N_2 and other products. How does it act to initiate the polymerization of ethylene?
10. The number of polymer chains formed using the initiator described in problem 9 is less than predicted from the consumption of initiator. In addition to polymers, some product having the formula C_8H_{18} is also formed. What is the mechanism of formation of this by-product?
11. How would the addition of the compound di-*n*-octyl phthalate to polyvinyl chloride affect the properties of the polymer?

$$\text{C}_6\text{H}_4\left(\overset{\displaystyle O}{\overset{\|}{C}}—OCH_2CH_2CH_2CH_2CH_2CH_2CH_2CH_3\right)_2$$

12. Rubber swells when in contact with hydrocarbon solvents. Why? Why does it not swell with water?
13. A highly cross-linked rubber is (less, more) rigid and is (easier, harder) to stretch than a moderately cross-linked rubber.
14. Is a synthetic rubber made from butadiene and styrene subject to attack by ozone? Why or why not?

15. How does the arrangement of polymer chains of nylon differ in a textile fiber and in a molded object?
16. Nylons made using diacids with longer hydrocarbon sequences between acid groups are (less, more) water resistant and (less, more) flexible than those made using shorter-chain diacids. Which nylon formulation would be more suitable for the bristles of a toothbrush? For the handle?
17. Draw the structure of a polymer prepared by reacting $NH_2CH_2CH_2NH_2$ with a 4-carbon diacid.
18. Chemists developed nylon as a substitute for silk and wool. What structure do you expect to be present in these protein fibers based on analogy to the structure of nylon?
19. Why is each of the following compounds unsuitable for use in making polyester for a textile fiber?
    (a) $HOCH_2(CH_2)_8CH_2OH$
    (b) Glycerol
    (c) Phthalic acid
20. When the triol glycerol replaces the diol ethylene glycol in making a polyester, the resulting polymer can no longer be melted. Why?
21. A cross-linked polystyrene polymer having $—CH_2N(CH_3)_3^+$ side groups can be used to separate (anions, cations) by ion exchange.
22. Why are Ca^{2+} and Mg^{2+} more tightly bound to cation-exchange resins than is Na^+?

CHAPTER

SEVENTEEN

BIOPOLYMERS

Living organisms have attained great variety while using only a relatively small number of chemical building blocks. Simple molecules combine in a variety of ways to form biopolymers. Cellulose that gives strength to wood, the enzymes that catalyze metabolic reactions, and the genetic material that governs inheritance are polymers formed from a small number of monomer units.

The sequence of monomer units is critical to the function of many biopolymers. Scientists have learned to read sequences both in proteins and in the nucleic acids that carry genetic blueprints for proteins. By studying these sequences, scientists are learning the molecular basis for genetic inheritance and for genetic diseases. This knowledge is leading to improvements in diagnosis and, in some cases, to new or improved therapies.

In this chapter, we will sample the variety of molecules found in living organisms and see how they combine to produce chemical and biological diversity.

SUGARS AND CARBOHYDRATES

Carbohydrates are sugars and their derivatives. Before the structures of sugars were known, their formulas [for example, $C_{12}H_{22}O_{11}$ or $C_{12}(H_2O)_{11}$] suggested the name "carbohydrate" or "hydrate of carbon."

Sugars are polyhydroxy aldehydes and ketones. Glucose is the most widely occurring simple sugar. It is a 6-carbon sugar having an aldehyde group. An isomeric sugar, fructose, has a ketone group. Because sugars have many OH groups, they are extremely soluble in water and may be difficult to recrystallize from the resulting syrups.

The carbon-oxygen double bond of a sugar can react with an —OH group to form a five- or six-membered ring. In solution, glucose is a mixture of two cyclic structures, α- and β-glucose, that interconvert through an acyclic form.

Dimers, trimers, and polymers are formed by further reaction at the oxygen atom of the aldehyde or ketone group. For example, sucrose or cane

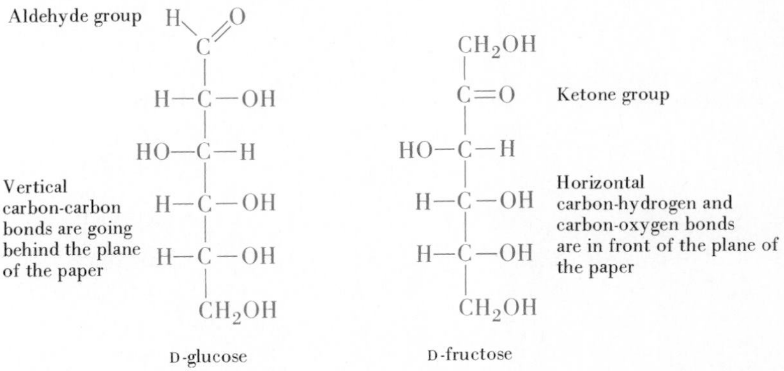

```
Aldehyde group  H    O
                 \  //
                  C                      CH2OH
                  |                        |
               H—C—OH                     C=O      Ketone group
                  |                        |
              HO—C—H                   HO—C—H
Vertical          |                        |        Horizontal
carbon-carbon  H—C—OH                   H—C—OH      carbon-hydrogen and
bonds are going   |                        |        carbon-oxygen bonds
behind the plane H—C—OH                 H—C—OH      are in front of the plane of
of the paper      |                        |        the paper
                CH2OH                    CH2OH

              D-glucose                D-fructose
```

FIGURE 17.1
Sugars are polyhydroxy aldehydes and ketones. The structures of two important sugars, D-glucose and D-fructose, are shown. The carbons shown to be bonded to four different groups are chiral. The geometry at these centers is fixed in each sugar, and sugars differing in geometry are isomers.

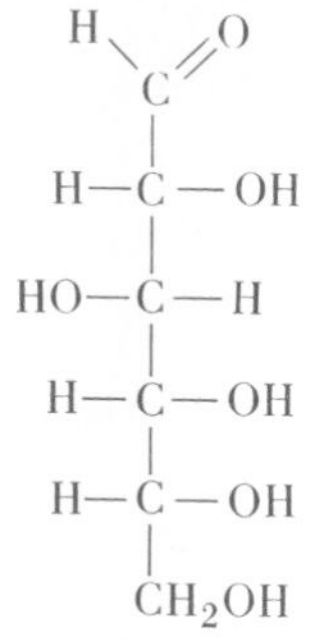

Recall that the aldehydes and ketones react as Lewis acids; intermolecular acid-base reactions give the cyclic forms of glucose

The O of this OH reacts with the C of the aldehyde group to form a six-membered ring

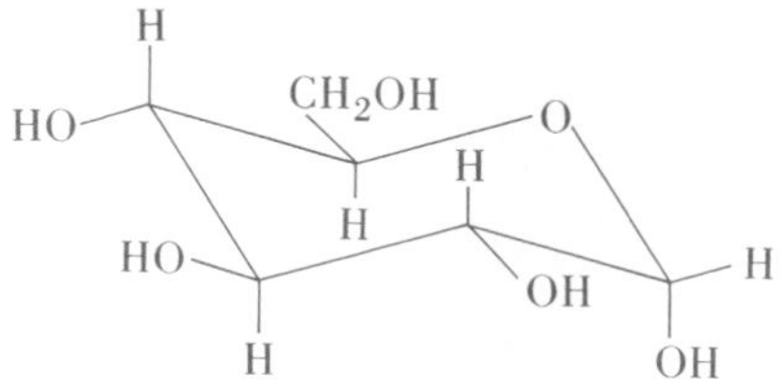

D-glucose

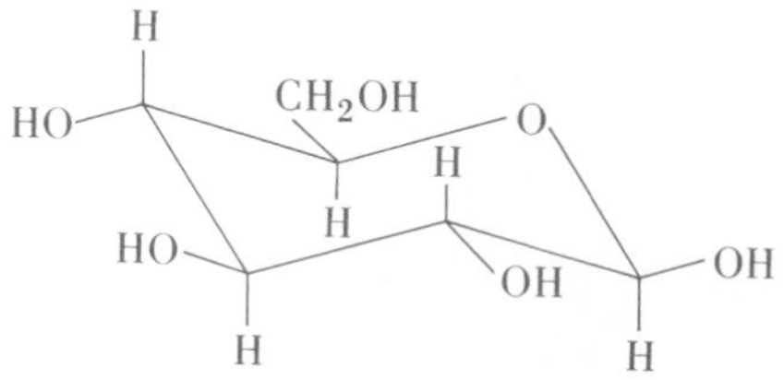

α-D-Glucose

β-D-Glucose

FIGURE 17.2
Glucose can exist as one of two cyclic isomers that can interconvert through an acyclic form. The cyclic isomers differ in their geometry at C-1, the former aldehyde carbon atom.

sugar is a dimer of glucose and fructose. The structures of lactose, the sugar in milk, and of maltose, obtained from starch, are also shown.

CELLULOSE AND STARCH

Cellulose and starch are both polymers of the sugar glucose, but cellulose contains β-glucose and starch contains α-glucose. Cellulose is a structural polymer in wood and cotton. Starch is an energy storage polymer from plants. (A closely related polymer, glycogen, stores energy in animals.) Hydrogen bonding contributes to the properties of each polymer. The different geometries of the α and β linkages and the different arrangements of polymer chains contribute to the very different properties of these polymers.

Cellulose chains are linear; they stack well and have strong interchain hydrogen bonds. Because polymer chains are close and parallel, many hydrogen bonds would need to be broken at the same time to separate chains. Hence, cellulose is insoluble in water even though it has many —OH groups.

Starch chains are wound in a helix. The —OH groups of starch are better exposed to interact with water than are those in cellulose. Some more linear starches dissolve completely; other more branched starches swell with water. These starches may be used as thickening agents, for example, in making gravy.

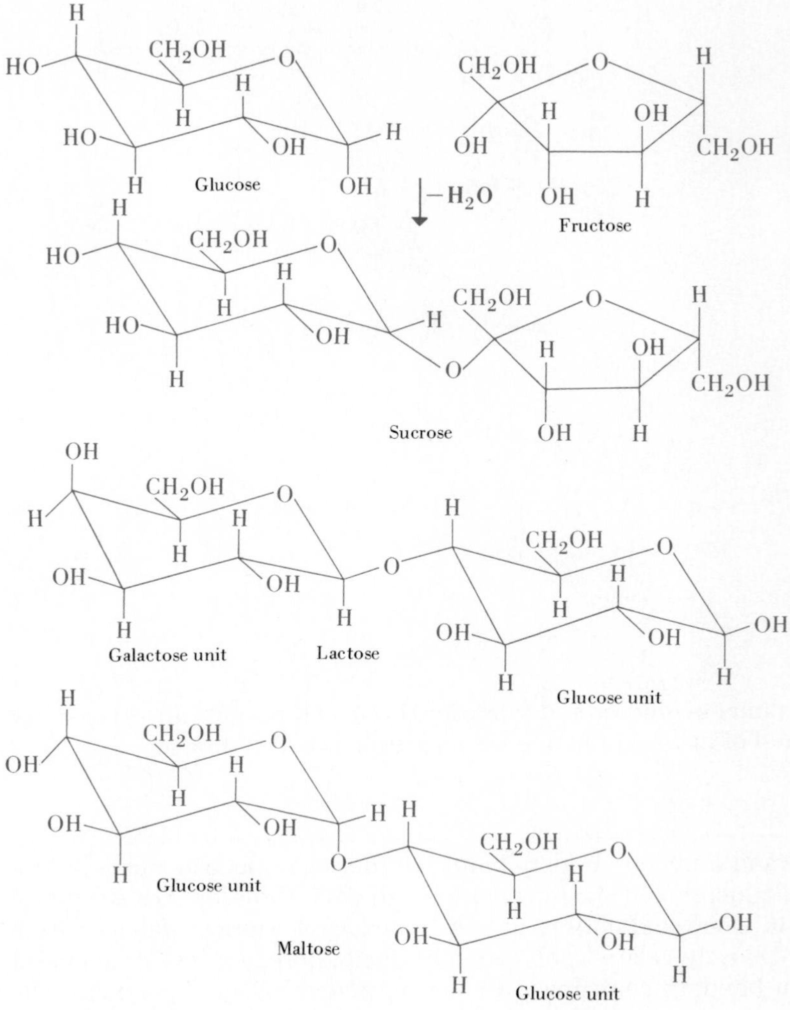

Glucose

$-H_2O$

Fructose

Sucrose

Galactose unit

Lactose

Glucose unit

Glucose unit

Maltose

Glucose unit

FIGURE 17.3
More complex sugars are formed from simple sugars by further reaction at the oxygen atom of an —OH group, bonded to the carbon of an aldehyde or ketone in the simple sugar.

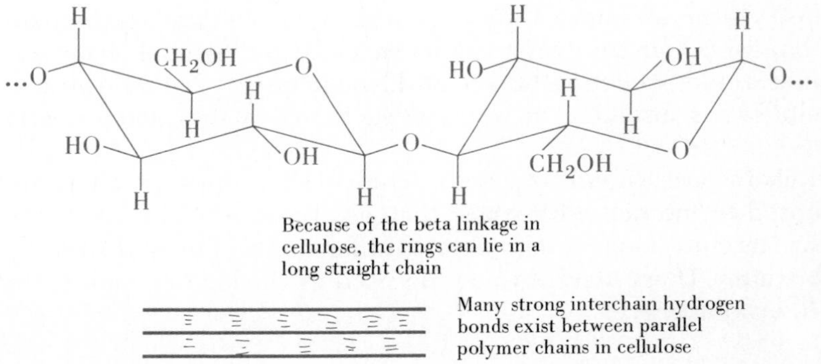

Because of the beta linkage in cellulose, the rings can lie in a long straight chain

Many strong interchain hydrogen bonds exist between parallel polymer chains in cellulose

FIGURE 17.4
Cellulose is a polymer of glucose that features long parallel chains and strong interchain hydrogen bonds.

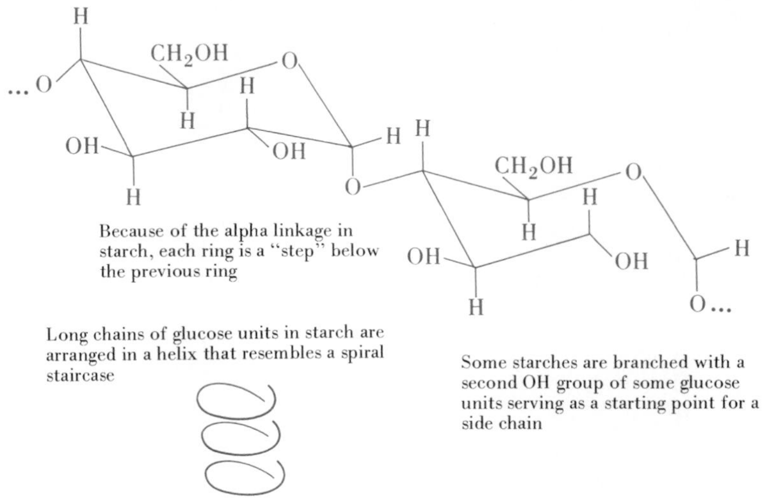

Because of the alpha linkage in starch, each ring is a "step" below the previous ring

Long chains of glucose units in starch are arranged in a helix that resembles a spiral staircase

Some starches are branched with a second OH group of some glucose units serving as a starting point for a side chain

FIGURE 17.5
Starch is a polymer of glucose and has a helical structure. These chains can unwind so that the —OH groups of starch can hydrogen bond to water.

A S I D E

CELLULOSES ARE MODIFIED FOR A VARIETY OF USES

Although cellulose is a strong, abundant polymer, it cannot be fabricated by methods used with synthetic thermoplastic resins because the molecular weight is too high and the interchain forces are too great. Chemists modify cellulose to prepare polymers with widely different properties. They lower the molecular weight of cellulose by treatment with sodium hydroxide. They chemically modify some of the —OH groups in cellulose to produce new polymers.

Cellulose nitrate, the product of the reaction of nitric acid and cellulose, was one of the first commercial synthetic polymers. Nitrate esters of cellulose known as gun cotton or cordite are used to make smokeless gunpowder for artillery shells. In this polymer, the oxidizing agent (nitrate groups) and the reducing agent (sugar units) are joined chemically. In the early days of the film industry, cellulose nitrate was used for the backing of films. The slow oxidation of this polymer has damaged or destroyed prints of many early films.

The rayon and acetate textile fibers are prepared by esterifying some —OH groups of cellulose. Esterification reduces interchain forces and increases solubility. The resulting polymers can be spun for monofilament applications. (The strength of a cotton thread is much less than that of its cellulose fibers; it depends instead on the frictional forces holding fibers together.)

Carboxymethylcellulose is a major food additive used in diet foods as an emulsifying agent. Because it has the β linkage of the cellulose chain, it is not attacked by the enzymes that break down starches. Made up of sugar units, carboxymethylcellulose has neither calories nor nutritional value.

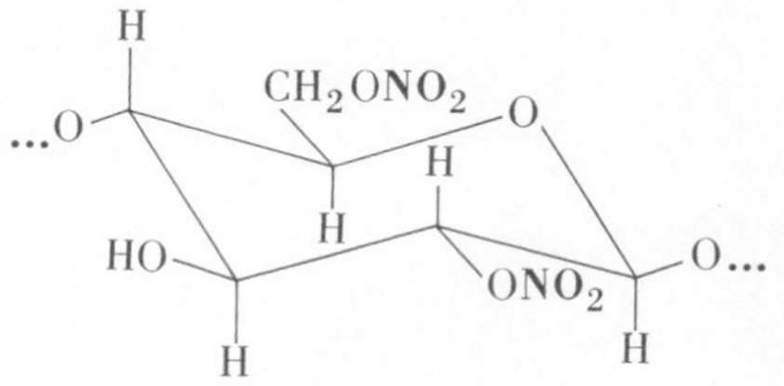

Cellulose nitrate chain

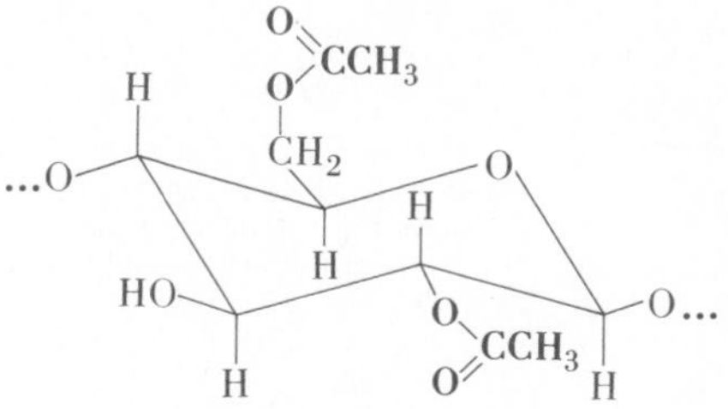

Cellulose nitrate is prepared by modifying some of the OH groups of cellulose by reaction with nitric acid. Cellulose nitrate has been used for billiard balls, as backing for photographic film, and as smokeless gunpowder.

Cellulose acetate chain

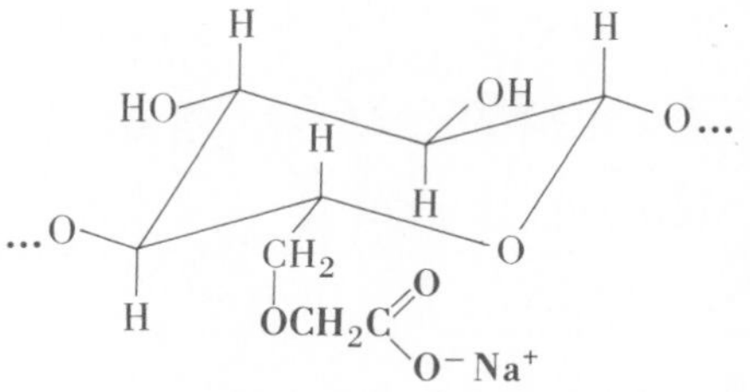

Cellulose acetate has fewer interchain hydrogen bonds than cellulose, so it can be dissolved and then spun to form monofilaments.

Sodium carboxymethylcellulose chain

Sodium carboxymethylcellulose has the beta linkage of the cellulose chain so it is not digested. It is used as a food additive in some diet foods.

FIGURE 17.6
A wide variety of polymers have been prepared by modifying some of the —OH groups of cellulose.

AMINO ACIDS FROM PROTEINS

When proteins are heated with aqueous solutions of strong acids, they cleave to give a mixture of *amino acids.* Each amino acid has an amine group and a carboxylic acid group attached to a carbon atom called the α carbon. Twenty-one amino acids that differ in the substituents attached at the α carbon atom are found in proteins. Substituent groups vary widely. Some are nonpolar; others are polar. Some are acidic; others are basic.

Amino acids are high-melting solids which are more soluble in water than in nonpolar organic solvents. They contain dipolar ions sometimes called zwitterions. In solution at a pH near 7, the dipolar form of the amino acid predominates. This ion can react either as an acid or as a base. At pH 1, amino acids have a net positive charge; at pH 13, they carry a negative charge.

Amino acids obtained from proteins are made up of chiral molecules. The chirality of amino acid building blocks is important for assembling efficient enzymes. If both right- and left-handed amino acids were joined in proteins, a much larger number of proteins could be formed. Proteins with the incorrect stereochemistry would be molecular debris. By using only L

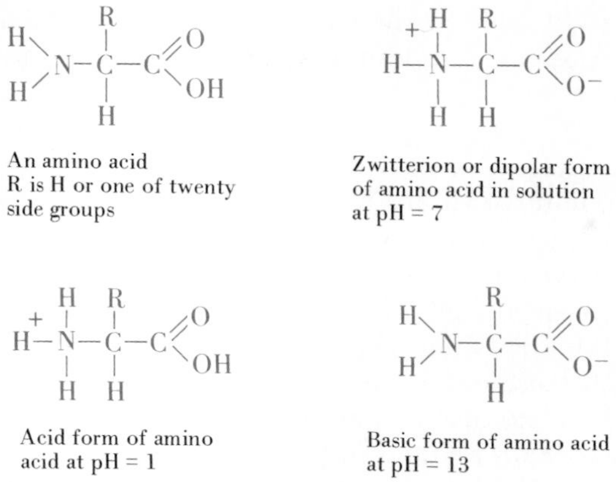

An amino acid
R is H or one of twenty side groups

Zwitterion or dipolar form of amino acid in solution at pH = 7

Acid form of amino acid at pH = 1

Basic form of amino acid at pH = 13

FIGURE 17.7
Amino acids are the building blocks of proteins. Because the carboxylic acid group is acidic and the amine group is basic, the form of an amino acid solution depends on pH.

isomers of amino acids, living organisms are efficient in assembling large protein molecules.

Because enzymes are composed of chiral amino acids, they can induce chirality in biochemical reactions. An achiral pencil fits equally well in either a right or a left hand. However, a chiral right glove will fit the right hand but not the left. Just as a garment worker can select a pattern for a

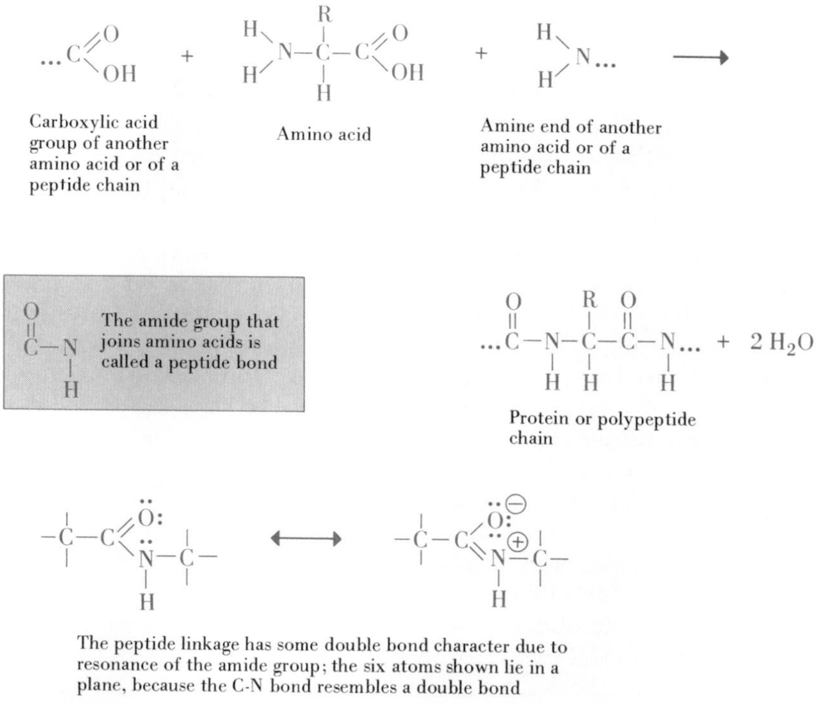

Carboxylic acid group of another amino acid or of a peptide chain

Amino acid

Amine end of another amino acid or of a peptide chain

The amide group that joins amino acids is called a peptide bond

$+ \ 2\,H_2O$

Protein or polypeptide chain

The peptide linkage has some double bond character due to resonance of the amide group; the six atoms shown lie in a plane, because the C-N bond resembles a double bond

FIGURE 17.8
Amino acids are joined together in proteins by amino linkages, also known as peptide bonds.

right-handed glove and make it from a flat piece of leather, an enzyme can produce an L amino acid from an achiral precursor.

PEPTIDE LINKAGES IN PROTEINS

Amino acids are joined together in proteins by amide bonds. These amide linkages are also called *peptide bonds*; consequently, proteins are called *polypeptides.* The peptide linkage of a protein is strong. The atoms bonded to carbon and nitrogen of the peptide bond are in the same plane because the C—N bond has some double-bond character. The chain of a protein is stiff and bends only at the two bonds to the α-carbon of each amino acid.

The proteins in silk and wool illustrate the two major types of regular structures found in protein molecules. In silk, the secondary structure is described as an antiparallel β-pleated sheet. Strong interchain hydrogen bonds link peptide groups in segments running in opposite directions. Small side-chain groups of amino acids extend above and below the folded sheet. The major protein in wool is arranged in α-helices. In an α helix, hydrogen bonding occurs between C=O and N—H that are located one turn or 3.6 residues apart. The side-chain groups of amino acids extend out from the helix. Hydrogen bonding of peptide groups favors the formation of both α-helices and β-pleated sheets in proteins. The β-pleated sheets, like those found in silk, and the α helices, like those found in proteins, are referred to as *secondary structures* in proteins.

ENZYMES

Enzymes (as well as polypeptide hormones) can be described as *informational macromolecules.* The sequence of amino acids is unique in each enzyme, and the shape and function of each enzyme is a consequence of its amino acid sequence. The sequence along the polypeptide chain is called the *primary structure* of the protein.

The shapes of a few crystalline proteins are known from x-ray diffraction studies. Heme, the protein in hemoglobin, has several segments of α helices; other proteins have few. Lysozyme is an example of an enzyme with segments of β-pleated sheet. Enzymes are folded in such a way that amino acids which are far apart on the polypeptide chain may be close together in space.

Proteins assume their shapes as they are being synthesized. The folding of proteins is, in part, a consequence of the solubilities of amino acid side chains. For highly folded, globular proteins, groups that interact strongly with water (acidic and basic groups as well as polar amide and alcohol groups with strong hydrogen bonding) are usually at the surface, and non-polar, hydrophobic side chains are in the interior.

Proteins found outside cells may be cross-linked. Cross-links form by the oxidation of thiol (—SH) groups in cysteine amino acids to give —SS— bonds. Such cross-links can perform the important function of holding separate peptide chains together in some enzymes and hormones. For example,

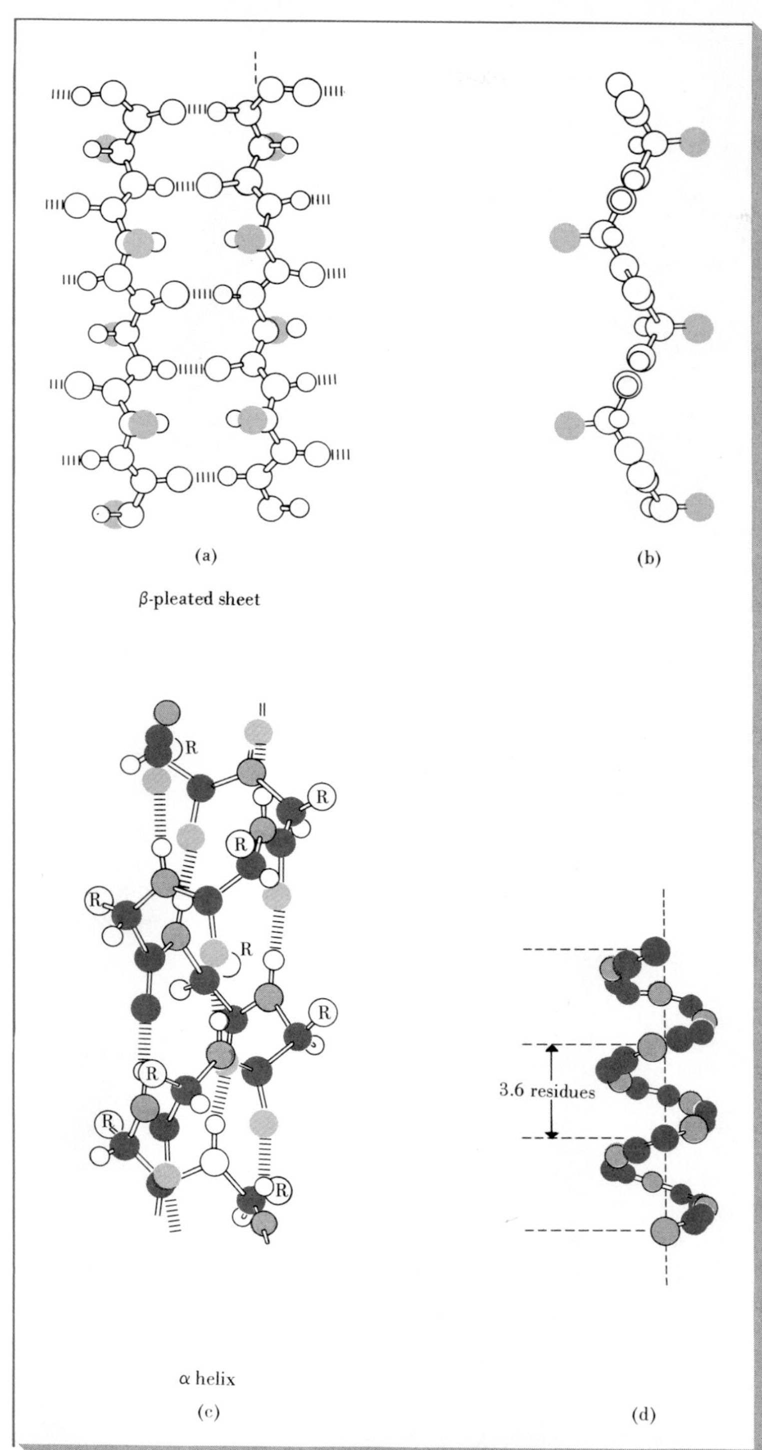

FIGURE 17.9
Two types of secondary structures found in proteins. *(a)* Top view of two chains arranged in a pleated sheet, showing the hydrogen-bond cross links between adjacent chains. The R groups are indicated in color. *(b)* An edge view showing the R groups extending out from the pleated sheet. *(c)* Ball-and-stick model of an α helix showing the intrachain hydrogen bonds. *(d)* The repeat unit is a single turn of the helix.

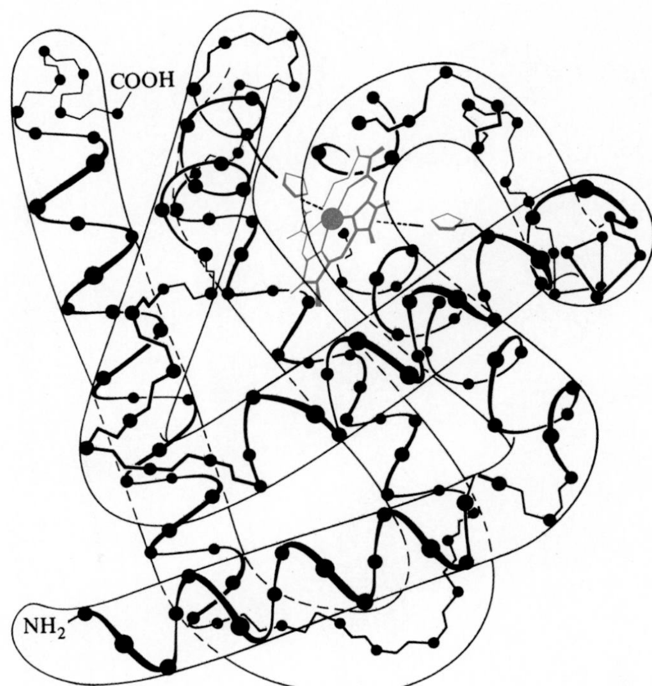

FIGURE 17.10
This diagram depicts the folding of the protein chain of myoglobin, an oxygen-binding protein closely related to hemoglobin. The planar heme ring of myoglobin appears in the upper middle of the drawing. (Adapted with permission from Richard E. Dickerson in "The Proteins," second edition, Volume II, H. Neurath, Editor, Academic Press, New York, NY, 1964.)

enzymes that digest proteins are synthesized as *zymogens* (inactive precursors larger than the active enzymes). After synthesis, zymogens are cross-linked by oxidation and then converted to active enzymes by the cleavage of ends or small loops. Resulting small changes in shape cause large increases in enzymatic activity.

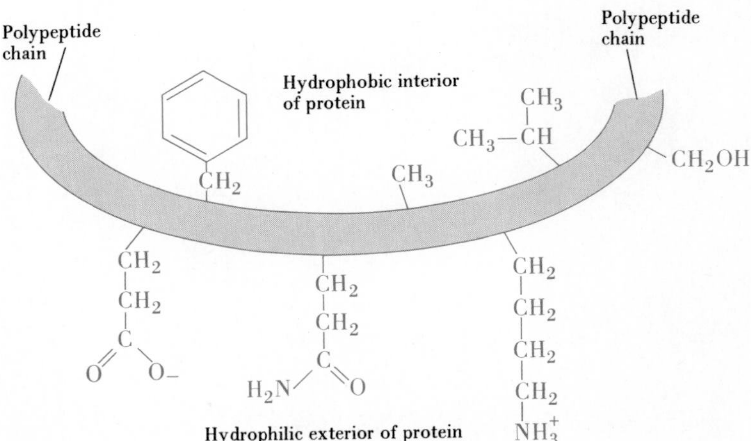

FIGURE 17.11
Proteins fold in such a way that the side chains of hydrophobic amino acids are in the interior and the side chains of charged or polar amino acids are on the outer surface.

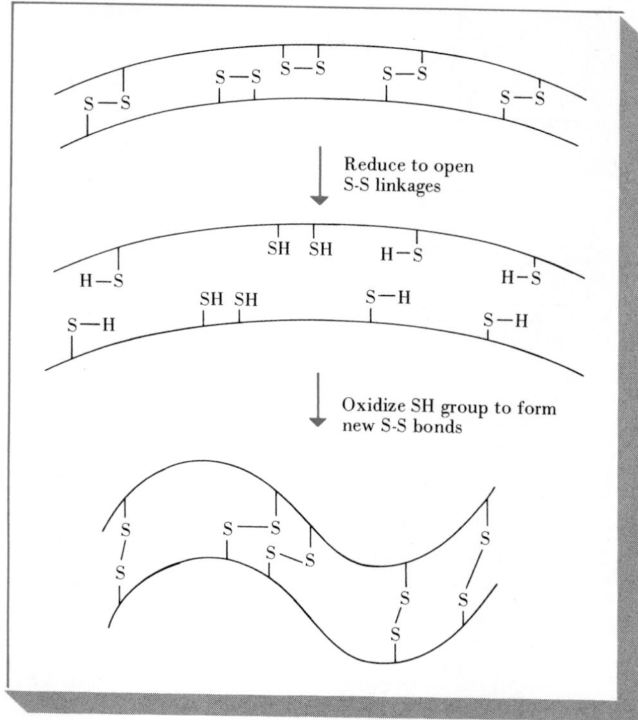

FIGURE 17.12
In proteins found outside of cells, the —SH group of cysteine amino acids are often oxidized to disulfide linkages. In a permanent, the disulfide groups bonding protein chains in hair are opened by reduction, the hair is set to change the arrangement of —SH neighbors, and new disulfide linkages are then formed by oxidation.

Permanents, used to modify the curl of hair, change the spacing of cross-links in hair proteins. First, a reducing solution cleaves sulfur-sulfur bonds of cross-links already present in the hair. The hair can then be curled or straightened to attain a new arrangement of neighboring —SH groups. Finally, an oxidizing agent forms new cross-links making the modified shape "permanent."

DETERMINING THE SEQUENCE OF AMINO ACIDS IN PROTEINS

The first determination of the sequence of amino acids in a protein, the hormone insulin, was completed in 1953. Since that time, methods for determining the sequence of amino acids in proteins have become faster and more sensitive, but important elements of strategy remain the same. A protein or a large polypeptide is cut into smaller segments using enzymes or chemical reagents. Small peptides are isolated and their amino acid content and sequence are determined.

Chemical and enzymatic reactions are used to determine the sequence of amino acids in these fragments. For example, the free amine group at one end of a peptide may be labeled by a reaction that enables it to be identified after the peptide bonds are cleaved. By using a variety of different methods to cut the polypeptide chain, overlapping sequences can be determined. This process is continued until one sequence best fits all the data.

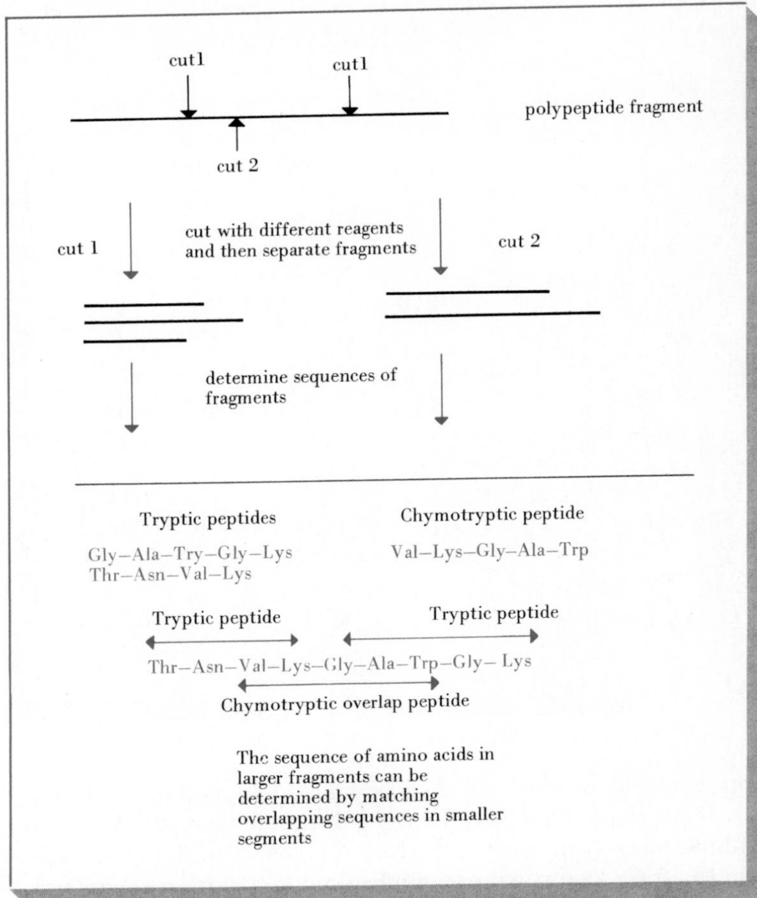

FIGURE 17.13
The general strategy for determining the sequence of amino acids in a protein.

The amino acid sequences of a large number of mutant hemoglobins have been determined. Most of them differ from normal hemoglobin by only a single amino acid. For example, the substitution of a nonpolar amino acid for a polar amino acid at a single position in the polypeptide chain converts normal hemoglobin to hemoglobin S, the protein found in persons

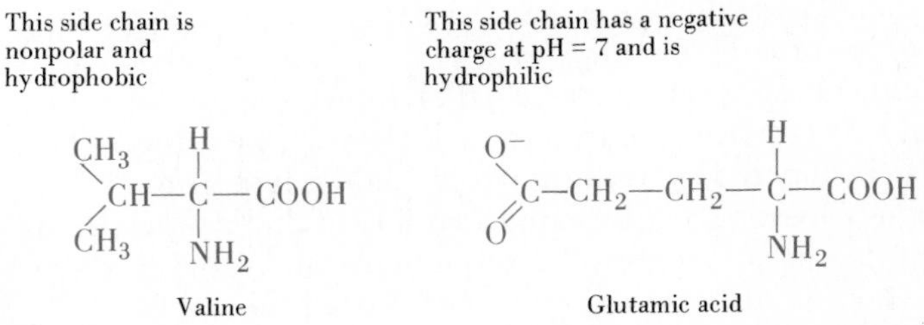

FIGURE 17.14
In the mutation responsible for sickle-cell anemia, valine is substituted for glutamic acid in position 6 of a hemoglobin chain. The mutated hemoglobin is less soluble in water and can precipitate, causing round, red blood cells to change to sickle-shaped cells that can aggregate and block capillaries.

This side chain is nonpolar and hydrophobic

$(CH_3)_2CH{-}CH(NH_2){-}COOH$

Valine

This side chain has a negative charge at pH = 7 and is hydrophilic

$^{-}O(O{=})C{-}CH_2{-}CH_2{-}CH(NH_2){-}COOH$

Glutamic acid

with the genetic disease, sickle-cell anemia. Hemoglobin S is less soluble than normal hemoglobin. If it crystallizes, red blood cells can become distorted and then clog capillaries.

A S I D E

MORPHINE IMITATES NATURAL PAINKILLING PEPTIDES

Alkaloids are amines isolated from plants. Most alkaloids are chiral, and many have strong physiological activity. For example, morphine obtained from opium poppies dulls pain. (Before the addictive nature of morphine had been discovered, morphine was widely used in medicine as a painkiller. Unfortunately this practice led to drug addiction among patients.)

The presence of alkaloids in plants and the molecular basis of their physiological activity are puzzling. Scientists do not yet know why plants should synthesize alkaloids, but they are beginning to understand why alkaloids produce such dramatic responses. Alkaloids bind to receptor sites present in the brain.

Morphine binds to the same pain-repressing site to which an endorphin binds. *Endorphins* are small peptides produced by the body to suppress pain. The body produces endorphins in response to injury, in response to the twirling needles used in acupuncture, and, perhaps, during the course of long exercise runs.

Morphine mimics the shape taken on by the endorphin. When the rigid morphine molecule is compared to the endorphin peptide, there are strong structural similarities. There are corresponding hydrophobic regions and corresponding groups with the potential to form hydrogen bonds. Morphine is shaped to bind to an endorphin-receptor site.

NUCELIC ACIDS AND GENETIC TRANSMISSION

Nucleic acids are polymers involved in the transmission and expression of genetic information. Nucleic acids are composed of a smaller number of monomers than are proteins. Nucleic acids can be cleaved to give phosphoric acid, a simple sugar, and a small number of organic bases. In DNA, ester linkages join phosphoric acid and deoxyribose, a 5-carbon sugar lacking the O in one —OH group. One of four organic bases is attached to each sugar unit. The four bases are represented by their initials A, T, C, and G.

X-ray diffraction patterns of DNA fibers show evidence of a helical structure. Working from scale molecular models, Watson and Crick formulated a double-helix structure for deoxyribonucleic acid, DNA, in 1953. The bases are in the center of the double helix, and two polymer strands formed from sugars and phosphates wind in opposite directions about the outside.

Base pairing provides the mechanism for recognition and copying. The pairing involves both base size and hydrogen bonds. Both AT and GC pairs

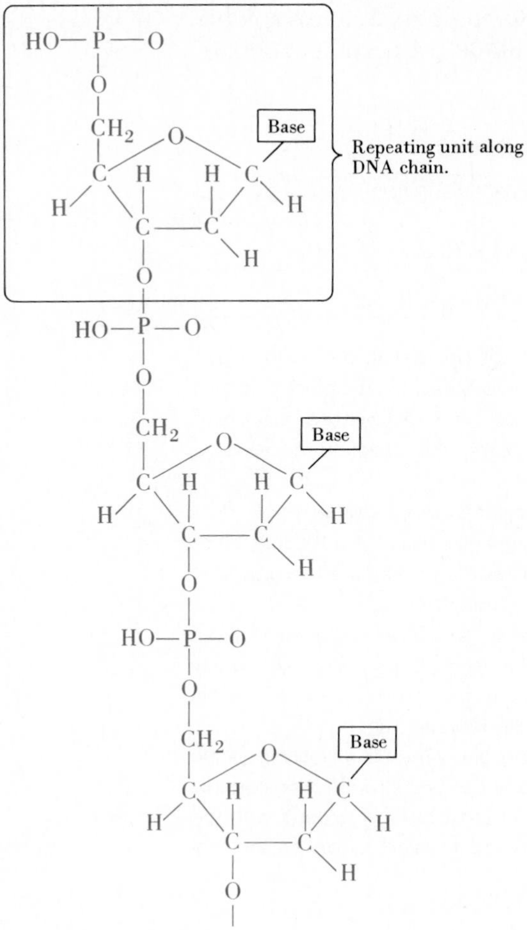

The structure of DNA

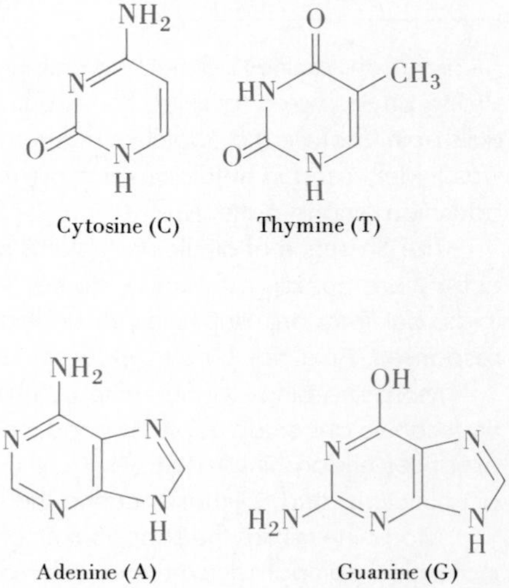

The bases in DNA

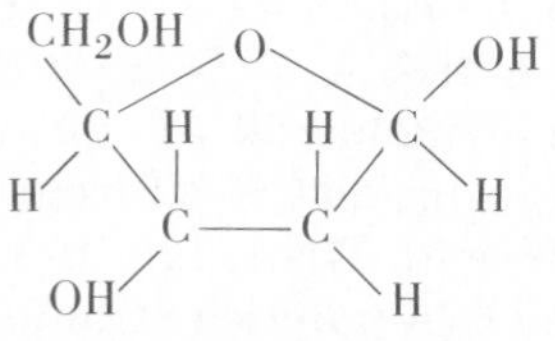

Deoxyribose, the sugar molecule present in DNA

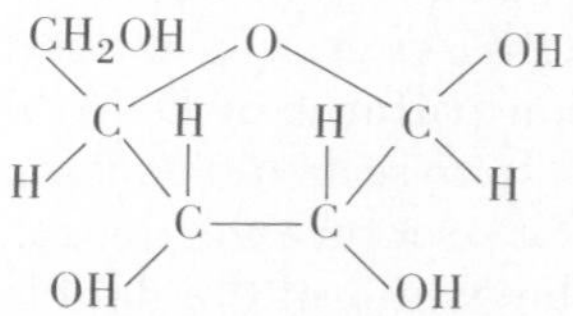

Ribose, found in RNA

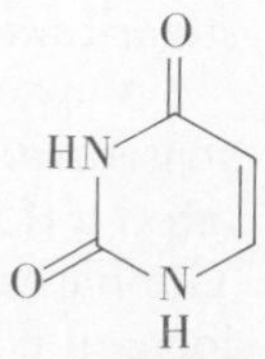

Uracil (U)

FIGURE 17.15
The nucleic acids DNA and RNA are involved in the storage and transcription of genetic information. Each of these molecules has a backbone built up from phosphoric acid and a 5-carbon sugar, and each has one of four bases attached to each sugar along the polymer backbone.

RNA resembles DNA but has a different sugar along its backbone and has uracil in place of thymine as one of its four bases.

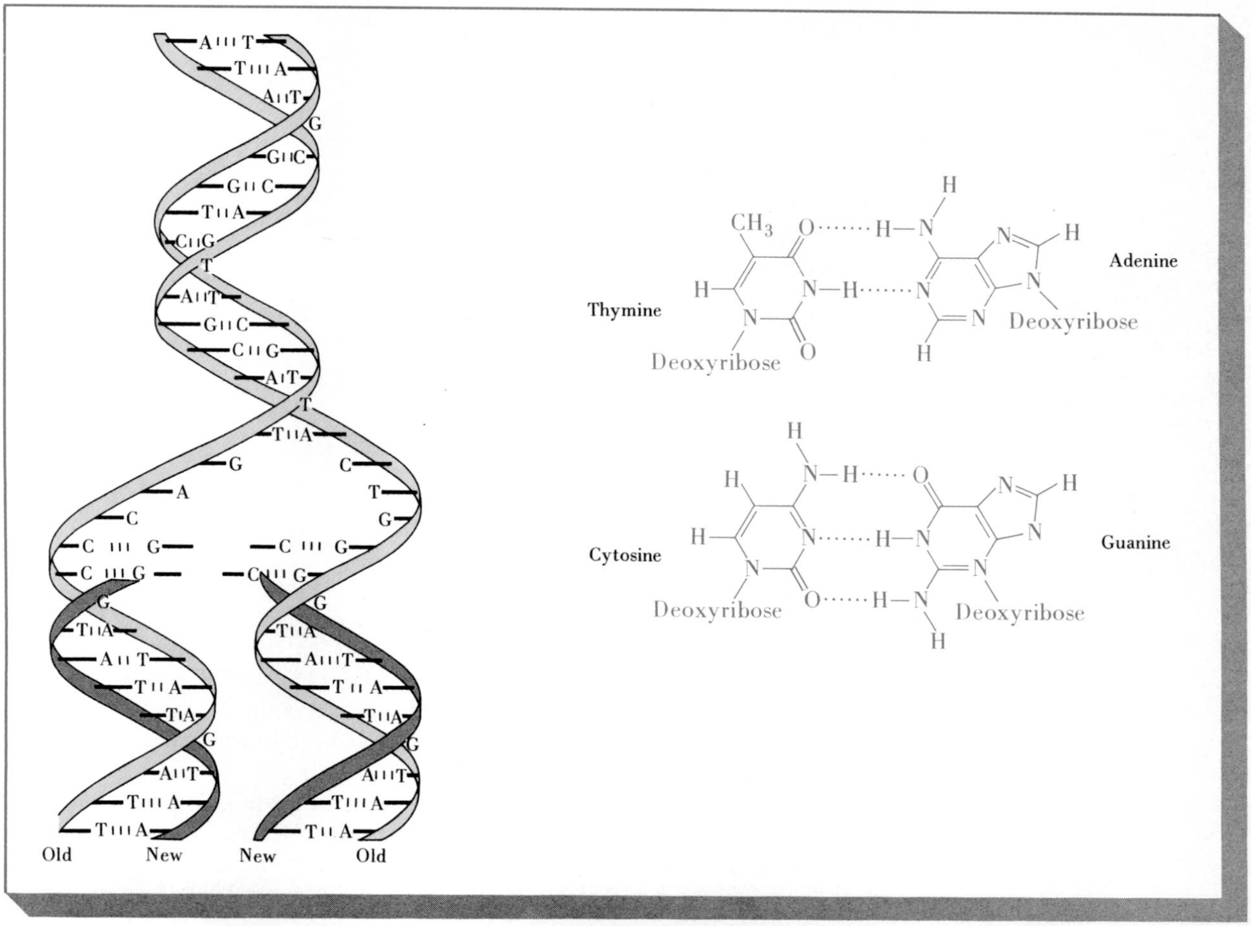

FIGURE 17.16
Base pairing provides a method for copying the information of DNA into new DNA.

span the distance between the backbone chains. A pair consisting of the bases A and G would be too large to fit in the double helix, and a pair between C and T would fail to bridge the gap. Hydrogen bonds favor AT and GC pairs. These bases pair with great fidelity. Adenine, A, on one strand pairs to a thymine, T, on the second strand; guanine, G, pairs with cytosine, C.

The Watson and Crick structure stimulated a rapid growth in the understanding of the molecular basis of genetics. Information is conveyed by the sequence of bases on a DNA polymer chain. Each strand carries the information needed to construct its complement. Strand separation and duplication provide for the transmission of genetic information from one generation to the next.

Genes are expressed in proteins. The phrase "one gene, one protein" once expressed the hypothesized correspondence. Genetic information flows from the linear DNA to the three-dimensional protein. The informa-

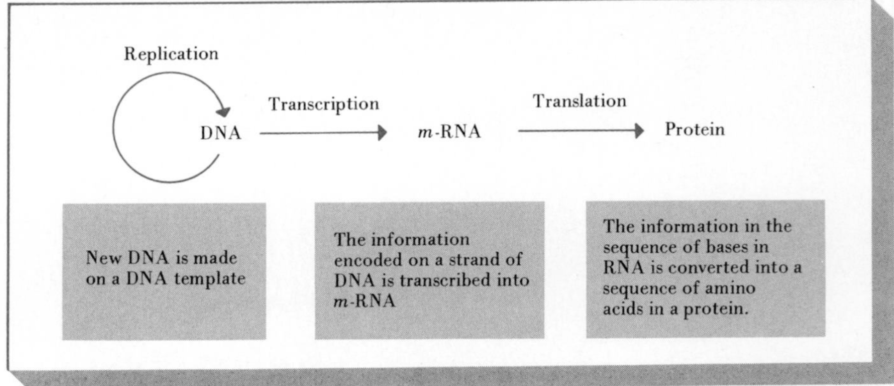

FIGURE 17.17 *The central dogma* describes the flow of genetic information from generation to generation through replication and the expression of that information in proteins by transcription and translation.

tion encoded in the sequence of letters (bases) along a strand of DNA prescribes which one of twenty-one amino acids is to be used at each stage in the construction of a protein.

Since protein synthesis occurs in the cytoplasm of cells and since DNA is located in the nucleus, a messenger molecule is required. Messenger RNA is synthesized in the nucleus and moves to the cytoplasm where it serves as a template for protein synthesis. Like DNA, RNA is a nucleic acid polymer. RNA differs from DNA in three ways. It is single-stranded, it has a ribose sugar in its backbone, and it has the base uracil rather than thymine as one of its four bases.

The model for genetic expression was named *the central dogma*. According to this model, information encoded in DNA is transcribed into mRNA and then translated into protein. Genetic mutations result in the synthesis of altered proteins and, since the enzymes that mediate metabolism are proteins, often in metabolic deficiencies.

By making short synthetic DNA segments with known base sequences and then using these segments to make peptides, scientists learned to read the genetic code. Information is encoded in base triplets such as GCA, CCG, etc. The number of different three-letter words that can be encoded using four letters is $4 \times 4 \times 4 = 64$. Start-and-stop signals together with a redundant code specifying amino acids have been determined. (Redundancy helps to protect messages from being mistranslated.)

A S I D E

READING THE SEQUENCES OF DNA AND RNA

A major revolution in the way we understand the mechanism of gene expression is under way. New experimental techniques have enabled scientists to read the base sequences of DNA and RNA in addition to determining the amino acid sequence of an

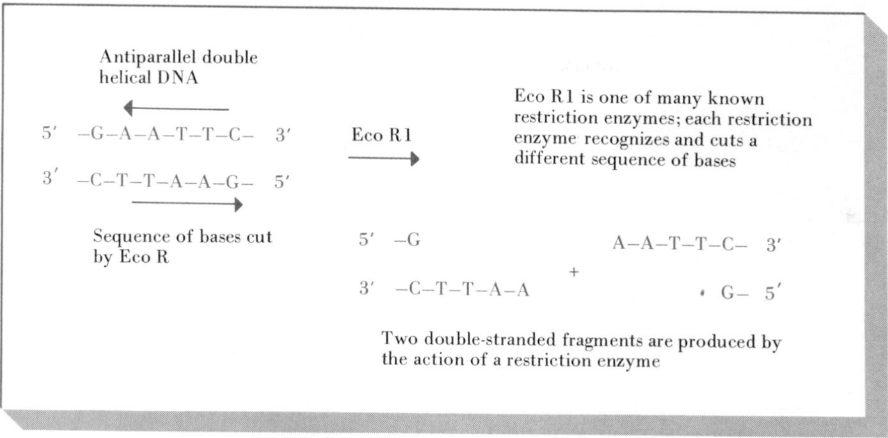

FIGURE 17.18
Double-stranded DNA is cut by restriction enzymes. Molecular biologists use these enzymes to make smaller fragments of DNA as an aid in determining the sequence of bases in DNA.

encoded protein. The simultaneous translation of DNA, of messenger RNA, and of the protein for a gene has given surprising results and is leading to fundamental changes in the way we understand the occurrence of genetic change.

In 1979, a Nobel prize was awarded for the discovery and characterization of restriction enzymes. These enzymes shred DNA of invading organisms while host DNA is protected. Restriction enzymes cut double-helical DNA at specific short sequences. By changing conditions, either double-stranded fragments or single-stranded fragments can be recovered from the reaction.

To cut intact DNA, restriction enzymes read information in the groove of the double helix rather than the base sequence of a single chain. With only four bases in DNA, the probability of finding a restriction sequence in any long piece of DNA is high. An organism having a restriction enzyme must protect its own DNA. Methylation enzymes are used to provide protection for an organism's own DNA. These enzymes can modify the bases in double-stranded DNA by attaching a methyl group. If one

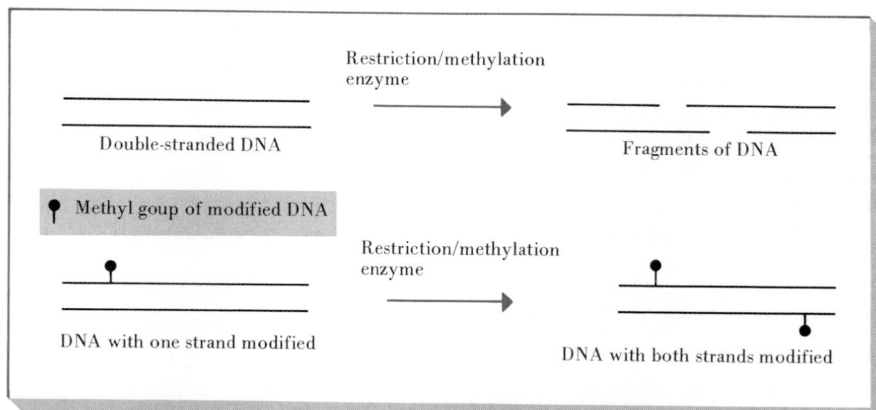

FIGURE 17.19
Restriction enzymes can recognize and destroy foreign DNA. Some restriction enzymes recognize their "own" DNA by the presence of methyl side groups. The same enzyme that cuts foreign DNA methylates the second strand of protected DNA. When that DNA separates and is duplicated during cell division, each of the progeny DNAs also carries protecting methyl groups.

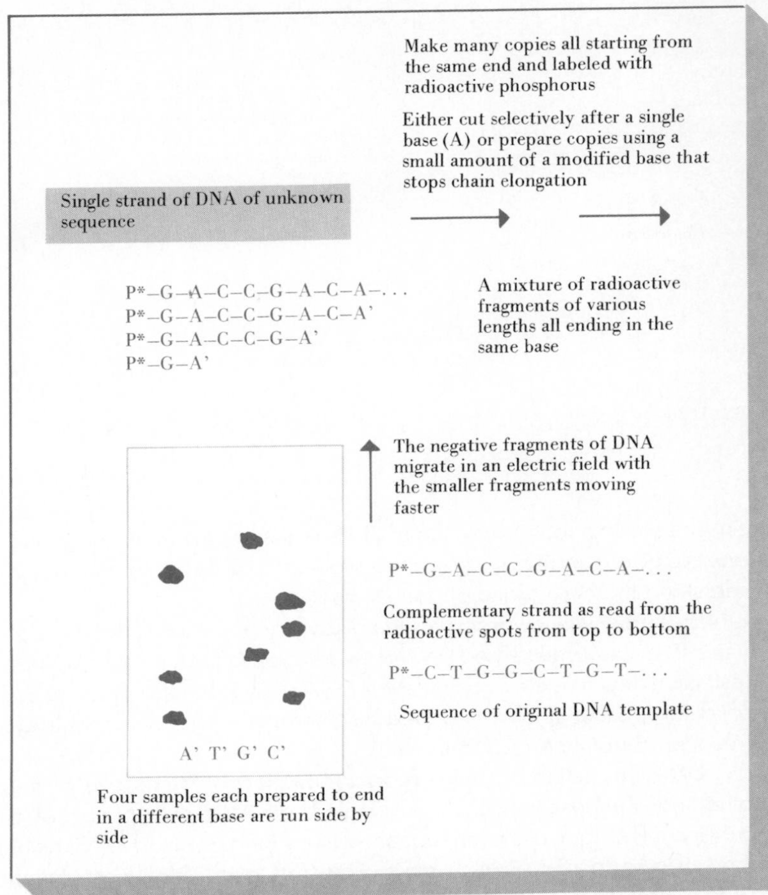

Make many copies all starting from the same end and labeled with radioactive phosphorus
Either cut selectively after a single base (A) or prepare copies using a small amount of a modified base that stops chain elongation
Single strand of DNA of unknown sequence
P*–G–A–C–C–G–A–C–A–. . .
P*–G–A–C–C–G–A–C–A'
P*–G–A–C–C–G–A'
P*–G–A'
A mixture of radioactive fragments of various lengths all ending in the same base
The negative fragments of DNA migrate in an electric field with the smaller fragments moving faster
P*–G–A–C–C–G–A–C–A–. . .
Complementary strand as read from the radioactive spots from top to bottom
P*–C–T–G–G–C–T–G–T–. . .
Sequence of original DNA template
A' T' G' C'
Four samples each prepared to end in a different base are run side by side

FIGURE 17.20
Scientists use a mixture of enzyme-catalyzed reactions and chemical reactions to make complementary copies of an unknown fragment of DNA. They separate these copies and use a radioactive marker to detect them. Using these methods, it is possible to read the sequence of bases in very large fragments of DNA.

strand of DNA is already methylated, these enzymes methylate the second strand. This chemical modification is done in a site along the groove rather than in a site used in base pairing.

To study genes, scientists cut the DNA with restriction enzymes. To determine the sequence of bases in each restriction fragment, a series of successively shorter copies, all of which originate at the same end, is prepared. Copies can be either complimentary DNA fragments or RNA fragments, depending on the enzymes used to synthesize them. Many copies of each fragment must be made so there will be enough material to detect. Detection is simplified by starting each copy with a radioactive phosphorus atom in the first phosphate group.

One approach is to build up successively longer copies that all end at the same base, say A. To do this, a small amount of a modified base A′ that stops the chain from growing is included in the mixture. Then all copies start with the same base and end at sites encoded for A. Alternatively, complete copies of the whole restriction fragment can be made and then nicked with a small amount of a chemical that selectively cleaves the polymer at A. Using either process, one can prepare a mixture of copies, all with a common origin and all ending at sites coded for A.

These chains can be separated according to length using an electric field and a supporting polymeric gel. The electric force moving a fragment through the gel is proportional to size, but the retarding frictional force increases faster with the size of the nucleic acid fragment. Hence, the smaller the fragment, the further it migrates. The positions of the nucleic acid fragments in the gel can be read by exposing the film to the decay of the radioactive phosphorus isotope. When separate samples ending in A, T, C, and G are run side by side, one can read the sequence of bases from the sequence of spots in the gel. Since the base pairing of copying is known, the template sequence can be inferred. For developing sequencing methods, Sanger and Gilbert shared a Nobel prize in 1980.

INFORMATIONAL MACROMAOLECULES—A LANGUAGE MODEL

The development of techniques to determine the sequence of proteins and nucleic acids has permitted a close examination of nucleic acid and protein structures. These studies show that protein and nucleic acid structural sequences can be treated literally as messages. The messages are expressed in a language with words, phrases, and other grammatical structures. The messages can be copied, modified, and transmitted. Some of the most provocative studies have resulted from textual comparisons of a DNA, the mRNA transcribed on it, and the translated protein. The results of some of these studies together with new questions raised by molecular biologists follow.

The first DNAs to be fully sequenced were those of small viruses. These viruses coded for only a few proteins. It came as a great surprise that the genetic information for some of these proteins was encoded in overlapping DNA segments. That a DNA phrase can be found in two protein sentences is evidence for complex punctuation. Alternative interpretations of a text must rely on a larger frame of reference than base triplets alone can furnish.

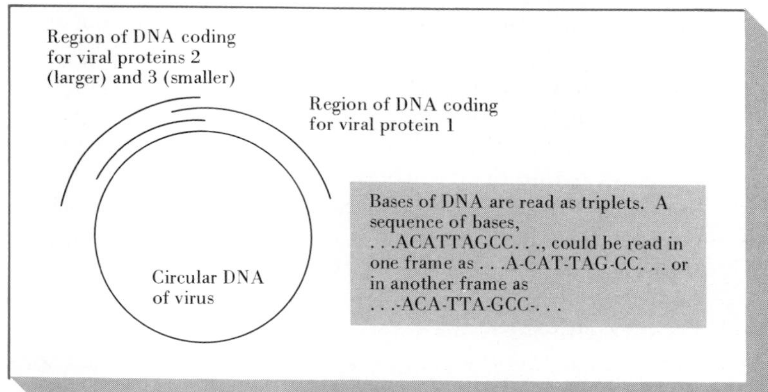

FIGURE 17.21
One of the first DNAs to be fully sequenced was that of SV40, a small virus. Scientists were surprised to learn that the DNA encoding three proteins found in the viral coat had regions of overlap. The reading frame for viral protein 1 differed from that of the other two proteins.

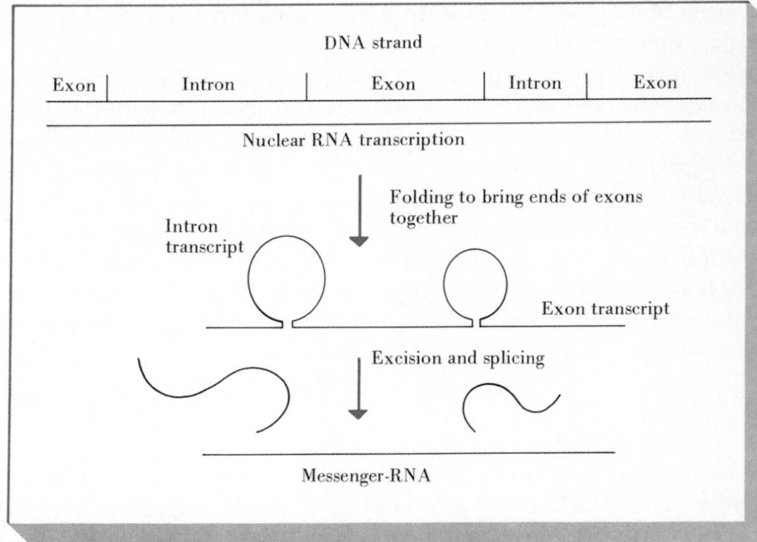

FIGURE 17.22
Many genes come in pieces. DNA expressed in proteins occurs in segments (exons); these segments are separated by intervening sequences (introns). The strand of DNA is copied completely; then it is edited to remove transcripts of introns and to complete the messenger RNA on which the protein is assembled.

When RNA transcripts of DNA genes from cell nuclei are examined, the results are striking. Far more RNA is synthesized on the DNA template than is necessary for the translation of the protein. The RNA in the nucleus differs from messenger RNA found in the cytoplasm where protein is synthesized. Nuclear RNA is edited to form the mRNA at sites in the nuclear membrane. The information to make most proteins is not encoded in a continuous segment of DNA. Genes are not continuous; they come in pieces. The DNA that is finally expressed in proteins occurs in segments known as exons that are separated by intervening sequences called introns.

It has been suggested that introns facilitate evolution. According to this model, genes evolve when gene fragments evolve. A given piece of DNA may be repeated several times in a long strand of DNA. Some of these repeated sequences, dormant in gene expression, may undergo mutation. If at a later time a modified fragment is reinserted into the active gene pool, a favorable mutation may be incorporated.

Examination of the protein segments encoded by each exon is underway for the genes encoding hemoglobin and lysozyme, proteins of known three-dimensional structure. The middle exon of the protein in hemoglobin encodes that portion of the protein that makes all contacts with the heme ring. Similarly, the active-site cleft of the enzyme lysozyme is encoded by a single exon. These findings suggest that exons encode functional segments of proteins.

Finally, for proteins secreted by a cell, it has been found that the protein synthesized on the mRNA is not identical to that found outside the cell. The protein precursor synthesized on the mRNA has an additional short sequence of amino acids called a leader sequence. This leader is excised as the protein crosses the cell membrane. It would appear that the language of gene expression includes shipping labels for export, perhaps even zip codes.

I would like to end this chapter on a personal note. I have always been intrigued by the philosophy of science. My early training was in chemistry and physics. I was exposed to the philosophy of the logical positivists and to an outlook influenced by the Heisenberg uncertainly principle and the probabilistic description of events furnished by quantum mechanics. The question "Why?" was not permitted; cause was reduced to an ordered sequence of events. When I studied biochemistry while on a sabbatical leave, I found that questions of purpose were assumed. Indeed, posing such questions led to productive scientific inquiry. This was another philosophical outlook coming from a scientific discipline with a different history.

QUESTIONS

1. Draw the structure of a 4-carbon sugar having a ketone functional group.
2. Which of the 3-carbon sugars is chiral?
3. The open-chain structure of the sugar ribose found in RNA is shown. Draw the open-chain structure of 2-deoxyribose, the sugar found in DNA.

```
H   O
 \ //
  C
  |
H—C—OH
  |
H—C—OH
  |
H—C—OH
  |
  CH2OH
```

4. Maltose is a disaccharide obtained from starch by the enzymatic cleavage of every second α-glycoside linkage. Draw the structure of maltose.
5. What enzymatic activity (of symbiotic microorganisms) enables termites to digest wood?
6. Is cellulose an informational macromolecule? Why or why not?
7. The formula of the open-chain form of glucose is $C_6H_{12}O_6$. What are the formulas of α- and β-glucose?
8. Why does the esterification of OH groups in cellulose reduce the attraction between polymer chains? Why would the extent of esterification greatly affect properties of the polymer?

9. The structure of glutamic acid is shown. What is the structure of this amino acid in solution at pH 7? What is the net charge on the ion?

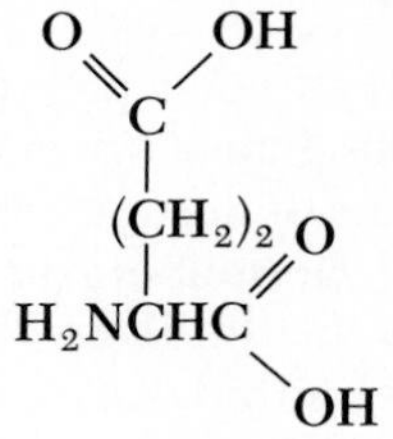

```
   O    OH
    \\ /
     C
     |
  (CH2)2 O
     |  //
  H2NCHC
        \
         OH
```

10. Two different dipeptides can be formed using alanine and glycine. Draw structures for each.

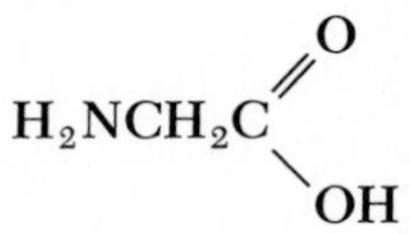

```
          O            CH3  O
         //             |  //
  H2NCH2C           H2NCHC
         \                 \
          OH                OH
      glycine          alanine
```

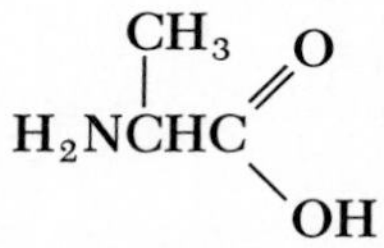

11. How many different dipeptides can be formed from the twenty-one amino acids found in proteins?
12. How many different tripeptides can be hydrolyzed to give alanine, serine, and lysine?
13. The amino acid lysine has a $-CH_2CH_2CH_2CH_2NH_2$ side chain. Draw the structure of lysine that would have a $+2$ charge. Draw the structure that has a -1 charge.
14. Draw a partial structure of a peptide showing only the six atoms that are coplanar.
15. Would proteins be effective as enzymes if there were no restrictions on their secondary structure and folding?
16. What is the complementary base sequence to AAGCT?
17. What base present in DNA must code with U in RNA since RNA is made on a DNA template?
18. For which of the following base pairs are the hydrogen-bonding attractions greatest, AT or GC?
19. Why is a minimum of three bases needed to encode each of the amino acids found in proteins?

CHAPTER

EIGHTEEN

SCIENCE AND SOCIETY

Chemistry, like other sciences, has grown rapidly in the last 200 years. Modern chemistry had its birth in the experiments and speculations of amateur scientists in the eighteenth century. It grew up during the industrial revolution. As scientific understanding grew, so did the applications of science to the refinement of earlier technologies and to the development of new ones. (In this chapter, chemistry provides examples of changes occurring in many sciences. At times the discussion will focus on chemistry, but much of the material is directed toward science in general.)

The roles of science in universities and in government have grown and changed. For example, from a late and modest beginning departments of chemistry have grown to rank among the larger departments, especially at the graduate level, of major universities. Government agencies fund the training of scientists and they employ scientists to assess the quality of air and water, to help make agriculture more productive, and to build new weapons.

Is rapid change inherent to science? What is the relationship between science and technology? How has the growth of one affected the other? What is the relation of science to the universities? To government? Are the forces for change still operating so that we might expect continued rapid change? If so, might we anticipate the directions of change? In this chapter, we shall explore these questions and consider some partial answers.

MEASURING THE GROWTH OF SCIENCE

One way to measure the rate of growth of science is to look at the increase in one of the products of science, articles and abstracts of articles. The curve for the graph of the number of scientific publications vs. time is characterized by a doubling time; for each six- to eight-year period, the number of articles has doubled. The growth in science shown by this graph is described as exponential. (The curve for the graph of radioactive decay vs. time is also exponential. It is characterized by a half-life, the time for one-half of a radioactive isotope to decay.) Exponential growth curves are observed if the number of science journals is plotted against time or if the number of abstracts is plotted against time. The history of chemistry has shown exponential growth; there has been an explosion of scientific knowledge.

New scientific disciplines have arisen; others have fallen into disfavor or neglect. Have important discoveries such as nuclear activity or antibiotics changed the rate of growth? Apparently not; the disciplines of science appear to have grown unchecked. Breaks in the growth curve are found only at the time of war, particularly World War II, when the resources of many nations were channeled into the war effort.

In recent times the changes accompanying growth have been dramatic. Less than one-quarter of the scientific information available now was known twenty-five years ago. What is known now will constitute less than one-half of the available scientific knowledge twenty years hence. Education is inherently incomplete and, in part, dated.

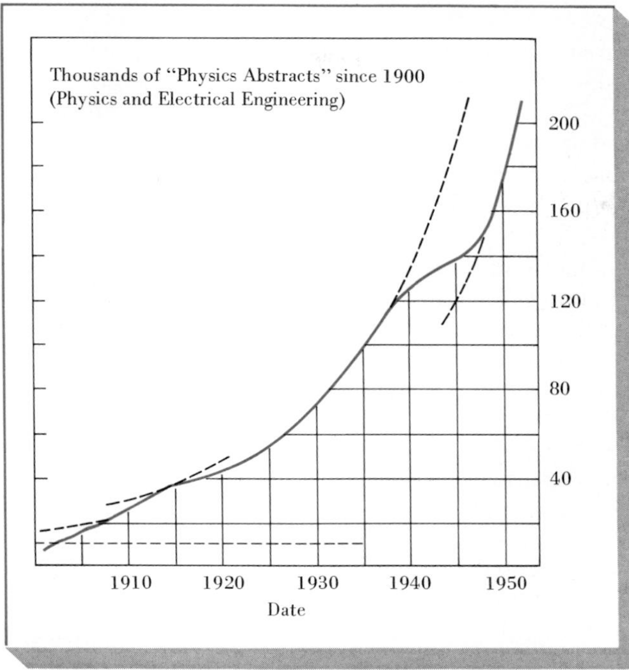

FIGURE 18.1
Total number of physics abstracts published since January 1, 1900. The full curve gives the total, and the broken curve represents the exponential approximation. Parallel curves are drawn to enable the effect of the wars to be illustrated. (From Derek J. de Solla Price, *Science Since Babylon*, New Haven: Yale University Press, 1961.)

However exponential growth in any area cannot continue indefinitely. Growth is limited by the available material and human resources. (No one believes that all people will be scientists before another century has passed, for example.) At some time the rate of growth must slow, and then it may begin to level.

We are living in a time when the rapid growth of science is slowing. As the rate of growth slows, competition for limited resources grows. The allocation of resources for science and among competing scientific disciplines increasingly becomes a political issue.

SCIENCE AND TECHNOLOGY

On a given day in the laboratory, one might attempt a reaction in the synthesis of a new organic compound. That night the same individual may boil eggs and make egg salad. Clearly, one activity is science and one is cooking. If the person quickly cools the eggs so that iron and sulfur in the yolks do not have time to react and form a green compound, is the person still cooking? Sharp distinctions between the activities of science and the applications of science can be difficult to maintain.

Science is the activity by which people seek knowledge of the world about them. Technology is the activity of applying knowledge to the production of goods and services. Applied science and engineering are terms used to describe the activities involved in making the knowledge of science applicable to the processes of technology. It can be useful to distinguish between applied science and basic science. Applied science has short-range, practical goals; basic science does not.

This text does not emphasize distinctions between basic or pure science and applied science; some other texts do. The author's reluctance to emphasize this distinction may stem from a tendency on the part of some people to use the words "pure" and "applied" as value words rather than as descriptive words. For some, pure science may be an activity of dreamers and eggheads while applied science is for those who show common sense and know-how. For others, pure scientists discover truth and beauty while applied scientists aid those who would despoil the environment.

The activities of basic science and of applied science are often intertwined. Efforts to find solutions to problems arising in technology can generate scientific knowledge, and increases in scientific knowledge can lead to improvements in old technologies and the creation of new ones. An effort to improve the properties of polyethylene led to the study of catalysis by transition-metal complexes. Efforts to improve explosives have led to increased knowledge of the kinetics of fast reactions. Or is it the other way around? Have designers of explosives benefited from the scientific database concerning rapid reactions?

Frequently the science of today becomes the technology of tomorrow. Nobel prizes were given to physicists and chemists who first studied the interactions of matter and radiant energy in a magnetic field. Decades later, many hospitals are equipped with instruments using such measurements to scan the soft tissues of the body. When did the use and study of nuclear magnetic resonance pass from the realm of science into technology? The magnetic resonance phenomenon was important. It provided knowledge of chemical structure not readily available by other means. The phenomenon was studied first on home-built instruments and then on commercially available instruments. Newer instruments were larger, more powerful, and faster. A new tool for probing nature had become available; another fire had been tamed.

THE RECRUITMENT AND TRAINING OF SCIENTISTS

As science has grown, the number of students being educated for careers in science has increased. Science has provided a major avenue for upward social mobility in the United States because recognition and advancement in science are tied to ability. People have entered science from a variety of economic and social backgrounds. (However, the numbers of women and of blacks entering chemistry remain low.)

As science has increased in complexity, the degree requirements to enter the profession have increased. Early in the twentieth century, most people beginning a career in science had either a bachelors or a masters degree. (Before that time, science was largely an avocation of amateurs with independent financial support.) Post-baccalaureate training has been extended, and now almost all scientists entering academic institutions or accepting research appointments in industry have a doctoral degree. Many have spent one or more years in post-doctoral research.

Conflicting demands for graduate and undergraduate education have created pressures within universities. Graduate education in science emphasizes research rather than courses. Students serve an apprenticeship in the laboratory of a research mentor. To successfully train graduate students for careers in science, a professor needs both time and money for research. Stipends for students, summer and often academic-year salaries for professors, and money for expendable supplies must come from research grants. And as scientific instruments have become more sophisticated, the costs of a vigorous research program have escalated rapidly.

Professors must publish results of research to maintain funding. Peer recognition and funding are tied to success in research. Excellence in graduate training is rewarded both professionally and financially. Indeed, at some universities the most widely recognized faculty may teach only a small fraction of their time and then only at the graduate level.

Quality education in elementary and secondary school and at the undergraduate level in college is important to maintaining a strong program in science, yet the profession and society reward research far more than teaching.

GOVERNMENT RESEARCH AND DEVELOPMENT

Much of the research and development in many areas of science are sponsored or performed by the federal government. The National Bureau of Standards, the United States Geological Service, the Weather Bureau, and laboratories of the Agriculture Department have had long-standing missions to advance knowledge and provide important information for industry and agriculture.

During World War II, the Manhattan project to develop the atomic bomb represented government involvement at a very high level. (At the same time in England, radar was being developed.) In addition, nylon and synthetic rubber were produced to replace silk and natural rubber, products no longer available from the Far East. Finally, the antibiotic penicillin was introduced to combat infections from wounds.

The government has continued a high level of involvement in science. The space program, the development of atomic energy, and the efforts to develop alternative energy sources mark areas where much of the scientific study has been federally sponsored to promote national interest. Still other forms of government science have included the monitoring of the workplace for safety, the advance of medicine, and the monitoring of the quality of the environment. Along with these peaceful applications of science, the government has continued to sponsor weapons research and development.

Science policy is made by the executive branch of the government and by the Congress in the budget-making process. (Fewer than 5 percent of the legislators have backgrounds in science.) In decisions to allocate money for research and development, competing needs for security, economic well-being, health care, and quality of the environment are sometimes in conflict. Choices made on the allocation of federal research efforts will do much to shape our future.

COMMUNICATION IN SCIENCE

Science has been characterized by open communication. A paper in a journal or a talk at a professional meeting is the traditional format by which scientists gain recognition among peers. Papers are particularly important; before being accepted for publication, they are reviewed by scientific peers active in research.

Journals have distinctive personalities. Some journals emphasize rapid communication; letters and brief reports of research in active fields are published with a minimum of delay. Other journals have interdisciplinary character; important papers of interest to scientists in more than one specialty are selected. Many journals become more specialized. The majority of papers are submitted to specialized journals that are read chiefly by members of a specialty. As a field grows, it may subdivide and spawn new, more specialized publications.

As more and more papers are written, less and less time may be available for reading them. Forms of communication have been developed to try to deal with the growing mass of scientific literature. One journal prints only the title pages of other journals to enable scientists to quickly scan the contents of many journals. An index gathers references to articles under the citations in footnotes found in articles. A scientist might use that index to find papers citing a method of interest or the work of a particular author. Finally, computers are used to search titles or abstracts for certain keywords. These methods are far from perfect, but no one has the time to read all articles or even all abstracts.

Personal contacts remain important for communication in science. Annual conferences may bring together participants in a field of research. Both lectures and informal conversations about avenues of research underway, enable participants to remain ahead of the literature. An important and active field of research may involve a relatively small number of scientists at a few universities and industrial or government laboratories all of whom know one another and have regular avenues of contact.

How is science communicated to nonscientists? The impact of science and technology on people's lives is great but, for many, it is a mysterious or magical process. Communication requires a scientifically literate public and a scientific community willing and able to communicate in language accessible to those not trained in science. Magazines are available to try to meet this need. Students completing this course might wish to read *Scientific American* and *Discover* to continue learning about the activities of science. Among newspapers, the *New York Times* publishes a section devoted to science each Tuesday.

Science is both national and international. Is it in the national interest to attempt to restrict the free flow of scientific information? Increasingly in the 1980s, some people in government question whether the free flow of scientific information is in conflict with national economic or defense interests.

FUNDING IN SCIENCE

The private sector of the economy has traditionally maintained substantial expenditures for research and development. Companies consider the development of new products and the savings provided by improved processes to be an important part of their future well-being. Patent protection allows companies developing new products time to profit from their inventions.

Within the private sector, distinctions are made between expenditures for long-range research and for developmental research. Companies spend more money for development than for basic research. The costs of finding ways to apply scientific ideas to the production of a product at a competitive price can be high. For example, small differences in the yields of a large-volume process may mean the difference between profit and loss, so that a substantial effort to maximize yields may be needed between the time a process is shown to be feasible and the time it becomes commercial. In the pharmaceutical industry, the costs of testing a potential new drug for safety and efficacy are very high.

University research is largely basic science. Published results from this research have been available to those who would improve existing technologies and develop new ones. Because of concern for the training of scientists to meet the needs of industry and government, university research is of interest to both the private and public segments of the economy.

World War II marked the beginning of a large and active role of the federal government in the support of research. Spending for science in the defense of the country was proven to be in the national interest. After the war, federal support for science continued and was seen to be in the nation's economic interest.

Federal support for research and the training of scientists continued after World War II. The navy provided funds for basic research at many universities. In 1948, the National Science Foundation (NSF) was established to support scientific research. At first, funding of the NSF grew slowly. After Russia launched a sputnik satellite in 1958, funds for science research and education grew rapidly. In the 1980s, funds have grown only slowly and many educational missions of the NSF have been cut.

The NSF has provided a model for the awarding of support for research. Research proposals are judged anonymously by scientists who are expert in the area of science covered in the proposal. Awards are made on the merits of the proposed research and on the record of successful research by the investigator. This method of awarding financial support has the general support of the scientific community; however, there are some criticisms for this method of funding. It is often difficult for young and untried investigators to find support for their research. Block grants were given to individual departments at some universities for distribution and administration. The active encouragement of early research efforts by the Research Corporation, a small private foundation, has helped to bridge this gap.

FUNDING FOR SCIENCE POSES POLITICAL QUESTIONS

Centers of academic excellence can have large favorable economic impacts on their communities and regions. High-technology industries grow up near universities. Examples include the industries located near Harvard and MIT near Boston, the industries in the "silicon valley" near Stanford and the University of California at Berkeley, and the industries in the research triangle near the University of North Carolina, Duke University, and North Carolina State University.

Who is served by elite centers of excellence? Would the wide dispersal of resources weaken the scientific effort everywhere? Because science has become a large social institution, decisions concerning science policy have entered the political arena. The awarding of research money to successful scientists at a relatively small number of major universities presents difficult political problems. Representatives from every state would like to see a larger share of federal support arrive in their state or district. Indeed, in the 1980s awards for the establishment of science facilities became a part of the pork barrel politics when the traditional peer review process was bypassed in making some awards directly to universities.

CHANGING RELATIONSHIPS BETWEEN INDUSTRY AND UNIVERSITIES

New relationships between universities and industry began appearing in the 1980s. In response to decreases in federal support, universities sought to gain dependable, long-range support for science elsewhere. In response to the shortened time to bring new technologies to market in the areas of microelectronics and biotechnology, industrial interest in the support of university research increased. To further complicate matters, in some areas the entrepreneurs were the very university scientists who had pioneered the new research.

There are potentials for both conflict and benefit in closer ties between specific industries and specific university laboratories. Universities are committed to the free flow of information. The work of scientists in university laboratories is published in journals. Industries are less open. Industrial research is often aimed at obtaining patents and profits. However, university research can provide both the scientific base and the trained personnel needed for industry. In addition, the flow of money from grants is often variable and almost always short-range. Dependable long-term funding offers stability to a program of research and to a faculty.

Negotiated agreements can insure the right of university researchers to freely publish results. In return for support, industries may receive early access to the results of research and the opportunity for their own scientists to visit and interact with university faculty.

THE FUTURE OF SCIENCE IN SOCIETY

Science has emerged as a large and important institution in society. Its interactions with other institutions in society are important. This has become increasingly obvious as limitations on resources have begun to limit the growth of science. Science is one institution among many that is competing politically for resources. Within science, disciplines and subdisciplines compete with one another. As it allocates resources, society is taking an ever larger role in determining the future directions of science.

Public awareness of the promise and limitations of science is called for. Science has grown so big and is so intimately connected to the economy, the quality of life, and to the future of this country and planet that decisions concerning science policy will have large impacts on the lives of all people. Resources to devote to science are limited, and decisions to support some science and not other science will be made and implemented by people not trained in science. We cannot afford the luxury of a scientifically illiterate or uninformed society.

APPENDIX

MEASUREMENTS AND CALCULATIONS

Chemists commonly measure quantities of chemicals in one of three ways. They measure the weight of solids and liquids using a balance. When working with gases, they measure volume, temperature, and pressure. The volume of a given gas at a fixed temperature and pressure can be converted to a weight. In many applications, chemists work with solutions and they frequently measure volumes of solutions with known concentration. From the volume and concentration measurements, the weight of a substance in solution can be calculated.

CHEMICAL EQUATIONS RELATE MOLES OF PRODUCTS AND REACTANTS

Because atoms are not measured directly, an interpretation of balanced equations in measured quantities is very desirable. A balanced equation is a statement about moles of elements and compounds in chemical reactions. Because reactions are often run using weights of reactants and products, it is often useful to interconvert moles and weights. Using the mole to express quantities in chemical reactions, a variety of relationships between reactants and products can be calculated.

SAMPLE CALCULATION

What weight of CO_2 is formed from 1000 g of C_8H_{18}?

$$2C_8H_{18} + 25O_2 \rightarrow 16CO_2 + 18H_2O$$

The balanced equation shows that 8 mol of CO_2 are formed for each mole of isooctane burned. To solve the problem, the weight of C_8H_{18} is converted into moles, the balanced equation is used to relate moles of product to moles of reactant, and, finally, the number of moles of CO_2 is converted into a weight.

$$\text{Molecular weight of } C_8H_{18} = 8 \text{ mol C} \frac{12.01 \text{ g}}{1 \text{ mol C}} + 18 \text{ mol H} \frac{1.008 \text{ g}}{1 \text{ mol H}} = 114.2 \text{ g}$$

$$\text{MW of } CO_2 = 1 \text{ mol C} \frac{12.01 \text{ g}}{1 \text{ mol C}} + 2 \text{ mol O} \frac{16.00 \text{ g}}{1 \text{ mol O}} = 44.01 \text{ g}$$

$$\text{Weight } CO_2 = \frac{1000 \text{ g isooctane}}{114.2 \text{ g/mol}} \frac{8 \text{ mol } CO_2}{1 \text{ mol isooctane}} \frac{44.01 \text{ g}}{1 \text{ mol } CO_2} = 3089 \text{ g}$$

GAS CALCULATIONS

Because gases expand or contract much more than liquids or solids, it is necessary to consider ways by which the quantities of gases might be measured. Pressure, volume, and temperature are measured more often than weight.

Boyle's Law

For a sample of a gas at constant temperature, the product of the pressure times the volume is found to be a constant. This relation is known as Boyle's law.

$$P\,V = \text{constant}$$

Pressure can be expressed using a variety of units. An atmosphere of pressure is equal to 760 torr, the pressure exerted by a column of mercury 760 mm high. A barometer reading of 720 torr is equal to 720/760 = 0.95 atm.

When the pressure of a gas changes at constant temperature, the volume of the gas at the new pressure can be related to the original volume and pressure by applying Boyle's law. At constant temperature, the product P_1V_1 at an initial set of conditions has the same value as the product P_2V_2 under new conditions. (The subscript 2 refers to the new conditions, and the subscript 1 refers to the initial conditions.)

$$P_1V_1 = P_2V_2$$

To solve for V_2, both sides of the equation are divided by P_2.

$$V_2 = \frac{P_1}{P_2}\,V_1$$

Note that a similar equation relates the new pressure of a gas to the initial pressure when the volume changes at constant temperature.

$$P_2 = \frac{V_1}{V_2}\,P_1$$

SAMPLE CALCULATION

What volume would 5000 L of helium in a balloon occupy if the pressure were decreased from 1.000 to 0.800 atm at constant temperature?

$$V_2 = V_1\,\frac{P_1}{P_2} = 5000\text{ L}\,\frac{1.000\text{ atm}}{0.800\text{ atm}}$$

$$V_2 = 6250\text{ L}$$

Charles' Law

At constant pressure, a gas expands when heated. A graph for gas volume vs. temperature at constant pressure is linear. Extrapolation of the plotted lines to very low temperatures gives the value $T = -273\,^\circ\text{C}$ for the intercept at $V = 0$. This relationship, known as Charles' law, is summarized in the following equation.

$$V = (\text{constant})\,[T\,(^\circ\text{C}) + 273]$$

If a new temperature scale is chosen, the equation for Charles' law can be simplified. On the kelvin scale, $T(\mathrm{K}) = T(^\circ\mathrm{C}) + 273$. Using the kelvin scale for temperature, Charles' law becomes:

$$V = (\text{constant})\,[T\,(\mathrm{K})]$$

$$\frac{V}{T} = \text{constant}$$

For a change of temperature at constant pressure, Charles' law can be used to relate the new volume to the initial volume and temperature, V_1 and T_1, and to the final temperature T_2.

$$\frac{V_2}{T_2} = \frac{V_1}{T_1}$$

$$V_2 = \frac{T_2}{T_1} V_1$$

SAMPLE CALCULATION

What volume would be occupied by 1.000 L of a gas if it were warmed from 20 to 30°C (293 to 303 K)?

$$V_2 = V_1 \frac{T_2}{T_1} = 1.000\ \mathrm{L}\ \frac{303\ \mathrm{K}}{293\ \mathrm{K}}$$

$$= 1.034\ \mathrm{L}$$

Note that Charles' law and Boyle's law are idealizations of the behavior of gases. All gases condense to liquids if cooled sufficiently at a high enough pressure.

A Combined Gas Law

A general equation that relates the volume of a gas at a new temperature and pressure to its initial volume, temperature, and pressure can be derived.

$$\frac{P_1 V_1}{T_1} = \frac{P_2 V_2}{T_2}$$

(At constant temperature, this formula reduces to Boyle's law; at constant pressure, it reduces to Charles' law.) By multiplying both sides of this equation by T_2/P_2, an expression for the new volume is obtained.

$$V_2 = \frac{P_1}{P_2} \frac{T_2}{T_1} V_1$$

Rather than try to remember subscripts in this equation, it is more useful to apply physical reasoning. If a gas is cooled, it contracts. In this case the ratio of initial and final temperatures should give a fraction smaller than 1. A gas also contracts if it is compressed. Again, the ratio of initial and final pressures should give a number smaller than 1. If a gas is heated, it expands. The initial volume should be multiplied by a ratio of temperatures greater than 1 to find the new volume. If the pressure on a gas is lowered, it expands. A ratio of pressures greater than 1 corresponds to expansion.

SAMPLE CALCULATION

A 400 mL sample of O_2 gas is collected at 22°C and 0.95 atm of pressure. What volume would the gas occupy at 0°C and 1.00 atm?

$$T\,(\mathrm{K}) = T\,(^\circ\mathrm{C}) + 273$$
$$T = 22^\circ\mathrm{C} + 273 = 295\ \mathrm{K}$$
$$V_2 = V_1 \frac{P_1}{P_2}\frac{T_2}{T_1}$$
$$= 400\ \mathrm{mL}\,\frac{0.95\ \mathrm{atm}}{1.00\ \mathrm{atm}}\,\frac{273\ \mathrm{K}}{295\ \mathrm{K}}$$
$$= 352\ \mathrm{mL}$$

The Molar Volume of Gases

At 1 atm and 273 K, the volume occupied by a mole of nitrogen, oxygen, or most other gases is 22.4 L. (Small differences exist.) This value can be used to calculate the number of moles of a gas from *P*, *T*, and *V* measurements or to calculate the volume of a given weight of a gas under standard conditions.

SAMPLE CALCULATION

What is the volume of 11 g of CO_2 gas at 273 K and 1.00 atm?

$$V = \text{moles } CO_2\,\frac{22.4\ \mathrm{L}}{\mathrm{mol}}$$
$$= \frac{11\ \mathrm{g}\ CO_2}{44\ \mathrm{g/mol}}\,\frac{22.4\ \mathrm{L}}{\mathrm{mol}}$$
$$= 0.25\ \mathrm{mol}\,\frac{22.4\ \mathrm{L}}{\mathrm{mol}}$$
$$= 5.6\ \mathrm{L}$$

SOLUTIONS

It is frequently convenient to measure quantities of chemicals by measuring volumes of solutions because it is faster and easier to measure volumes of solutions than to weigh substances. To express the concentration of solutions, chemists often use units of molarity (moles/L, M). For example, a 1.00 M solution of NaCl contains 1.00 mol of NaCl/L of solution. To prepare this solution, a chemist would dissolve 1.00 mol of NaCl in water and then dilute the solution to a final volume of 1.00 L. If the concentration of a solution is known, the measured volume can be used to calculate the number of moles of the dissolved substance.

SAMPLE CALCULATION

What is the concentration of a solution of NaOH (FW = 40.0) if 10.0 g of NaOH is dissolved in water and diluted to 500 mL?

$$\begin{aligned} C &= \frac{\text{mole}}{\text{L}} = \frac{10.0 \text{ g NaOH}/40.0 \text{ g/mol NaOH}}{0.500 \text{ L}} \\ &= \frac{0.250 \text{ mol NaOH}}{0.500 \text{ L}} \\ &= 0.50\ M \end{aligned}$$

SAMPLE CALCULATION

What weight of NaOH is required to make 2.0 L of a 0.20 M solution?

$$C = \frac{\text{mole}}{\text{volume}}$$

$$\text{Moles} = C\ V$$

$$\begin{aligned} \text{Moles}_{\text{NaOH}} &= \frac{0.20 \text{ mol}}{\text{L}}\ 2.0 \text{ L} = 0.40 \\ \text{Weight}_{\text{NaOH}} &= (\text{moles}_{\text{NaOH}})\ (\text{FW}_{\text{NaOH}}) \\ &= 0.40 \text{ mol}\ \frac{40 \text{ g}}{1 \text{ mol}} = 16 \text{ g} \end{aligned}$$

ANSWERS

TO SELECTED QUESTIONS

CHAPTER 1

1. There are only a small number of planets against a background of a very large number of stars.
3. Gases that are very soluble in water cannot be collected over water.
5. Water cannot be an element.
7. 44 g.
9. Limestone has been decomposed into simpler (lighter) substances.
11. Samples A and B have the same composition by weight.
13. The properties of compounds differ from the properties of the elements combined.
15. The ratio of the masses of oxygen combined with 70 g of iron in the two compounds is $30/20 = 3/2$.
17. With the same fixed weight of hydrogen, the ratio of the weights of carbon in the two compounds is $4.89/1.63 = 3/1$.
19. No, Dalton did not say that an atom of an element had the same properties as the element.
21. If two compounds are formed from a pair of elements and if atoms combine in small whole number ratios, then with a fixed number of atoms of one element, the numbers of the second element would be a small, whole number ratio. Since all atoms of an element have the same weight, the same fixed number of atoms of the first element would have the same weight in the two compounds. The ratio of the weights of the second element would be in the same small, whole number ratio as the ratio of atoms.

CHAPTER 2

1. HO_2.
3. UF_3, UF_6.
5. (*a*) 5; (*b*) 6, 3; (*c*) 2.
7. 50.
9. Correct formulas are required to determine atomic weights.
11. 65.4, Zn.
13. (*a*) 64.1; (*b*) 80.1; (*c*) 56.1; (*d*) 136.1.
15. Families—similar properties and formulas of compounds; rows—increasing atomic weights (unless family assignment indicated a reversal of atomic weight or the presence of a gap for a yet undiscovered element).
17. (*a*) AsH_3; (*b*) GeH_4; (*c*) SeH_2; (*d*) HI.
19. Eka boron is scandium (Sc), and eka aluminum is gallium (Ga).

CHAPTER 3

1. $Ba^{2+} + 2e^- \rightarrow Ba$.
3. No, the radiation observed was the beta decay of radioactive "daughters" of uranium 238.
5. Some atoms are not "indestructible."
7. Radioactivity is associated with a particular chemical element.

9. With only a small number of drops, the largest common divisor might not correspond to the charge on a single electron. For example, all the drops might have an even number of electrons.
11. The fraction of alpha particles passing directly through a foil would be similar, for the nucleus of all atoms is very small. In a collision with a lighter nucleus, an alpha particle would be deflected less.
15. $^{238}_{92}U$ has 146 neutrons; $^{14}_{6}C$ has 6 protons and 8 neutrons; $^{35}_{17}Cl$.
17. In Cl^-, the same positive nucleus is surrounded by one more electron than in an atom of Cl. The increase in repulsion between negative electrons causes the anion to be larger.
19. (*a*) $^{222}_{86}Rn$; (*b*) $^{4}_{2}He$; (*c*) $^{218}_{84}Po$; (*d*) $^{214}_{82}Pb$.
21. $^{30}_{15}P \rightarrow {}^{30}_{14}Si + {}^{0}_{+1}e$.

CHAPTER 4

1. Longest, blue; shortest, red.
3. Neon with eight outer electrons.
5. As minor constituents of air, the physical properties of the noble gases were masked by the properties of nitrogen and oxygen. Unlike other elements, they were not found combined in solid compounds that could be isolated and studied.
7. The value of $\left(\frac{1}{2^2} - \frac{1}{n^2}\right)$ increases and approaches $\frac{1}{4}$ as n gets larger.
9. The smallest value of $\left(\frac{1}{2^2} - \frac{1}{n^2}\right)$ is $\frac{5}{36}$ when $n = 3$. As n becomes large, the value in parentheses approaches $\frac{1}{4}$.
11. 50.
15. A wave.
17. (*a*) 1; (*b*) 3; (*c*) 5; (*d*) 7; (*e*) 9.

19.

Atomic Number	Element	Protons	Inner Electrons	Kernel Charge	Outer Electrons
4	Be	4	2	+2	2
12	Mg	12	10	+2	2
20	Ca	20	18	+2	2
8	O	8	6	+6	6
16	S	16	10	+6	6
34	Se	34	18	+6	6

21. O 2*p*; Ca 4*s*; Fe 3*d*; Li 2*s*; Ag 4*d*; Hg 5*d*; U 5*f*.

CHAPTER 5

1. −3, −1, +2, +3.
3. LiI, K_2O, CaS, $BaCl_2$, AlF_3.

5. Lattice with Mg^{2+} and O^{2-} ions.
7. $Ba^{2+} > Sr^{2+} > Ca^{2+} > Mg^{2+}$.
9. CsI has the largest cation and the largest anion, so the attractive forces between the +1 cation and the −1 anion should be weaker.
11. KNO_3, Li_2CO_3, $MgSO_4$, NH_4Br.
14. (*b*)

```
    H   H             H
    |   |             |   ..
H — C — C — H     H — C — O :
    |   |             |   |
    H   H             H   H
```

(*c*)

```
H       H        H
 \     /          \      ..
  C = C            C = O
 /     \          /      ..
H       H        H
```

(*f*)

```
    H
    |                 ..
H — O :+          H — O :−
    |                 ..
    H
```

16. With a filled shell of electrons and a high kernel charge, neon cannot readily lose electrons to form a cation, gain electrons to form an ion, or share electrons to a form covalent bond.
18. Linear, triangular pyramid, tetrahedral, bent.

CHAPTER 6

1. KI, BaS, and AlF_3 are ionic.
3. The difference in electronegativity between Ag and F is greater than the differences between Ag and the other halogens. Ionic bonding is favored when there is a large difference in electronegativity.
5. BF_3 is planar with boron at the center of an equilateral triangle formed by the fluorines.
7. Nitrogen gas consists of diatomic molecules with weak intermolecular forces. Diamond is a giant covalent molecule.
9. Ammonia molecules form strong hydrogen bonds to water molecules; methane molecules do not.
11. If a water molecule was linear, it would not have a dipole moment and would not solvate ions as well.
13. HF is a very polar molecule, and C_8H_{18} is not. Polar molecules can solvate ions to help overcome lattice forces.

CHAPTER 7

1.

```
    ..         ..        ..             ..
H — Cl: + H — O : → H — O+ — H + :Cl:−
    ..         |         |              ..
               H         H
```

3. Basic, a weak base.
5. 0.10; 0.010; 0.0010.

7. 110 mL.
9. $2OH^- + CO_2 \rightarrow CO_3^{2-} + H_2O$.
11. (*a*) and (*c*) $H_3O^+ + NH_3 \rightarrow H_2O + NH_4^+$; (*c*) is a buffer.
13. (*a*) and (*b*) PH_3.
15. $HAc + HCO_3^- \rightarrow Ac^- + H_2O + CO_2$.

CHAPTER 8

1. (*a*) +5, +5, +3; (*b*) +6, +4, +2, −2; (*c*) 0, +2, +4, +7; (*d*) −1, −1, +1, +7.
3. Oxidized, reduced.
5. Ease of oxidation Cu > Ag > Au.
7. (*a*) +3; (*b*) nitric acid; (*c*) Cl^-.
9. The outer electron in I^- is farther from a + 7 kernel than the outer electron in Cl^-.
11. Cl^-, NH_4^+.
13. More current at constant voltage.
15. $2SO_2 + O_2 \rightarrow 2SO_3$; $H_2O + SO_3 \rightarrow H_2SO_4$.
17. (*a*) $NaCl + H_2SO_4$; (*b*) electrolysis; (*c*) electrolysis.

CHAPTER 9

1. 20 min.
3. To find the fraction of a half-life that has elapsed, it is necessary to know both the original and the final amount of uranium 238 present. Lead 206 is a measure of the uranium 238 that has decayed.
5. The carbon 14 is undergoing decay both when a tree is alive and after it dies.
7. Finely divided NaCl dissolves faster than large crystals because it has a larger surface area.
9. Liquids drain more rapidly when air entering through a vent replaces the liquid.
11. The platinum catalyzes the addition of H_2 to double bonds.
13. Simultaneous three-way collisions are rare events.
15. Energy-producing reactions occur more slowly at lower temperatures.
17. They slow the rate of reproduction.

CHAPTER 10

1. Only part of the energy in food is converted into work.
3. 3.74 cal.
5. More than (hydrogen-bonded water is more ordered in the liquid state).
7. Water has a vapor pressure.
9. Gases mix; the direction of spontaneous change is toward greater disorder.

11. Helium is produced by alpha-decay reactions. It is easier to recover helium from natural gas in which its concentration is greater than in the atmosphere.
13. CO, N_2, NO, SO_2 are among the gases produced by the redox reactions of gunpowder. The exothermic reaction heats the gases causing them to expand.
15. When a bottle is opened, CO_2 vapor escapes lowering the concentration of CO_2 over the liquid, and then CO_2 leaves the liquid to replenish the depleted vapor.

CHAPTER 11

3. Natural gas contains smaller amounts of sulfur and nitrogen compounds than most petroleums.
5. Oxidation of the oil and reaction with water would have produced small amounts of corrosive sulfuric acid.
7. Less refining is necessary.
15. The energy produced in a nuclear reactor can be modified fairly rapidly by inserting or withdrawing control rods.
17. The dangerous radiation produced by short-lived isotopes took this long to greatly diminish.
19. Concerns for the proliferation of nuclear weapons.
21. (*a*) and (*c*) because uranium 235 and not uranium 238 would have undergone fission.

CHAPTER 12

1. (*d*).
3. Fe^{2+} is harder to reduce than Cu^{2+}; Ag is harder to oxidize than Cu.
5. $CuS + O_2 \rightarrow Cu + SO_2$; $SnO_2 + C \rightarrow Sn + CO_2$.
7. Reactions between two solids are slow due to the limited area of contact between them.
9. Yes, it is important to know what is present.
11. A coating of tough, adhering, impenetrable MgO protects the surface of magnesium.
13. Most of the aluminum in scrap aluminum is already reduced, and a major cost in producing aluminum is the reduction of aluminum oxide.
15. As in the case of Al^{3+}, the high positive charge on Fe^{3+} pulls electrons from surrounding water molecules causing them to be more acidic.
17. Impurities that are both easier to oxidize and harder to oxidize are removed in the electrolytic purification of copper because the process includes both an oxidation step and a reduction step. Because the impurities present in aluminum ores (particularly iron compounds) contain metals easier to reduce than aluminum, these impurities would be reduced along with aluminum and would weaken the protective oxide coating on aluminum.
19. Germanium.

CHAPTER 13

1. -6; triangular with three bridging oxygens.
3. $CaCO_3 + H_3O^+ \rightarrow Ca^{2+} + HCO_3^- + H_2O$.
5. $2HCO_3^- \rightarrow H_2CO_3 + CO_3^{2-}$ $\quad H_2CO_3 \rightarrow H_2O + CO_2$. Loss of CO_2 shifts the acid-base equilibrium toward H_2CO_3 and CO_3^{2-}. An increase in the concentration of CO_3^{2-} shifts the precipitation reaction to the right. $Ca^{2+} + CO_3^{2-} \rightarrow CaCO_3$.
7. The acid H_3O^+ attacks O, and the base F^- attacks Si.
9. The energy required to make a pound of glass is far less than the energy required to produce a pound of aluminum, yet the transportation costs in collecting each would be nearly the same.
11. When heated, $MgCO_3$ decomposes to MgO and CO_2. Basic MgO reacts with the silicates of clay to produce the glassy surface.
13. K^+ is bound to anionic particles in soils by strong attractions between opposite charges. K^+ cannot wash out but it can be exchanged for other cations.
15. The pH of runoff water can be greatly different than the pH of precipitation because of acid-base reactions occurring with basic groups present in soil.
17. Clay swells with water, and the water tends to facilitate the slippage of one layer past another. More water would be taken up by the clay during frequent light rains than during a single heavy rain.
19. Cs^+ most resembles K^+, and it would be taken up by plants. To the extent that K^+ is not permanently taken up in tissue as Ca^{2+} is in bone, Cs^+ is less a threat than Sr^{2+}.

CHAPTER 14

1. 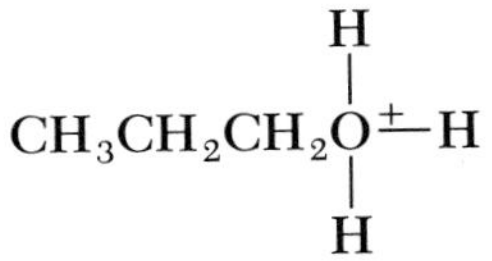 $CH_3CH_2CH_2\overset{+}{O}H_2$ (O bonded to two H: $CH_3CH_2CH_2\text{–}O^+(\text{–H})\text{–H}$) $\qquad CH_3CH_2CH_2\ddot{\underset{..}{O}}\!:^-$
3. $CH_3CH_2CH_2OCH_3 \qquad CH_3CH_2OCH_2CH_3$

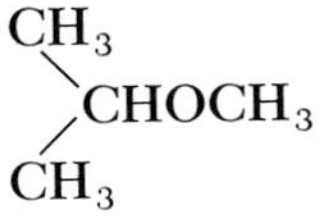

   $(CH_3)_2CHOCH_3$ (drawn as CH₃ and CH₃ branches on CHOCH₃)
5. (*a*)(*a*′); (*b*)(*c*′); (*c*)(*b*′)
7. (*a*) $(CH_3)_2CHOCH(CH_3)_2$ $\qquad$ (*b*) $CH_3CH_2CH(OH)CH_2CH_3$

   (*c*) $(CH_3)_2CHCH(CH_3)_2$ $\qquad$ (*d*) $H_2C{=}CHCH_2CH_3$ (drawn with H, H on one carbon and H, CH_2CH_3 on the other)

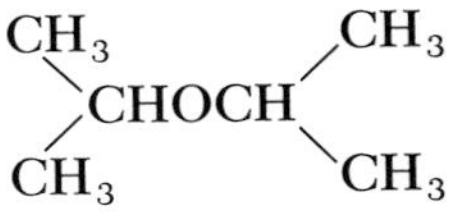

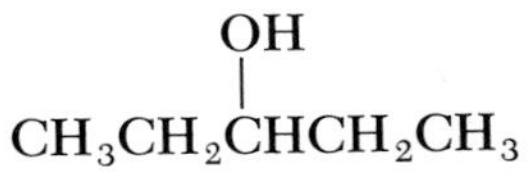

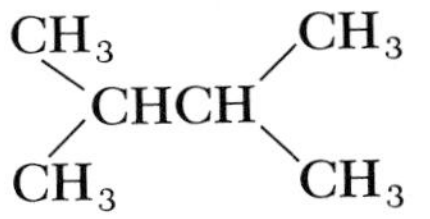

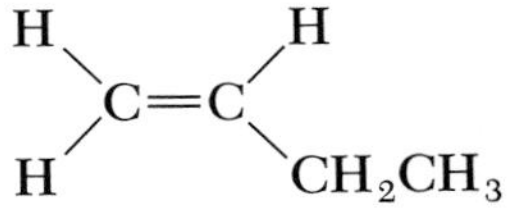

9. 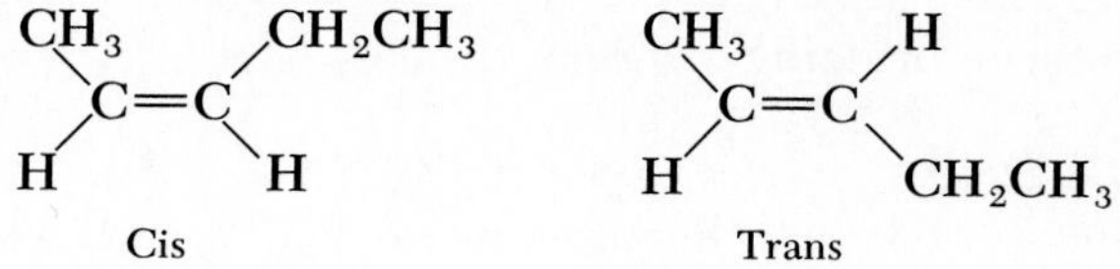 Cis: $CH_3(H)C{=}C(H)CH_2CH_3$ Trans: $CH_3(H)C{=}C(H)CH_2CH_3$

Cis Trans

11. 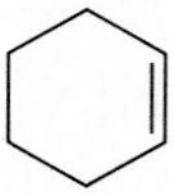 (cyclohexene) C_6H_{10}

13. 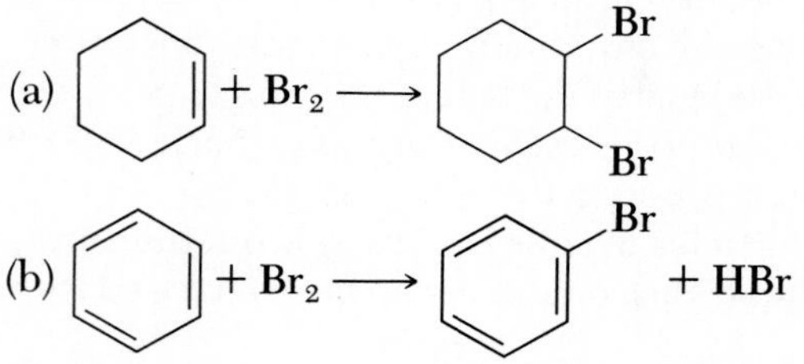 (a) cyclohexene $+ Br_2 \longrightarrow$ 1,2-dibromocyclohexane (Br, Br)

(b) benzene $+ Br_2 \longrightarrow$ bromobenzene (Br) $+ HBr$

15. More refining steps are required to produce high-octane compounds, branched alkanes and aromatic hydrocarbons, than to produce low-octane, linear alkanes. Because some energy is required for each refining step, there are losses in each step.
17. Some of the NO_2 produced at ground level or at low levels in the atmosphere dissolves in water and returns to the earth before ever reaching the ozone layer.

19. (a) 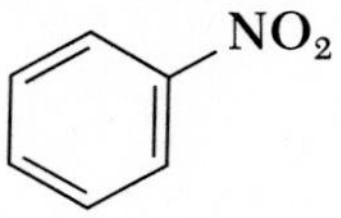 nitrobenzene (NO_2) (b) 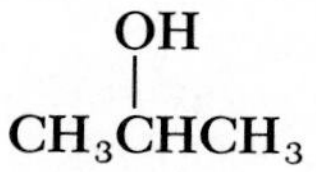 $CH_3CH(OH)CH_3$

CHAPTER 15

1. Methyl *t*-butyl ether for use as an antiknock additive for gasoline sold in Europe is manufactured in Saudi Arabia from petroleum produced there. Petrochemicals have a much greater value than crude petroleum.

3. 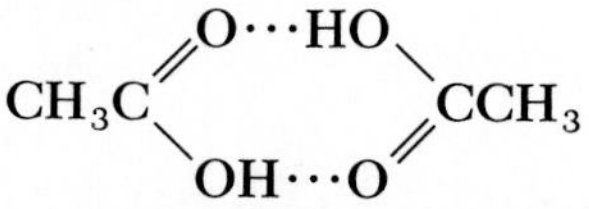 $CH_3C(=O\cdots HO)(-OH\cdots O=)CCH_3$ (acetic acid dimer: CH_3C with $O\cdots HO$ above and $OH\cdots O$ below, linked to CCH_3)

Strong hydrogen bonds exist between acetic acid molecules in the vapor state.

5. (*a*) $CH_3C(=O)H$ (*b*) $CH_3C(=O)CH_2CH_3$

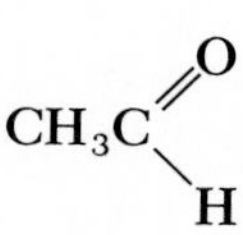

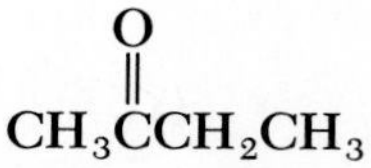

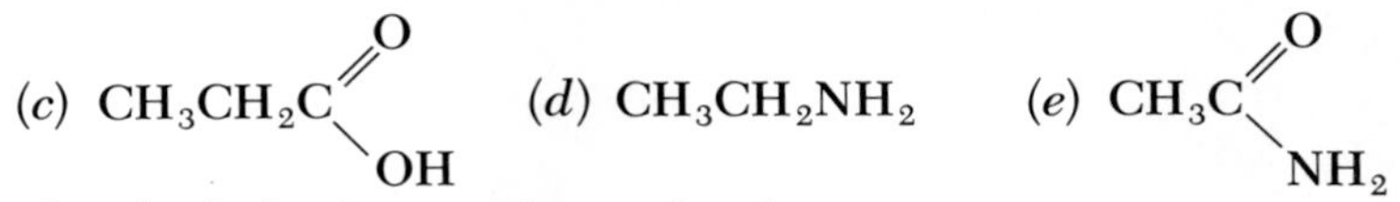

(*c*) $CH_3CH_2C(=O)OH$ (*d*) $CH_3CH_2NH_2$ (*e*) $CH_3C(=O)NH_2$

7. The alcohol 2-butanol has a chiral center at C-2.
9. Unless the soap contains a large hydrophobic group, it will not form micelles capable of carrying off grease.
11. Ions are not solvated by hydrocarbons so they cannot pass through the hydrocarbon interior of a lipid bilayer.
13. (*a*) Butanoic acid (butyric acid); (*b*) propanol (propionaldehyde); (*c*) 3-pentanone; (*d*) ethyl methanoate (ethyl formate); (*e*) ethanamide (acetamide).
15. 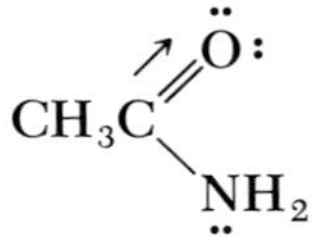 $CH_3C(\nearrow \ddot{O}:)(\ddot{N}H_2)$ Electrons are pulled toward the oxygen making nitrogen more positive and less able to share its lone pair of electrons.
17. (*a*) Ketone and alcohol, carboxylic acid; (*b*) amine and aldehyde, amide.

CHAPTER 16

1. 16,400, 4.
3. 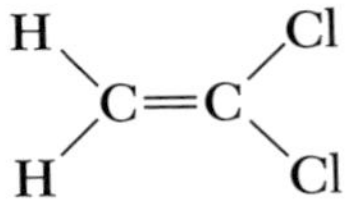 $H_2C=CCl_2$ (H, H on one carbon; Cl, Cl on the other)
5. (*c*).
7. Linear polyethylene chains stack better than the chains in branched polyethylene; and so, high-density polyethylene has stronger crystalline regions.
9. $C_4H_9—N=N—C_4H_9 \rightarrow 2C_4H_9\cdot + N\equiv N$. This compound decomposes to form free radicals that initiate polymerization.
11. Di-*n*-octyl phthalate is a plasticizer that causes polyvinyl chloride to be more flexible.
13. More, harder.
15. The chains are more nearly parallel in a textile fiber.
17. 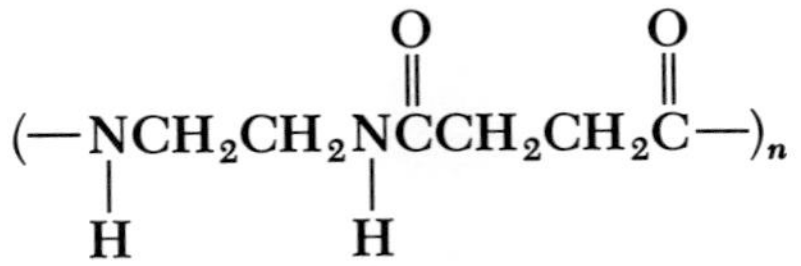 $(—\underset{H}{N}CH_2CH_2\underset{H}{N}\overset{O}{\overset{\|}{C}}CH_2CH_2\overset{O}{\overset{\|}{C}}—)_n$
19. (*a*) A polyester made with a long-chain diol would have too little crystallinity and strength. (*b*) Reaction of a diacid with glycerol gives a rigid, 3-dimensional thermosetting resin that cannot be spun to form fibers. (*c*) A linear polyester made using phthalic acid would not stack well to form parallel chains, so it would lack the strength to form a good fiber.
21. Anions.

CHAPTER 17

1. $HOCH_2C(=O)CH(OH)CH_2OH$

3. $O{=}C(H)CH_2CH(OH)CH(OH)CH_2OH$

5. No, there is a single monomer.
7. $C_6H_{12}O_6$, $C_6H_{12}O_6$.
9. $H_3\overset{+}{N}CH(CH_2CH_2COO^-)COO^-$, -1
11. $21 \times 21 = 441$.
13. $H_3\overset{+}{N}CH(CH_2CH_2CH_2CH_2\overset{+}{N}H_3)COOH$ $\qquad$ $H_2NCH(CH_2CH_2CH_2CH_2NH_2)COO^-$
15. No, for without a fixed shape, the binding of substrates and the catalysis of reactions would not take place.
17. Adenine.
19. If only two bases were required, $4 \times 4 = 16$ possible pairs are possible. This is not enough to code for all the amino acids found in proteins. With three bases, there are a total of $4 \times 4 \times 4 = 64$ possible combinations.

GLOSSARY

acid a substance that increases the hydronium ion concentration in aqueous solution; a proton donor
acid, Lewis a molecule or ion that can act as an electron-pair acceptor
acid-base reaction transfer of a proton from a proton donor to a proton acceptor
acid-base reaction, Lewis formation of a covalent bond by electron-pair donation
acid rain acidic precipitation resulting from the emission of sulfur dioxide and nitrogen oxides
acidic (aqueous) solution a solution that contains a greater concentration of H_3O^+ than OH^- ions; pH < 7
activation energy the energy that reactants must have for reaction to occur
addition reactions addition of atoms or groups to a double or triple bond
alcohol ROH, where R is an alkyl group
aldehyde RCHO, where R is an alkyl or an aryl group
alkali metal a column 1 metal
alkaline earth metal a column 2 metal
alkanes hydrocarbons with only single bonds
alkenes hydrocarbons with double bonds
alkyl group group containing one less hydrogen than an alkane
alkyl halides RX, where R is an alkyl group and X is a halogen atom
alkynes hydrocarbons with triple bonds; acetylenes
alloy intimate mixture of two or more metals or metals plus nonmetals in a substance that has metallic properties
α decay (alpha decay) emission of an α particle by a radioactive nuclide
α particle (alpha particle) a dipositive helium ion, He^{2+}
amalgam alloy of mercury and another metal
amide $RCONH_2$, where R is an alkyl or an aryl group

amine RNH_2, R_2NH, R_3N, where R is an alkyl or an aryl group
anhydrous denoting the absence of water
anion negatively charged ion
anode electrode at which oxidation occurs
aqueous solution a solution in water
aromatic hydrocarbons benzenelike hydrocarbons
aryl group group having one less hydrogen than an aromatic hydrocarbon
asymmetric atom an atom bonded to four different groups
atmosphere all of the gases that surround the earth
atom smallest particle of an element that can participate in a chemical reaction
atomic mass unit one-twelfth the weight of one carbon-12 atom; the unit of atomic and molecular weights
atomic number the number of protons in the nucleus of each atom of an element
atomic orbital the space in which an electron with a specific energy is most likely to be found
atomic weight the average weight of the atoms of the naturally occurring element relative to one-twelfth the weight of an atom of carbon-12
Avogadro's law equal volumes of gases, measured at the same temperature and pressure, contain equal numbers of molecules
Avogadro's number the number of atoms in exactly 12 g of carbon-12; 6.022×10^{23}
base a substance that in aqueous solution increases the hydroxide ion concentration; a proton acceptor
base, Lewis a molecule or ion that can act as an electron-pair donor
basic (aqueous) solution a solution that contains a greater concentration of OH^- ions than H_3O^+ ions; $pH > 7$
battery two or more voltaic cells combined to provide electric energy
β decay (beta decay) electron emission by a radioactive nuclide
β particle (beta particle) high-speed electron produced by a nuclear reaction
binary compound compound containing only two elements
binding energy (nuclear) the energy required to decompose a nucleus into its component nucleons
boiling point temperature at which a liquid boils (bubbles of vapor form throughout the liquid)
bond length average distance from nucleus to nucleus in a stable compound
breeder reactor a nuclear reactor that produces more fissionable atoms than it consumes
buffer solution a solution that resists changes in pH when small amounts of strong acid or strong base are added
calorimeter a device for measuring the heat given off or absorbed in a chemical reaction
carbohydrate a polyhydroxy aldehyde, a polyhydroxy ketone, or a compound derived from these subunits

carbonyl group C=O
carboxylate acid RCOOH, where R is an alkyl or an aryl group
carboxylate ion $RCOO^-$, where R is an alkyl or an aryl group
catalyst a substance that increases the rate of a reaction but may be recovered from the reaction unchanged
cathode electrode at which reduction occurs
cathode rays streams of electrons flowing from the cathode to the anode in a gas discharge tube
cation a positively charged ion
ceramic a material made from clay that has been hardened by firing at high temperature
chain reaction a reaction taking place by a series of repeated reaction steps
changes of state interconversions between the solid, liquid, and gaseous states
chemical bond a force that acts strongly enough between two atoms to hold them together
chemical equation symbols and formulas representing the total chemical change that occurs in a chemical reaction
chemical equilibrium a dynamic equilibrium in which the amounts of species present do not change with time
chemical formula the symbols for the elements combined in a compound and the subscripts which indicate how many atoms of each element are present
chemical kinetics the study of reaction rates and reaction mechanisms
chemical properties properties that can only be observed in chemical reactions
chemical reaction process in which at least one substance is changed in composition and identity
chemistry the study of matter and of its transformations
chirality the property of having nonsuperimposable mirror images
cis-trans isomers compounds with identical groups arranged in different ways on either side of a double bond or rigid ring
collision theory a reaction that occurs when particles collide, only a very small portion of which collisions result in reaction
combustion a process or instance of burning; a high-temperature reaction with oxygen that releases heat and light
compound a substance of definite composition in which two or more elements are chemically combined
condensation movement of molecules from the gaseous phase to the liquid phase
continuous spectrum radiation at all wavelengths
covalent bond bond based upon electron-pair sharing; the attraction between two atoms that share electrons
critical mass smallest mass that will support a self-sustaining nuclear chain reaction
cross-linking bonding between adjacent chains in a polymer

crystalline solid substance in which the atoms, molecules, or ions have a characteristic, regular, and repetitive three-dimensional arrangement
deionized water water with almost all ions removed
density mass per unit volume
diatomic molecule a molecule made of two atoms
diffraction scattering of light by regular array of lines or points
dipole moment measure of the polarity of a chemical bond or molecule
doping addition of impurities to semiconducting elements
double bond two electron pairs shared between the same two atoms
electrochemistry study of oxidation-reduction reactions that either produce or utilize electric energy
electrode a conductor through which electric current enters or leaves a conducting medium
electron negatively charged subatomic particle
electronegativity a measure of the ability of an atom to attract electrons to itself
element a substance composed of only one kind of atom
emission spectrum the spectrum of radiation emitted by a substance
enantiomers mirror-image isomers
endpoint the point at which chemically equivalent amounts of reactants in a titration have been combined
entropy a measure of the randomness or disorder of a system
enzyme a protein that catalyzes a biological reaction
equilibrium state of balance between opposing changes
ester RCOOR, where R's are alkyl or aryl groups
eutrophication process in which a lake grows rich in nutrients and fills with organic sediment and aquatic plants
excited state state of energy reached by electrons when an atom or molecule has absorbed sufficient extra energy
exothermic process a process that releases heat
family the elements in a single vertical column in the periodic table; also called a group
faraday the amount of electricity represented by 1 mol of electrons; 96,500 C
flotation concentration of metal-bearing mineral in a froth of bubbles that can be skimmed off
fossil fuels natural gas, coal, and petroleum
free radicals reactive species that contain an odd number of electrons
frequency the number of wavelengths passing a given point in unit time
functional group a group that contributes a characteristic chemical behavior to the molecule
gas-discharge tube a glass tube that can be evacuated and in which there are sealed electrodes
Gay-Lussac's law of combining volumes a law that in a chemical reaction, the ratios of the volumes of the gases involved, measured at the same temperature and pressure, are small, whole numbers
gene a portion of DNA that codes for a specific protein

glass an amorphous solid formed when a mixture of silica and other compounds is melted and then cooled rapidly
greenhouse effect the effect of warming by absorption and reemission of radiation
ground state lowest-energy state of an electron in an atom
half-life the time it takes for one-half of the nuclei in a sample of a radioactive isotope to decay
halides binary compounds of halogen atoms with other elements
halogen a column 17 element
hard water water that contains dipositive ions that form precipitates with soap or upon boiling
heat capacity the amount of heat required to raise the temperature of a given amount of material by 1°C
Heisenberg uncertainty principle the precept that it is impossible to know simultaneously both the exact momentum and the exact position of an electron
heterogeneous reaction a reaction between substances in different phases
homogeneous reaction a reaction between substances in the same gaseous or liquid phase
hydrocarbon a compound composed of carbon and hydrogen
hydrogenation addition of hydrogen to a molecule
hydrogen bond unusually strong dipole-dipole attraction between a group in which hydrogen is bonded to N, O, or F and a lone pair of electrons on another atom or ion
hydronium ion H_3O^+
hydroxide ion OH^-
hydroxyl group $—OH$
indicator an organic acid or base that changes color when it reacts with hydronium ion or hydroxide ion
inhibitors substances that slow down a catalyzed reaction
initiation production of the first reactive intermediate in a chain reaction
inorganic chemistry the chemistry of all the elements and their compounds, with the exception of hydrocarbons and hydrocarbon derivatives
intermolecular forces the forces of attraction or repulsion between individual molecules
ion an atom or group of atoms that has a net positive or negative charge
ion exchange replacement of one ion by another
ionic bonding bonding based on the attraction between positive and negative ions
isomers compounds that differ in structure but have the same molecular formula
isotopes atoms with different mass numbers but the same atomic number
kernel the nucleus and the inner shell electrons of an atom
ketone $R_2C{=}O$
lattice energy energy liberated when gaseous ions combine to give a crystalline, ionic substance

law statement of an empirical relation between phenomena that is always the same under the same conditions

law of conservation of mass the law that in chemical reactions, matter is neither created nor destroyed

law of definite proportions the law that in a pure compound, elements are combined in definite proportions by weight

law of multiple proportions the law that if two elements combine to form more than one compound, a fixed weight of element A will combine with two or more different weights of element B so that the different weights of B are in the ratio of small whole numbers

leaching a process of extracting a soluble compound from insoluble material

Le Chatelier's principle the precept that if a system in equilibrium is subjected to a stress, a change that will offset the stress will occur in the system

line spectrum radiation at only certain wavelengths

lone pair pair of outer electrons not involved in bonding

mass an intrinsic property; the quantity of matter in a body

mass number the sum of the number of protons and the number of neutrons in a nucleus

melting point temperature at which the solid and liquid phases of a substance are in equilibrium

metallic bonding the attraction between positive metal ions and surrounding freely mobile electrons

metallurgy all aspects of the science and technology of metals

mineral a naturally occurring substance with a characteristic range of chemical composition

mixture any combination of two or more substances in which the substances combined retain their identity

molarity moles of dissolved substance per liter of solution

mole an Avogadro's number of a chemical compound or species

molecular weight sum of the atomic weights of the total number of atoms in the formula of a chemical compound

molecule smallest particle of a pure substance that has the composition of that substance and is capable of independent existence

natural radioactivity decay of radioactive isotopes found in nature

net ionic equation an equation that shows only the species involved in the chemical change

neutral (aqueous) solution a solution that contains equal numbers of H_3O^+ and OH^- ions

neutralization the reaction of an acid with a base

neutron fundamental, subatomic particle that carries no charge has a mass almost the same as the mass of the proton

nitrogen fixation the formation of compounds from molecular nitrogen

noble gas a column 18 element

nonpolar bond a bond in which electrons are shared equally between two atoms

normal boiling point boiling point at 1 atm pressure
normal hydrocarbon chain an unbranched chain
***n*-type semiconductor** negative electrons are the majority of the current carriers; contains a donor impurity
nuclear-binding energy energy that would be released in the combination of nucleons to form a nucleus
nuclear fission splitting of a heavy nucleus into two lighter nuclei of intermediate mass number
nuclear fusion combination of two light nuclei to give a heavier nucleus
nuclear reactions reactions that result in changes in atomic number, mass number, or energy state of nuclei
nuclear reactor equipment in which nuclear fission is carried out at a controlled rate
nucleon a proton or a neutron
nucleus a central region in the atom that is very small by comparison with the total size of the atom and in which all of the mass and positive charge of the atom are concentrated
octet rule atoms tend to combine by gain, loss, or sharing of electrons so that each atom holds or shares eight outer electrons
olefins hydrocarbons with double bonds; alkenes
optical isomerism occurrence of pairs of molecules of the same molecular formula that rotate plane-polarized light equally in opposite directions
optically active compound that which rotates the plane of vibration of plane-polarized light
ore a mixture of minerals from which a particular metal or several metals can profitably be extracted
organic chemistry the chemistry of carbon compounds and their derivatives
osmosis the passage of solvent molecules through a semipermeable membrane and from a more dilute solution into a more concentrated solution
oxidation any process in which oxidation number increases
oxidation number a number that represents the positive or negative character of atoms in compounds found by a set of rules
oxidation-reduction reactions reactions in which oxidation and reduction occur together; also called redox reactions
oxides binary compounds of oxygen
oxidizing agent an atom, molecule, or ion that can cause another substance to undergo an increase in oxidation number
peptide bond (linkage) the amide linkage joining two amino acids
period a horizontal row of elements in the periodic table
periodic law the physical and chemical properties of the elements are periodic functions of their atomic numbers
periodic table a chart showing all the elements arranged in columns with similar chemical properties
petroleum cracking a process of breaking large molecules into small molecules, usually in the gasoline range

petroleum isomerization conversion of straight-chain alkanes into branched-chain alkanes

petroleum reforming conversion of noncyclic hydrocarbons to aromatic compounds

pH $-\log [H_3O^+]$

photoelectric effect the giving off of electrons by certain metals when light shines on them

photon a single quantum of radiant energy (1 $h\nu$)

photosynthesis reaction of carbon dioxide and water in green plants to give carbohydrates

physical properties properties that can be exhibited, measured, or observed without changing the composition and identity of a substance

polar bond a covalent bond in which electrons are shared unequally

pollutant an undesirable substance added to the environment, usually by the activities of earth's human inhabitants

polyatomic ions ions that incorporate more than one atom

polymer a large molecule made of many units of the same structure linked together

polypeptide a polymer formed from amino acids joined by peptide linkages

positron particle identical to an electron in all properties except charge, which is $+1$

precipitate solid that forms during a reaction in solution

pressure force per unit area

products new substances produced in a chemical reaction

propagation production of the species that initiate further steps in a chain reaction

proton fundamental, subatomic particle with a positive charge equal in magnitude to the negative charge of the electron

***p*-type semiconductor** a semiconductor in which positive holes are the majority of the current carriers; contains an acceptor impurity

pure substance a form of matter that has identical physical and chemical properties regardless of its source

quantized measurements measurements restricted to quantities that are multiples of the basic unit, or quantum, for a particular system

quantum numbers numbers that specify the amounts of energy of a system such as an electron in an atom

radiation energy traveling through space

radioactive isotopes isotopes that decay spontaneously

radioactivity spontaneous emission by unstable nuclei of particles, or of electromagnetic radiation, or of both

rate constant the proportionality constant between the rate and the reactant concentrations

rate-determining step the slowest step in a reaction mechanism

rate equation an equation that gives the relationship between the reaction rate and the concentration of the reactants

reactants substances that are changed in a chemical reaction

reaction mechanism the pathway from reactants to products
reaction rate speed with which products are produced or reactants are consumed in a particular reaction
reagent any chemical used to bring about a desired chemical reaction
redox reactions oxidation-reduction reactions
reducing agent an atom, molecule, or ion that can cause another substance to undergo a decrease in oxidation number
reduction any process in which oxidation number decreases
resonance hybrid actual molecular structure of a molecule or ion which is not adequately described by a single Lewis structure
reversible reaction a chemical reaction that can proceed in either direction
salts neutral ionic compounds composed of the cations of bases and the anions of acids
saponification hydrolysis of an ester by a base
saturated hydrocarbons hydrocarbons containing only single covalent bonds
silica SiO_2
silicates compounds containing metal cations and silicon-oxygen groups
single bond a bond in which two atoms are held together by sharing two electrons
slag molten mixture of minerals
smelting a process for producing a free metal from its ore
solubility the amount of solute that can dissolve in a given amount of solvent
solution homogeneous mixture of the molecules, atoms, or ions of two or more substances; a single-phase mixture
solvation interaction of a solute molecule or ion with solvent molecules
solvent the component of a solution usually present in the larger amount
stable isotopes isotopes that do not decay spontaneously
states of matter the gaseous state, the liquid state, and the solid state
steel alloys of iron that contain carbon (up to 1.5%) and usually other metals as well
strong acid an acid that is virtually 100% ionized in dilute aqueous solution to produce hydronium ions
strong base a base that is completely dissociated in dilute aqueous solution to produce hydroxide ions
subatomic particle a particle smaller than the smallest atom
theory unifying model or set of assumptions that explains phenomena and laws based on observation
thermodynamics the study of energy and its transformations
titration measurement of the volume of a solution of one reactant that is required to react completely with a measured amount of another reactant
triple bond three electron pairs between the same two atoms
unsaturated hydrocarbons hydrocarbons with double or triple covalent bonds

unsaturated solution a solution that can still dissolve more solute under the given conditions

voltaic cell a cell that generates electric energy from a spontaneous redox reaction

wavelength the distance between two peaks or troughs on adjacent waves

weak acid a compound or ion that reacts with water only to a small extent to produce hydronium ions

weak base a compound or ion that reacts with water only to a small extent to produce hydroxide ions

weight the force exerted on a body by gravity

TABLE OF THE ELEMENTS

Name	Symbol	Atomic number	Name	Symbol	Atomic number
Actinium	Ac	89	Molybdenum	Mo	42
Aluminum	Al	13	Neodymium	Nd	60
Americium	Am	95	Neon	Ne	10
Antimony	Sb	51	Neptunium	Np	93
Argon	Ar	18	Nickel	Ni	28
Arsenic	As	33	Niobium	Nb	41
Astatine	At	85	Nitrogen	N	7
Barium	Ba	56	Nobelium	No	102
Berkelium	Bk	97	Osmium	Os	76
Beryllium	Be	4	Oxygen	O	8
Bismuth	Bi	83	Palladium	Pd	46
Boron	B	5	Phosphorus	P	15
Bromine	Br	35	Platinum	Pt	78
Cadmium	Cd	48	Plutonium	Pu	94
Calcium	Ca	20	Polonium	Po	84
Californium	Cf	98	Potassium	K	19
Carbon	C	6	Praeseodymium	Pr	59
Cerium	Ce	58	Promethium	Pm	61
Cesium	Cs	55	Protactinium	Pa	91
Chlorine	Cl	17	Radium	Ra	88
Chromium	Cr	24	Radon	Rn	86
Cobalt	Co	27	Rhenium	Re	75
Copper	Cu	29	Rhodium	Rh	45
Curium	Cm	96	Rubidium	Rb	37
Dysprosium	Dy	66	Ruthenium	Ru	44
Einsteinium	Es	99	Samarium	Sm	62
Erbium	Er	68	Scandium	Sc	21
Europium	Eu	63	Selenium	Se	34
Fermium	Fm	100	Silicon	Si	14
Fluorine	F	9	Silver	Ag	47
Francium	Fr	87	Sodium	Na	11
Gadolinium	Gd	64	Strontium	Sr	38
Gallium	Ga	31	Sulfur	S	16
Germanium	Ge	32	Tantalum	Ta	73
Gold	Au	79	Technetium	Tc	43
Hafnium	Hf	72	Tellurium	Te	52
Helium	He	2	Terbium	Tb	65
Holmium	Ho	67	Thallium	Tl	81
Hydrogen	H	1	Thorium	Th	90
Indium	In	49	Thulium	Tm	69
Iodine	I	53	Tin	Sn	50
Iridium	Ir	77	Titanium	Ti	22
Iron	Fe	26	Tungsten	W	74
Krypton	Kr	36	Unnilhexium	Unh	106
Lanthanum	La	57	Unnilpentium	Unp	105
Lawrencium	Lr	103	Unnilquadium	Unq	104
Lead	Pb	82	Uranium	U	92
Lithium	Li	3	Vanadium	V	23
Lutetium	Lu	71	Xenon	Xe	54
Magnesium	Mg	12	Ytterbium	Yb	70
Manganese	Mn	25	Yttrium	Y	39
Mendelevium	Md	101	Zinc	Zn	30
Mercury	Hg	80	Zirconium	Zr	40